Advances in Industrial Control

More information about this series at http://www.springer.com/series/1412

Juš Kocijan

Modelling and Control of Dynamic Systems Using Gaussian Process Models

Springer

Juš Kocijan
Department of Systems and Control
Jožef Stefan Institute
Ljubljana
Slovenia

and

Centre for Systems and Information
Technologies
University of Nova Gorica
Nova Gorica
Slovenia

Additional material to this book can be downloaded from http://extras.springer.com.

ISSN 1430-9491 ISSN 2193-1577 (electronic)
Advances in Industrial Control
ISBN 978-3-319-21020-9 ISBN 978-3-319-21021-6 (eBook)
DOI 10.1007/978-3-319-21021-6

Library of Congress Control Number: 2015954342

Springer Cham Heidelberg New York Dordrecht London

Printed on acid-free paper

Springer International Publishing AG Switzerland is part of Springer Science+Business Media (www.springer.com)

Series Editors' Foreword

The series *Advances in Industrial Control* aims to report and encourage technology transfer in control engineering. The rapid development of control technology has an impact on all areas of the control discipline. New theory, new controllers, actuators, sensors, new industrial processes, computer methods, new applications, new philosophies…, new challenges. Much of this development work resides in industrial reports, feasibility study papers and the reports of advanced collaborative projects. The series offers an opportunity for researchers to present an extended exposition of such new work in all aspects of industrial control for wider and rapid dissemination.

For classical low-level industrial PID controllers a widespread tuning paradigm is based on nonparametric tuning methods pioneered by Ziegler and Nichols in the 1940s. The transition from analogue to digital controller technology led to new tuning flexibilities and, inspired by techniques like Åström and Hägglund's relay experiment, autotuning or push-button controller tuning became a reality; however, the field of more advanced control methods based on nonparametric models has not enjoyed the success achieved by methods that use a parametric model. Whether this is a feature of the lack of flexibility or variety of nonparametric models or whether it is because the research focus has been lacking is an open question but there is a utilitarian attractiveness about the idea of being able to design and tune advanced controllers without having to pursue a preliminary detailed parametric model identification exercise. This becomes even more of an advantage if the process to be controlled is a nonlinear system where nonlinear system identification creates its own difficulties for control design.

Some solution methods to advanced nonlinear system control design are those that use fuzzy-logic models or neural-network models. On the other hand, such routes also involve the selection and tuning of parameters that are embedded in the model, and it might be argued that these are also parametric methods. A class of nonparametric methods that may have the potential to become "mainstream" is that based on Gaussian process models. A Gaussian process model is a nonparametric probabilistic model. It is a Gaussian random function characterised by its mean and covariance function. The outputs of Gaussian process models are normally distributed random variables. Professor Juš Kocijan has been working in this field for some time and feels that a sufficient body of work has been published in the literature to be able to present a coherent framework for the use of Gaussian process models in control systems design and industrial applications. His thoughts have been captured in this *Advances in Industrial Control* monograph entitled *Modelling and Control of Dynamic Systems Using Gaussian Process Models*.

The monograph opens with a short introductory chapter and illustrative example of Gaussian process model regression using data generated by the simple nonlinear function, $y = z^3$. Chapters 2 and 3 present more thorough details of system identification using Gaussian models and on how to incorporate structural and prior information into the models. Throughout these chapters, and indeed the complete monograph, careful referencing of the appropriate source literature enables the reader to develop a fuller appreciation of the subject's development. In Chap. 4 the reader is treated to an introductory overview of how Gaussian process models can be used in advanced control system design. Professor Kocijan is careful to stress the developing nature of this field of research and points out the advantages, disadvantages and open questions that still exist with this class of methods. Chapter 5 summarises many of these areas for the interested reader. The monograph closes with a chapter that contains three diverse case studies to demonstrate the use of Gaussian model techniques in realistic industrial and environmental problems.

Modelling and Control of Dynamic Systems Using Gaussian Process Models is an unusual entry to the *Advances in Industrial Control* monograph series. So, many of the monographs present in the series are reports of "finished" work, but Prof. Kocijan's monograph reports the current status of this nonparametric method and seeks to inspire researchers and engineers to join the evolving development of future work and applications, solving open questions and creating implementation guidelines for industrial applications; it is a welcome and interesting contribution to the series.

Michael J. Grimble
Michael A. Johnson

Preface

We are living in an era of rapidly developing technology. Dynamic systems control is not a new methodology, but it is heavily influenced by the development of technologies for sensing and actuating devices, data storage and communication. All these advances create new opportunities for the control based on data-driven models.

Systems control, especially systems control design, relies on mathematical models. These can be developed from an understanding of the underlying relations in the systems or from measurement data. The term for data-driven modelling used in the control community is system identification.

Increasingly complex systems have to be controlled, and this means that we are dealing with increasingly complex models. Examples of such systems are coming from the fields of biological systems, environmental systems, transportation networks, energy grids and others. This increased complexity triggers a strong need for new methods that deal with data in a scalable and robust way. System identification methods are usually based on statistical analyses, resulting in parametric or nonparametric mathematical models that can be used for systems analyses and control design. System identification methods are well established in the control community.

Large data sets provided with ever-better sensing devices require new identification methods, and this book is written with purpose of demonstrating a type of nonparametric model, coming mainly from the machine-learning community, for use in the applications of the engineering community. On the other hand, small data sets where larger amount of data is required for system identification is another border situation met in practice that also needs appropriate attention.

The particular aim of the book is to describe how Gaussian process (GP) models can be used for the design of dynamic systems control. System identification based on GP models is an integral part of the control design and its description as such also forms an integral part of this book. Using GP models for system identification is a relatively recent approach, where the research activities are very lively. Consequently, this book cannot give a complete picture of the application field, rather it attempts to provide an overview of the current situation and opens up directions for further applications as well as further research options.

The book is intended to open up new horizons for engineers and researchers in academia and industry and all those who are dealing with, or who are interested in, new developments in the field of system identification and control. It addresses issues, some of which are of more interest to engineers and others to researchers. The emphasis is on guidelines for working solutions and on practical advice for their implementation. The emphasis is not on the theoretical background of GP models, for which other books exist, nor does it describes the basics of nonlinear systems identification; instead it shows the potential of the selected modelling method for use in dynamic-systems identification and control design. It is not written in a theorem/proof style. The mathematics is kept to a minimum. The emphasis of the book is on an intuitive understanding of the topic and on providing guidelines and case studies to facilitate practical applications.

An introductory course in nonlinear system identification, probability, statistics and computation intelligence methods is considered as the most appropriate prerequisite for reading this book.

The book was inspired by many years of research and involvement in applications using GP models. I thought it necessary to convey my experience and fascination with the topic to the audience in an integrated form. Since my first introduction to GP modelling more than a decade ago, I see GP modelling as a very handy tool for engineers working with dynamic systems and my wish is that this book conveys this fact.

The content of the book was developed over a few years, but the last three were focused on the book preparation itself. Research on this topic is very dynamic and I am well aware of the fact that I was not able to include everything that relates to GP models, system identification and control design. However, I hope I have enough content to convey the main features of using GP models when modelling dynamic systems and design control system with these models.

The book consists of five chapters that lead the reader from a basic understanding of GP models, via their application for system identification, to the use of the obtained models for system control. Real-life examples are presented to illustrate the explained concepts in the last chapter.

I wish to express my gratitude to the many people I have interacted with during my research on the topic and the writing of this book. I wish to thank Martin Stepančič, Kristjan Ažman, Dejan Petelin, Matej Gašperin, Bojan Likar, Djani Juričić, Stanko Strmčnik, and all my other colleagues from the Department of Systems and Control at the Jožef Stefan Institute. In addition, I would like to thank Alexandra Grancharova, Gregor Gregorčič, Jan Přikryl, Roderick Murray-Smith, David Murray-Smith, Douglas Leith, Bill Leithead, Agathe Girard, Carl Rasmussen, Joaquin Quiñonero Candela, Keith Thompson, Keith Neo and many others. They all helped me by proofreading draft material, or discussed the topic with me, or had a strong influence on my research.

I am very grateful to the Jožef Stefan Institute and the University of Nova Gorica for providing a highly stimulating research environment and for giving me the freedom to write this book. Thanks to Springer for all their support.

I would like to thank Miroslav Štrubelj for assisting me with drawing the figures and Paul McGuiness for improving the English language of the book.

Finally, I would like to thank my entire family for all their support throughout the years it has taken me to write this book.

Ljubljana
March 2015

Juš Kocijan

Contents

Symbols and Notation

$\sim$	Distributed according to, for example: $z \sim \mathcal{N}(\mu, \sigma^2)$
$\mathbf{k}^{\mathrm{T}}$	The transpose of vector $\mathbf{k}$
$\|\|\mathrm{x}\|\|_{\mathbf{A}}$	The weighted Euclidean norm
$C(\mathbf{z}_i, \mathbf{z}_j)$	Covariance function evaluated at $\mathbf{z}_i$ and $\mathbf{z}_j$
$\mathcal{D}$	Data set: $\mathcal{D} = \{(z_i, y_i) \| i = 1, \ldots, N\}$
D	Dimension of the input space $\mathcal{Z}$
$\Delta u(k)$	Difference between two time-consecutive samples $\Delta u(k) = u(k) - u(k-1)$
δ_{ij}	Kronecker delta, $\delta_{ij} = 1$ iff $i = j$ and 0 otherwise
δ	Dirac delta function that is zero everywhere except at zero, with an integral of one over the entire real line
E	Expectation
$f(\mathbf{z})$	Latent function values at the inputs $\mathbf{z}$, or mapping between the inputs $\mathbf{z}$ and the output
$\phi(\mathbf{z})$	Fixed basis function of the regression vector $\mathbf{z}$
$\mathcal{GP}$	Gaussian process: $f \sim \mathcal{GP}(m_f(\mathbf{z}), C_f(\mathbf{z}_i, \mathbf{z}_j))$, the function f is distributed as a Gaussian process with the mean function $m_f(\mathbf{z})$ and the covariance function $C_f(\mathbf{z}_i, \mathbf{z}_j)$
$\mathbf{I}$	The identity matrix
J	Cost function
$\|\mathbf{K}\|$	Determinant of the $\mathbf{K}$ matrix
$\mathbf{K}$	$N \times N$ covariance matrix for the (noisy) output values or Gram matrix for independent homoscedastic noise, $\mathbf{K} = \mathbf{\Sigma}_f + \sigma_n^2 \mathbf{I}$
$\mathbf{k}(\mathbf{z}^*)$	Vector, short for $C(\mathbf{Z}, \mathbf{z}^*)$, when there is only a single test case
$k(\mathbf{z}_i, \mathbf{z}_j)$	Kernel function evaluated at $\mathbf{z}_i$ and $\mathbf{z}_j$
k	Sampling instant
$\kappa(\mathbf{z}^*)$	Autocovariance of the test input data
ℓ	The value of the logarithm of the evidence or the marginal likelihood function
$\ln(z)$	Natural logarithm (base e)

λ	Penalty term in the minimum variance cost function
$m(\mathbf{z})$	The mean function of a Gaussian process
μ_y	The mean value of the random variable y with a Gaussian probability distribution, $\mu_y = E(y)$
$\mu(\mathbf{z})$	The mean value of the output Gaussian probability distribution at the input data $\mathbf{z}$
N	The number of input data (identification points)
N_h	MPC prediction and optimisation horizon
N_u	MPC control horizon
$\mathcal{N}(\boldsymbol{\mu}, \boldsymbol{\Sigma})$	Gaussian or normal distribution with the mean vector $\boldsymbol{\mu}$ and the covariance matrix $\boldsymbol{\Sigma}$
$\mathbb{N}$	The integer numbers
ν	The noise signal
$\mathcal{O}(N)$	The time complexity of an algorithm that quantifies the amount of time by an algorithm that depends on the number of input data points N
P	The coincidence point
$p(y\|z)$	The probability (density) of the conditional random variable y given z
$\mathbb{R}$	The real numbers
$\boldsymbol{\Sigma}_f$	The covariance matrix for the (noise-free) f values
σ_n^2, σ_ν^2	The noise variance
σ_y^2	The variance of the random variable y with a Gaussian probability distribution, $\sigma_y^2 = \mathrm{var}(y)$
$\sigma^2(\mathbf{z})$	The variance of the output Gaussian probability distribution at the input data $\mathbf{z}$
T_s	The sampling period
$\boldsymbol{\theta}$	The vector of hyperparameters (parameters of the covariance function)
$u(k), y(k)$	Samples of the measurements for the signals u and y at the time instance k
$\mathcal{U}, \mathcal{X}$	The set of (constrained) input values and state values
V^*	The minimum value of the cost function J
x and x(k)	The vector of states in the state-space model
$y\|z$	The conditional random variable y given z
y^*	The prediction of output at the input data $\mathbf{z}^*$
$\hat{y}$	The prediction of output
$\hat{y}(k)$	The prediction at the time instant k
$\hat{y}(k+j)$	The j-step-ahead prediction at the time instant k
$\mathcal{Z}$	The space of regressors
$\mathbf{Z}$	The $D \times N$ matrix of regressors $\{\mathbf{z}_i\}_{i=1}^N$
$\mathbf{z}_i$	The ith regression vector
z_{di}	The dth regressor of the ith regression vector

Acronyms

ALGPDP	Gaussian process dynamic programming with active learning
ANN	Artificial neural network
ARD	Automatic relevance determination
ARMA model	Autoregressive and moving-average model
BIC	Bayesian information criterion
CPU	Central processing unit
DE	Differential evolution
DTC	Deterministic training conditional
EA	Evolutionary algorithms
ELLE	Extended local linear equivalence
FDI	Fault detection and isolation
FITC	Fully independent training conditional
FSGP model	Fixed-structure Gaussian process model
GA	Genetic algorithms
GP	Gaussian process
GP-NAR model	Gaussian process nonlinear autoregressive model
GP-NARX	Gaussian process nonlinear autoregressive model with exogenous input
GP-NFIR model	Gaussian process nonlinear finite-impulse response model
GP-NMPC	Gaussian-process-based nonlinear model predictive control
GP-NOE model	Gaussian process nonlinear output-error model
GPDP	Gaussian process dynamic programming
GPGPU	General-processing graphics processing unit
GPU	Graphics processing unit
IMC	Internal-model control
IV	Instrumental variable
LFM	Latent force model
LMGP model	Local models incorporated into a Gaussian process model
LMN	Local model network
LOO-CV	Leave-one-out cross-validation
LPD	Log predictive-density error

LPV model	Linear-parameter-varying model
MAP	Maximum a posteriori
MCMC	Markov-chain Monte Carlo
mp-NLP	Multi-parametric nonlinear programming
MPC	Model predictive control
MRSE	Mean relative square error
MSE	Mean-squared error
MSLL	Mean standardised log loss
MVM	Matrix-vector multiplication
NARMAX	Nonlinear autoregressive and moving-average model with exogenous input
NARX	Nonlinear autoregressive model with exogenous input
NBJ model	Nonlinear Box–Jenkins model
NFIR model	Nonlinear finite impulse-response model
NMPC	Nonlinear model predictive control
NOE model	Nonlinear output-error model
PCA	Principal component analysis
PFC	Predictive functional control
PILCO	Probabilistic inference and learning for control
PITC	Partially independent training conditional
PSO	Particle swarm optimisation
PWL	Piecewise linear
RRMS	Relative-root-mean-square error
SMSE	Standardised mean-squared error
SoD	Subset of data
SoR	Subset of regressors
SPGP	Sparse pseudo-input Gaussian processes
SSGP	Sparse spectrum Gaussian process regression
VBL	Velocity-based linearisation

Chapter 1
Introduction

This book deals with the topic of nonlinear system identification using Gaussian process (GP) models and their role in the design of the control of dynamic systems. We believe that this method offers great potential in dynamic-systems modelling as well as in engineering practice. GP modelling has also been used for dynamic-systems modelling since the end of the twentieth century.

In order to describe a variety of systems, different kinds of models are used. The fundamental property of a model is that it uses a system's observations to form a pattern that possesses the same properties as the observed system. Various forms of models exist, but in the case of modelling dynamic systems in the framework of systems theory we are mainly interested in mathematical models.

Mathematical models of dynamic systems are used for prediction, control design, fault detection, etc. These models can be further divided by the type of modelling. One example is the white-box models, which are obtained from theoretical modelling based on physical, chemical or other basic principles. The alternative is black-box models, which are obtained from experimental modelling or identification. In this case, the structure of the mathematical model and the corresponding parameters are identified from the system's input and output data. When the structure is known, partly known, or presumed, and the parameters are optimised, the combination of both, i.e. the theoretical and the experimental, modelling methods results in grey-box models.

Model structures are often, especially in engineering practice, presumed to be linear. The identification of linear systems is a well-established engineering tool. However, the presumption of linearity is usually not correct, although it is acceptable under certain operating conditions. In this case, when a more detailed model for a wider operating range is necessary, then nonlinear models are required. Artificial neural networks, fuzzy models, local model networks, wavelets, support vector machines and many other methods exist for modelling of nonlinear systems. The identification of nonlinear systems is incomparably more complex than the

J. Kocijan, *Modelling and Control of Dynamic Systems Using Gaussian Process Models*, Advances in Industrial Control,
DOI 10.1007/978-3-319-21021-6_1

identification of linear systems. Nonlinearity can express itself in many different ways, and the number of possible structures increases tremendously, as does the complexity of the model. The problems that arise with this increase are, for example, the non-transparency of the models and the increase in their computational complexity. The identification of nonlinear systems and the application of these models for design are still fields of active research. On the other hand, these are also fields of interest for engineering practice.

Various sub-disciplines of mathematics, like statistics, go hand in hand with computer science, for example, with machine learning, as well as with other scientific fields. This offers an abundance of methods for modelling of systems among these methods, as well as for modelling of dynamic systems. Nevertheless, some of these methods have never found their way into engineering practice, regardless of their attractive properties. The reasons for this cannot be easily generalised, but the difficulties associated with using these methods and the lack of interpretation when it comes to solving engineering problems are certainly significant factors.

The GP model belongs to the class of black-box models. GP modelling differs from most other black-box identification approaches in that it does not try to approximate the modelled system by fitting the parameters of the selected basis functions, but rather it searches for the relationship among the measured data. The model is composed of input–output data that characterises the behaviour of the modelled system and the covariance function that describes the relation of the output data with respect to the input data. The prediction of the GP model output value is given as a normal distribution, expressed in terms of the mean and the variance. The mean value represents the most likely output value, and the variance can be interpreted as a measure of its confidence. The obtained variance, which depends on the amount and the quality of the available identification data, is important information that distinguishes the GP models from other methods.

The GP model has been known for a long time in the field of geostatistics, where the method was named 'kriging' by Krige [1]. It was first used to solve a regression problem in the late 1970s by O'Hagan [2], and it gained popularity within the machine-learning community in the late 1990s. This was initially due to the research of Neal [3], who showed the relationship between GP models and artificial neural networks. It continued with the research of Rasmussen [4], who placed GP modelling within the Bayesian probability framework. The research of Gibbs [5], Williams [6], and many others followed. More about the history of GP modelling development can be found in [7].

GP models are used for regression, where the model output is continuous, but may be also extended for classification, where the model output is classified into a finite, discrete number of sets. This book deals with regression models in accordance with its aim of dynamic-systems identification. In this book, we are taking the position of a practical user of nonlinear, black-box models. The book complements other books where GP models have been described in detail from the theoretical point of view, for example, [7, 8].

The application of GP models for dynamic-systems identification was initiated within the European research project MAC [9]. The publications sharing the results of this project are among the first publications where GP models are used for dynamic-systems identification and the simulation of the obtained dynamic models.

The obtained research results indicated that GP models might be attractive in engineering practice for the following reasons:

- they are simple to use in system identification due to the relatively small number of design decisions, like the selection of input regressors and the covariance function;
- they work well with a relatively small number of identification data, which is sometimes the case in identification, e.g. the identification of the dynamics in a region away from equilibrium;
- it is possible to include various sorts of prior knowledge about the modelled process in the model, e.g. linear local models, static characteristics, etc.;
- they work well although the identification output data is noisy;
- the model prediction contains a measure of confidence, which can be exploited in many ways for system identification, control system design and model-based fault detection.

In the following section, a brief introduction to GP models is given. This introduction will be upgraded with more details, especially about the structure, training and validation necessary for dynamic systems modelling in Chap. 2.

1.1 Introduction to Gaussian-Process Regression

A GP model is a probabilistic, nonparametric model for the prediction of output-variable distributions. Its use and properties for modelling are thoroughly described in, e.g., [7, 8]. Here, only a brief description, necessary to introduce the concepts used in this book, is given. More details about GP models, especially those dealing with the application of GP models for modelling of dynamic systems, are given in Chap. 2.

This section is divided into some preliminaries and a basic introduction to GP regression.

1.1.1 Preliminaries

Some of the fundamental topics necessary for further understanding, which are briefly discussed here, are regression, kernel methods and the Bayesian approach to modelling.

Introduction to Regression Problem

Let us take a look at the following *regression* problem. Given some noisy values, for example, measurements of a dependent variable y at certain values of the variables described with vector $\mathbf{z}$, what is the best estimate of the dependent variable for a new value of the variables that y depends upon?

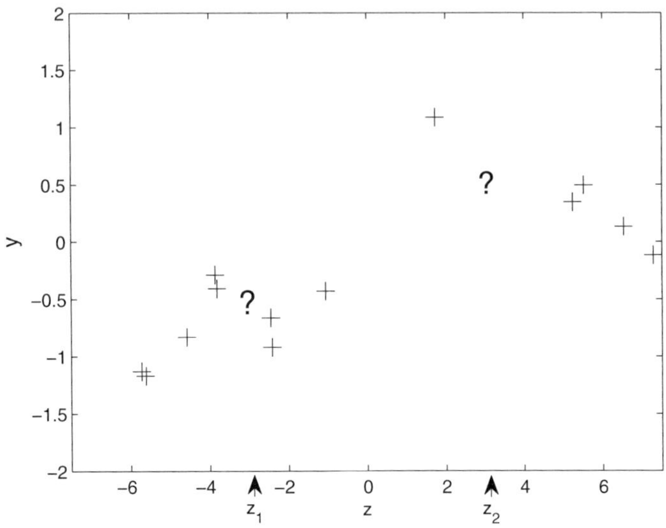

Fig. 1.1 Given 13 data points, what are the function values at z_1 and z_2?

We can formulate this problem as follows. Let us consider the following description:

$$y = f(\mathbf{z}) + \nu, \tag{1.1}$$

where $f(\mathbf{z})$ is an underlying function and ν is the noise. We are looking for an estimate of $f(\mathbf{z})$ that will enable us to find the prediction $\hat{y}$ at any new value $\mathbf{z}$.

The illustration of this problem for the case when z is a scalar for two values of the variable z is given in Fig. 1.1.

Frequently, this kind of problem is solved so that at first some mathematical structure is selected, based on prior knowledge about the system or some other assumptions. Then the parameters of this mathematical structure are optimised for the available measurements so that the model predictions fit the available measurements of the dependent variable. Elements of the vector $\mathbf{z} \in \mathbb{R}^D$, i.e. $z_i : i = 1, \ldots, D$ are called *regressors* and the vector $\mathbf{z}$ is called the *regression vector*. More about regression as one of the frequently used statistical tools can be found in general statistical reference books, e.g., [10, 11], or, for example, in reference books on pattern recognition, e.g., [12].

Kernel Methods

A frequent approach in modelling with regression is to approximate the nonlinear function between the regression vector $\mathbf{z}$, representing the system's input data, and y, representing the output data, also called an input–output mapping function, with a set of basis functions that can be described with a vector of basis functions $\phi(\mathbf{z}) = [\phi_1(\mathbf{z}), \ldots, \phi_i(\mathbf{z}), \ldots]^{\mathrm{T}}$. These functions have parameters that are optimised during the learning phase using training data, i.e. measurement data selected to be used for modelling. These data are discarded after the learning phase.

On the other hand, *kernel methods* [12] are a class of algorithms that are used for modelling relations, i.e. mapping functions, in data. The idea of kernel methods is to solve the modelling problem using some similarity function over pairs of data points. A direct relation between the data is sought.

Instead of introducing a set of basis functions, a localised *kernel function* is defined directly on the data and can be used to make predictions for new input vectors of data, given the observed training set of data. The kernel function or kernel is a general name for a function k of two arguments mapping a pair of inputs $\mathbf{z}_i \in \mathcal{Z}, \mathbf{z}_j \in \mathcal{Z}$ into $\mathbb{R}$ [7]. The training data is therefore kept and used during the prediction phase.

The kernel function $k(\mathbf{z}_i, \mathbf{z}_j)$ can be actually derived from the vector of basis functions $\phi(\mathbf{z})$, with the arguments that are the regression vectors $\mathbf{z}_i$ and $\mathbf{z}_j$. The regression vectors represent, for example, the dynamic system's inputs sampled at different time instants:

$$k(\mathbf{z}_i, \mathbf{z}_j) = \phi(\mathbf{z}_i)^{\mathrm{T}}\phi(\mathbf{z}_j). \tag{1.2}$$

The kernel function is a metric that measures the similarity of any two data vectors in the input space. The value of the kernel function describes the relation between two data vectors in the input space and not the value of the data vectors themselves. Since the kernel function can be used to make model predictions, it can be interpreted that the prediction depends on the training data directly.

The property of kernel methods is that the computation of an explicit, nonlinear mapping function between the input and output data is avoided and replaced with the training data. We obtain the identification of the mapping in the space where the number of parameters to be optimised is smaller. The parameters for the optimisation of the kernel function are called hyperparameters so as to be distinguished from those of the basis functions.

The kernel function can be any function of two regression vectors with the condition that it can be formulated as an inner product, also called a dot product, of basis-functions vectors in some, possibly also infinite, space of regressors as described by Eq. (1.2). This also means that the kernel function can be any function that generates a symmetric, semi-positive, square matrix named the Gram matrix $\mathbf{K}$ [12]. Therefore, if $k_{i,j} = [K_{i,j}]$, then $k_{i,j} = k_{j,i}$ for every $1 \leq i, j \leq N$ and $\mathbf{c}^{\mathrm{T}}\mathbf{K}\mathbf{c} \geq 0$ for every $\mathbf{c} \in \mathbb{R}^N$.

Let us introduce the case where the mapping between the input and output data is possible only with an infinite number of basis functions. Such modelling is realisable in practice, if the model can be obtained using a kernel method. In this case it is not necessary to know the basis functions themselves, it will do if the properly selected kernel function can be formulated as the inner product. An example of such a kernel function is a Gaussian kernel

$$k(\mathbf{z}_i, \mathbf{z}_j) = \exp(-\|\mathbf{z}_i - \mathbf{z}_j\|^2/2l^2) = \lim_{N\to\infty} \sum_{m=1}^{N} \phi_m(\mathbf{z}_i)\phi_m(\mathbf{z}_j), \tag{1.3}$$

where l^2 represents the lengthscale parameter. The kernel function described with Eq. (1.3) replaces infinite dimensionality that is required when formulated as an inner product.

There are numerous forms of kernel functions used by different kernel methods like support vector machines, e.g., [13], principal component analysis (PCA) method, e.g., [12] and others. More details about kernel methods can be found in, e.g., [12].

Bayesian Approach to Modelling

From a probabilistic perspective we model the predictive distribution $p(y|\mathbf{z})$ because this quantifies our uncertainty about the value y for each value $\mathbf{z}$. From this conditional distribution the predictions of the random variable y are made for any new $\mathbf{z}$. The Bayesian treatment as an established probabilistic method is a base for some practical techniques for addressing the question of model complexity, reduces the level of overfitting and provides the framework for treating uncertainties.

Gaussian-process modelling uses the ideas of the Bayesian approach to modelling, which is based upon the expression of knowledge in terms of a probability distribution. It seeks to maximise the probability of a model, given some data. More about the Bayesian approach to modelling, especially in the context of dynamic system identification, can be found in, e.g., [14].

Bayes' theorem (Appendix A) connects a prior belief expressed as a prior probability distribution with a posterior probability distribution. The *posterior*, i.e. posterior probability distribution, is inferred from combining the information present in the *prior*, i.e., the prior probability distribution, with that of the information gained from the data known as *likelihood*:

$$\text{Posterior} = \frac{\text{Likelihood} \times \text{Prior}}{\text{Evidence}}. \tag{1.4}$$

The *evidence*, also known as the *marginal likelihood*, is acting as a normalising constant to ensure that the probability sums to unity.

Let us illustrate Bayes' modelling on an example of function modelling from the input measurements $\mathbf{Z} = [\mathbf{z}_1, \mathbf{z}_2, \dots \mathbf{z}_N] \in \mathbb{R}^{D \times N}$ and a vector of the corresponding output measurements $\mathbf{y} = [y_1, y_2, \dots y_N]^{\mathrm{T}} \in \mathbb{R}^N$ forming the data $\mathcal{D} = \{(\mathbf{z}_i, y_i)|i = 1, \dots, N\} = \{(\mathbf{Z}, \mathbf{y})\}$. We are looking for a function f to describe the input–output relation between the data pairs $\mathbf{z}_i, y_i$. The Bayesian modelling of function $f = f(\mathbf{z}, \mathbf{w})$, where random variables $\mathbf{w}$ are the function parameters, can be pursued with Bayes' theorem.

The goal is to infer a probability distribution over the model parameters that is conditional on the data:

$$p(\mathbf{w}|\mathcal{D}) = \frac{p(\mathcal{D}|\mathbf{w})p(\mathbf{w})}{p(\mathcal{D})}, \tag{1.5}$$

where the probability density functions are as follows: $p(\mathcal{D}|\mathbf{w})$ is the likelihood, $p(\mathbf{w})$ is the parameters' prior, $p(\mathcal{D})$ is the evidence and $p(\mathbf{w}|\mathcal{D})$ is the parameters' posterior. The posterior distribution can be seen as the product of the likelihood and the prior, and captures everything we know about the parameters. Bayesian model inference in general does not contain any optimisation procedure.

The Gaussian process regression framework [7, 12] joins the idea of the kernel methods with the Bayesian modelling methods.

GP modelling not only utilises the principles of Bayesian modelling, as we will see in the continuation, but also its vocabulary. For example, the expression 'prior', which expresses prior belief, is often used not only for a prior probability distribution, but also generally for various prior beliefs and knowledge. In general 'prior', means before the observations have been made [15].

1.1.2 Gaussian-Process Regression

Before GP regression is explained, let us take a look at what GP is.

Gaussian Process

The *Gaussian process* is a generalisation of the Gaussian or normal probability distribution, where the sample function generated over time $\{f(\mathbf{z}_1), f(\mathbf{z}_2), \ldots, f(\mathbf{z}_N)\}$ has the property that any finite set of function values is jointly normally distributed. The random variables represent the value of the random function $f(\mathbf{z})$ at the location $\mathbf{z}$.

GPs can be viewed as a collection of random variables $f(\mathbf{z}_i)$ with a joint Gaussian distribution for any finite subset. GP is therefore a stochastic process containing random variables with a normal probability distribution. It is a Gaussian random function, fully described by its mean function and covariance function.

$$p(f(\mathbf{z}_1), \ldots, f(\mathbf{z}_N)|\mathbf{z}_1, \ldots, \mathbf{z}_N) = \mathcal{N}(m_f, \boldsymbol{\Sigma}_f), \tag{1.6}$$

where
$m_f = E(f(\mathbf{z}_1), f(\mathbf{z}_2), \ldots, f(\mathbf{z}_N))$ and $\boldsymbol{\Sigma}_f = \mathrm{cov}(f(\mathbf{z}_1), f(\mathbf{z}_2), \ldots, f(\mathbf{z}_N))$, and $p(f(\mathbf{z}_1), \ldots, f(\mathbf{z}_N)|\mathbf{z}_1, \ldots, \mathbf{z}_N)$ means the joint probability density function of $f(\mathbf{z}_1), \ldots, f(\mathbf{z}_N)$ given $\mathbf{z}_1, \ldots, \mathbf{z}_N$.

Note that no distinction is made in this book between the notation for random variable and the values that the random variable can take. This simplifies the notation, provided that the interpretation is clear from the context.

GP Model

Let us return to the regression problem from the previous section, this time to solve it with GP regression.

There are two perspectives from which GP modelling can be seen. The first perspective is the input–output function modelling perspective, called the function-space view in [7], and it can be explained as follows.

Rather than claiming that the input–output function $f(\mathbf{z})$ relates to some models of a specific mathematical structure, a GP is used to represent $f(\mathbf{z})$ indirectly, based directly on the data. In the case of GP regression, instead of parameterising the function $f(\mathbf{z})$, initial knowledge or belief about the function, called a prior, is placed directly on the space of all possible functions $f(\mathbf{z})$. All the candidate functions for $f(\mathbf{z})$ represented with GP can represent the nonlinear mapping from the input $\mathbf{z}$ to the output y. The GP prior represents the modeller's beliefs about the mapping, for example, smoothness assumptions.

This prior is combined with the likelihood of the identification, also training, set of N observed input–output data pairs, $\{(\mathbf{z}_i, y_i)|i = 1, \ldots, N\}$, to provide us with the posterior distribution for the model predictions by applying Bayes' theorem. We will explain the model-selection procedure in more detail in Chap. 2.

As the function f is represented, i.e. modelled, with a random function, which is a Gaussian process, the model is called the *Gaussian process model*. The GP model represents a mapping between the deterministic input data $\mathbf{z}$ and the corresponding random variable with normal distribution at the output $y = f(\mathbf{z})$.

A GP model is therefore completely described by its mean function m_f and covariance function C

$$\begin{aligned} m_{f_i}(\mathbf{z}_i) &= E(f(\mathbf{z}_i)) \\ C(\mathbf{z}_i, \mathbf{z}_j) &= E((f(\mathbf{z}_i) - m_{f_i})(f(\mathbf{z}_j) - m_{f_j})) = \text{cov}(f(\mathbf{z}_i), f(\mathbf{z}_j)). \end{aligned} \tag{1.7}$$

Any finite set of values from random function $f(\mathbf{Z}) = [f(\mathbf{z}_1), \ldots, f(\mathbf{z}_N)]^{\mathrm{T}}$ is jointly Gaussian distributed and can be written as

$$p(f(\mathbf{Z})) = \mathcal{N}(m_f, \boldsymbol{\Sigma}_f). \tag{1.8}$$

The function that is GP is denoted as

$$f(\mathbf{Z}) \sim \mathcal{GP}(m_f, \boldsymbol{\Sigma}_f). \tag{1.9}$$

If we do not have prior information about the mean function of the GP model, then we set $m_f \equiv \mathbf{0}$. In general, the mean values can be removed from the mean function, i.e., $m_f \equiv \mathbf{0}$, and added later if necessary, see Sect. 2.4.4.

A covariance matrix $\boldsymbol{\Sigma}_f$ can then be generated from evaluating the covariance function given all the pairs of measured data. The elements Σ_{ij} of the covariance matrix $\boldsymbol{\Sigma}_f$ are covariances between the values of the functions $f(\mathbf{z}_i)$ and $f(\mathbf{z}_j)$ corresponding to the arguments $\mathbf{z}_i$ and $\mathbf{z}_j$:

$$\Sigma_{ij} = \text{cov}(f(\mathbf{z}_i), f(\mathbf{z}_j)) = C(\mathbf{z}_i, \mathbf{z}_j). \tag{1.10}$$

This means that the covariance between the random variables that represent the outputs, i.e. the functions of the arguments numbers i and j, equal the function called the *covariance function* between the arguments numbers i and j.

From the second perspective, GP modelling is viewed as a kernel method using Bayesian inference. A covariance function is in fact a kernel function. The relation in Eq. (1.10) derives from a property of the kernel methods [7]. Note that the covariance function is denoted with $C(\cdot)$ in this book so as to distinguish it from the kernel functions used with other methods, even though it plays exactly the same role. The other reason is to distinguish the covariance function from the time instant of a sampled system, which is denoted by k.

Another property of GP regression that is common to the kernel methods is that more emphasis is given to the data itself, rather than to the selection of the model structure, which is common to methods using basis functions for modelling.

Any function $C(\mathbf{z}_i, \mathbf{z}_j)$ can be a covariance function, providing that it generates a positive, semi-definite, covariance matrix $\boldsymbol{\Sigma}_f$. The covariance function $C(\mathbf{z}_i, \mathbf{z}_j)$ can be interpreted as a measure of the correlation between the function values $f(\mathbf{z}_i)$ and $f(\mathbf{z}_j)$. For systems modelling, it is usually composed of two main parts:

$$C(\mathbf{z}_i, \mathbf{z}_j) = C_f(\mathbf{z}_i, \mathbf{z}_j) + C_n(\mathbf{z}_i, \mathbf{z}_j), \tag{1.11}$$

where C_f represents the functional part and describes the unknown system we are modelling and C_n represents the noise part and describes the model of the measurement noise which can be derived based on the Gaussian likelihood [16].

For the noise part in Eq. (1.11), it is most common to use the covariance function that has a constant value that is different from zero in the case of the same arguments, which is presuming white Gaussian noise. The choice of the covariance function for the functional part in Eq. (1.11) also depends on the stationarity of the stochastic process. A *stationary process* is a stochastic process whose joint probability distribution does not change when shifted in time or space. Assuming the stationarity of the data set, the most commonly used covariance function is the squared exponential covariance function. Some other possible choices for the covariance functions [7] are the Matérn class of covariance functions, exponential, rational quadratic, etc. In the case of assuming a non-stationary data set, the polynomial covariance function can be used. These covariance functions will be discussed in Sect. 2.3.3 in the next chapter.

For example, the composite covariance function composed of the squared exponential covariance function for the functional part and the constant covariance for the noise is therefore

$$C(\mathbf{z}_i, \mathbf{z}_j) = \sigma_f^2 \exp\left[-\frac{1}{2}\sum_{d=1}^{D} w_d (z_{id} - z_{jd})^2\right] + \sigma_n^2 \delta_{ij}, \tag{1.12}$$

where w_d, σ_f and σ_n are the *hyperparameters* of the covariance function, D is the input dimension, and $\delta_{ij} = 1$ if $i = j$ and 0 otherwise. δ_{ij} is called the Kronecker delta function. The name hyperparameters originates from the kernel methods, because a covariance function is a kernel function.

Briefly About Modelling

We continue to consider the system

$$y = f(\mathbf{z}) + \nu \tag{1.13}$$

with the white Gaussian noise $\nu \sim \mathcal{N}(0, \sigma_n^2)$, with the variance σ_n^2 and the vector of regressors $\mathbf{z}$ from the operating space $\mathbb{R}^D$. The estimate of the function f is uncertain due to finiteness of measurements y and the presence of noise. The Bayesian framework enables us to express this uncertainty with probability distributions, requiring the concept of distributions over functions.

Within this framework, we have $[y_1, \ldots, y_N]^{\mathrm{T}} \sim \mathcal{N}(0, \mathbf{K})$ with

$$\mathbf{K} = \mathbf{\Sigma}_f + \sigma_n^2 \mathbf{I}, \tag{1.14}$$

where $\mathbf{\Sigma}_f$ is the covariance matrix for the noise-free f of the system described with Eq. (1.13) and $\mathbf{I}$ is the $N \times N$ identity matrix. The elements of $\mathbf{\Sigma}_f$ are defined as in Eq. (1.10).

After the data for modelling is collected $\mathcal{D} = \{(\mathbf{z}_i, y_i) | i = 1, \ldots, N\} = \{(\mathbf{Z}, \mathbf{y})\}$, and prior belief in the form of the mean function and the covariance function is selected, then the posterior GP model is inferred. It is not the aim of this section to explain the GP model identification or training in detail, rather to point out some of its specifics.

Following the Bayesian modelling framework we are looking for the posterior distribution over f, which for the given data $\mathcal{D}$ and hyperparameters $\boldsymbol{\theta}$ is

$$p(f|\mathbf{Z}, \mathbf{y}, \boldsymbol{\theta}) = \frac{p(\mathbf{y}|f, \mathbf{Z}, \boldsymbol{\theta}) p(f|\boldsymbol{\theta})}{p(\mathbf{y}|\mathbf{Z}, \boldsymbol{\theta})}, \tag{1.15}$$

where $p(\mathbf{y}|f, \mathbf{Z}, \boldsymbol{\theta})$ is the likelihood, $p(f|\boldsymbol{\theta})$ is the function f prior for the given hyperparameters $\boldsymbol{\theta}$, $p(\mathbf{y}|\mathbf{Z}, \boldsymbol{\theta})$ is the evidence and $p(f|\mathbf{Z}, \mathbf{y}, \boldsymbol{\theta})$ is the posterior distribution over f.

The implementation of Bayesian inference requires evaluation of several integrals which may be analytically intractable. One solution to the problem of intractable expressions is to adopt some sort of analytical or numerical approximation.

One of the more efficient methods for approximation is estimating the hyperparameters with the maximisation of the evidence. More details about the evidence or marginal-likelihood maximisation will be explained in Sect. 2.4.1.

The prior distribution over the function f prior is set as a GP model and also the posterior is a GP model:

$$f \sim \mathcal{GP}(m_f, \mathbf{K}). \tag{1.16}$$

The posterior GP model is described with

$$\mu(\mathbf{z}^*) = E_f(f(\mathbf{z}^*)|\mathbf{Z}, \mathbf{y}, \boldsymbol{\theta}) \tag{1.17}$$

$$\sigma^2(\mathbf{z}^*) = \mathrm{var}_f(f(\mathbf{z}^*)|\mathbf{Z}, \mathbf{y}, \boldsymbol{\theta}), \tag{1.18}$$

where $\mathbf{z}^*$ is an arbitrary vector, which is called the validation or test input data.

Prediction

After the model is identified, the regression task is to predict a new output estimate $\hat{y}$, denoted here as y^*, of the GP model at the input data $\mathbf{z}^*$ using the relation $y^* = f(\mathbf{z}^*)$. As we presume that the hyperparameter estimates $\hat{\boldsymbol{\theta}}$ after modelling are known, we will no longer condition on them for notational convenience.

The overall problem of prediction in the Bayesian framework corresponds to inferring the posterior $p(f(\mathbf{z}^*))$ and making the prediction probability distribution $p(y^*|\mathcal{D}, \mathbf{z}^*)$ of the new output estimate y^*, given the training data $\mathcal{D}$ and a new input data $\mathbf{z}^*$. Due to the nature of our prior, the predictive distribution of y^* can be obtained from the application of conditional probability instead of applying the Bayes' theorem. The conditional probability can be interpreted as [17]

$$p(y^*|\mathcal{D}, \mathbf{K}, \mathbf{z}^*) = \frac{p([\mathbf{y}^{\mathrm{T}}, y^*]^{\mathrm{T}}|\mathbf{K}, \mathbf{Z}, \mathbf{z}^*)}{p(\mathbf{y}|\mathbf{K}, \mathbf{Z})}. \tag{1.19}$$

Using Eq. (A.10) from Appendix A the prediction probability distribution function is

$$p([\mathbf{y}^{\mathrm{T}}, y^*]^{\mathrm{T}}|\mathcal{D}, \mathbf{K}, \mathbf{z}^*) = \frac{1}{(2\pi)^{\frac{N+1}{2}}|\mathbf{K}_{N+1}|^{\frac{1}{2}}} e^{-\frac{1}{2}([\mathbf{y}^{\mathrm{T}}, y^*]\mathbf{K}_{N+1}^{-1}[\mathbf{y}^{\mathrm{T}}, y^*]^{\mathrm{T}})}. \tag{1.20}$$

At this point, we show how the covariance matrix $\mathbf{K}_{N+1}$ is updated through the introduction of a new input data. For the collection of random variables $\{y_1, \ldots, y_N, y^*\}$ we can write:

$$\mathbf{y}, y^* \sim \mathcal{N}(\mathbf{0}, \mathbf{K}_{N+1}) \tag{1.21}$$

with the covariance matrix

$$\mathbf{K}_{N+1} = \begin{bmatrix} \mathbf{K} & \mathbf{k}(\mathbf{z}^*) \\ \mathbf{k}^{\mathrm{T}}(\mathbf{z}^*) & \kappa(\mathbf{z}^*) \end{bmatrix}, \tag{1.22}$$

where $\mathbf{y} = [y_1, \ldots, y_N]^{\mathrm{T}}$ is an $N \times 1$ vector of training targets, $\mathbf{k}(\mathbf{z}^*) = [C(\mathbf{z}_1, \mathbf{z}^*), \ldots, C(\mathbf{z}_N, \mathbf{z}^*)]^{\mathrm{T}}$ is the $N \times 1$ vector of covariances between the training input data and the test input data, and $\kappa(\mathbf{z}^*) = C(\mathbf{z}^*, \mathbf{z}^*)$ is the autocovariance of the test input data.

We can write

$$p(y^*|\mathcal{D}, \mathbf{K}, \mathbf{z}^*) = \frac{|\mathbf{K}|^{\frac{1}{2}}}{(2\pi)^{\frac{1}{2}}|\mathbf{K}_{N+1}|^{\frac{1}{2}}} e^{-\frac{1}{2}([\mathbf{y}^{\mathrm{T}}, y^*]\mathbf{K}_{N+1}^{-1}[\mathbf{y}^{\mathrm{T}}, y^*]^{\mathrm{T}} - \mathbf{y}^{\mathrm{T}}\mathbf{K}^{-1}\mathbf{y})}. \tag{1.23}$$

Finally, the prediction probability distribution from Eq. (1.23) is Gaussian and can be written as a normal distribution with the following mean and variance [5, 7]:

$$E(y^*) = \mu(\mathbf{z}^*) = \mathbf{k}^{\mathrm{T}}(\mathbf{z}^*)\mathbf{K}^{-1}\mathbf{y} \tag{1.24}$$

$$\mathrm{var}(y^*) = \sigma^2(\mathbf{z}^*) = \kappa(\mathbf{z}^*) - \mathbf{k}^{\mathrm{T}}(\mathbf{z}^*)\mathbf{K}^{-1}\mathbf{k}(\mathbf{z}^*). \tag{1.25}$$

The vector $\mathbf{k}^{\mathrm{T}}(\mathbf{z}^*)\ \mathbf{K}^{-1}$ in Eq. (1.24) can be interpreted as a vector of smoothing terms that weight the training output data $\mathbf{y}$ to make a prediction at the test point $\mathbf{z}^*$. For the stationary covariance function like Eq. (1.12), the following holds: if the new input data is far away from the data points, the term $\mathbf{k}^{\mathrm{T}}(\mathbf{z}^*)\ \mathbf{K}^{-1}\ \mathbf{k}(\mathbf{z}^*)$ in Eq. (1.25) will be small, so that the predicted variance $\sigma^2(\mathbf{z}^*)$ will be large. The regions of the input space where there is few data or there is corruption with noise are, in this way, indicated by a higher variance.

The solution to the problem with a scalar input value z depicted in Fig. 1.1 is illustrated in Fig. 1.2. The smoothing of the noisy points is illustrated in Fig. 1.3, where the predicted mean values smooth the noisy measurements denoted by crosses. For a clearer presentation a GP model posterior in three dimensions is depicted in Fig. 1.4.

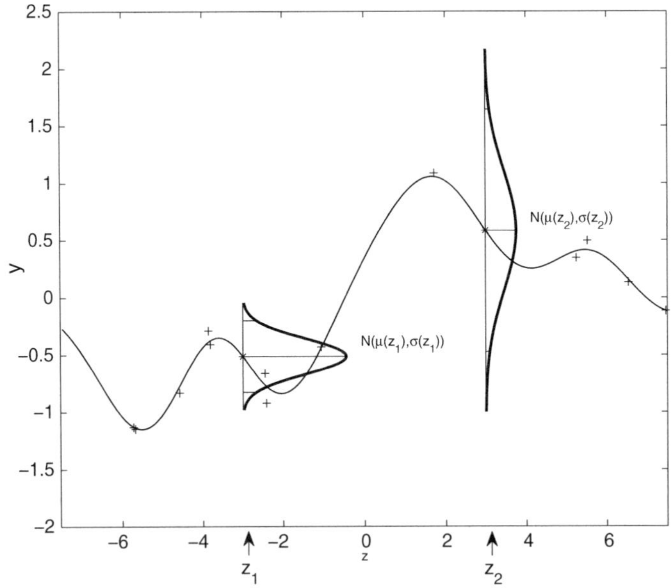

Fig. 1.2 Gaussian prediction at the new points z_1 and z_2, conditioned on the training points denoted by *crosses*. The training points are taken from the function denoted by the full line with the addition of some noise

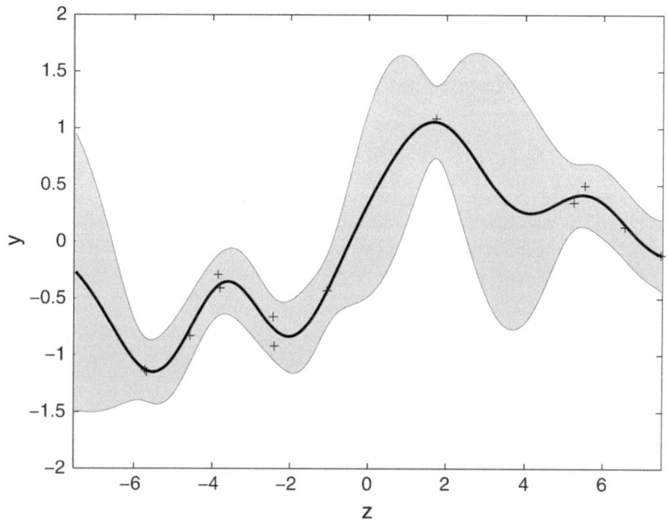

Fig. 1.3 Using GP models for regression: in addition to the mean value (prediction), we obtain a 95 % confidence interval for the underlying function $f(z)$ (shown in *grey*)

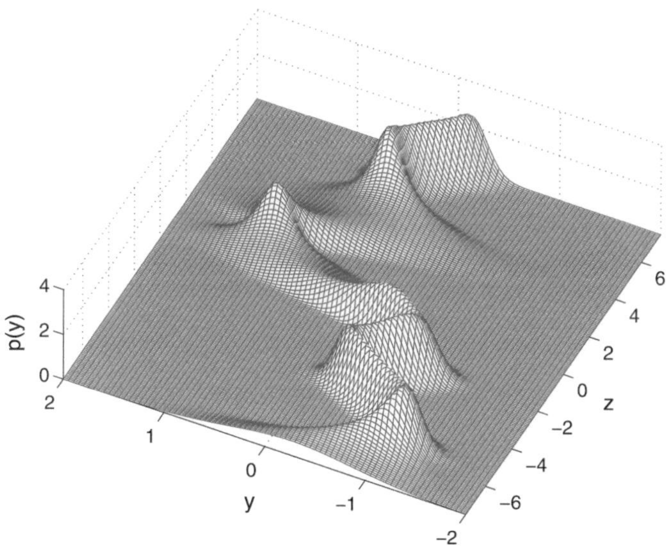

Fig. 1.4 Presentation of the GP model posterior where the normal predictive distribution is determined for every input data in the interval $z \in [-7.5, 7.5]$

The described GP regression is just a walk through the procedure of GP modelling, which will be discussed and explained in the following chapter. However, before this an illustrative example will demonstrate the use of the explained procedure.

Example 1.1 Nonlinear function modelling

This example illustrates the application of modelling with a GP model on a regression problem with a one-dimensional input data. The function to be modelled is $f(z)$:

$$y = f(z) + \nu = z^3 + \nu, \tag{1.26}$$

where z is the independent variable in the interval $z \in [-1.5, 2]$ and ν is the white noise with the normal distribution, mean value $\mu_\nu = 0$ and variance $\sigma_\nu^2 = 0.01$, which is added to the function values. Nine non-uniformly distributed training pairs containing input–output data points are sampled from the nonlinear function described with Eq. (1.26) for the corresponding independent variable. The function without noise $f(z)$ and the training points are depicted in Fig. 1.5.

The used covariance function in Eq. (1.12) is composed of the squared exponential covariance function with a single input variable for the functional part and a constant function, representing the white noise, for the noise part:

$$C(z_i, z_j) = \sigma_f^2 \exp\left[-\frac{1}{2}w(z_i - z_j)^2\right] + \sigma_n^2\delta_{ij}. \tag{1.27}$$

The prior mean function is set to 0. The details of the hyperparameters' selection with the evidence maximisation are avoided here for the sake of simplicity. The obtained hyperparameter values after optimisation are: $\sigma_f^2 = 1.57$, $w = 13.4$ and $\sigma_n^2 = 0.01$.

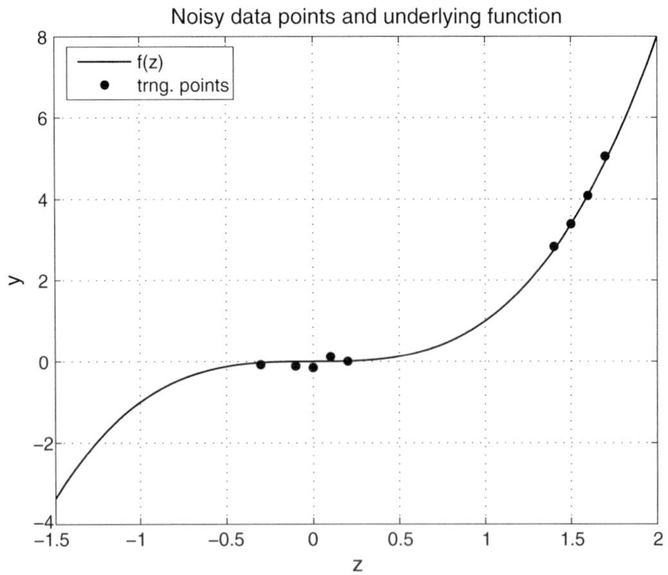

Fig. 1.5 Training data points and underlying function to be modelled

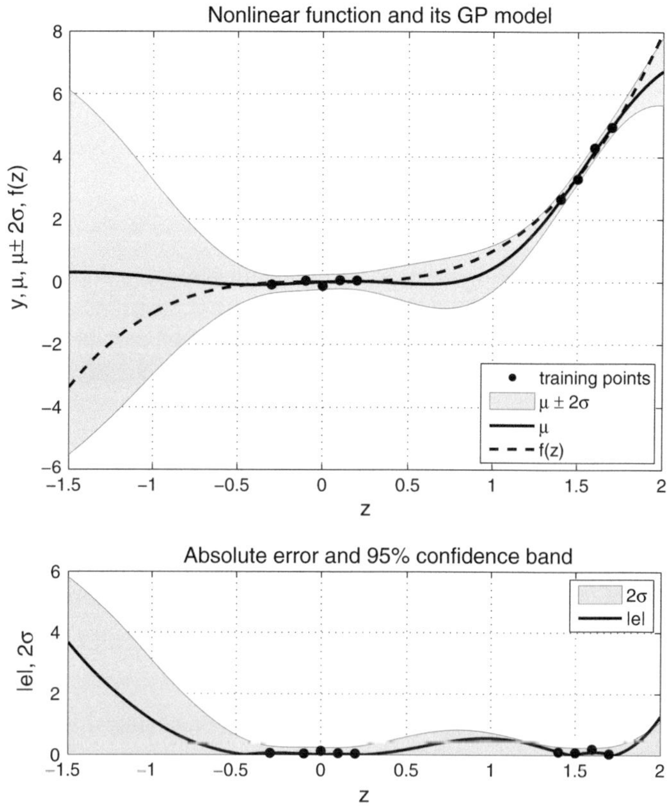

Fig. 1.6 Comparison of the original function and GP model identified on nine data points. While the dots in the *upper figure* represent training target points, the dots in the *lower figure* represent the absolute differences between the training target points and the function to be modelled, which in fact are absolute values of noise for the target points

The results of modelling are given in Fig. 1.6, from which it is clear that the model approximates well to the function $f(z)$ in the intervals $-0.3 < z < 0.2$ and $1.4 < z < 1.7$, which are well populated with training points. The model also predicts relatively well in the interval $0.2 < z < 1.4$, which lies between the previously mentioned intervals. However, the variance of the model prediction in this interval is higher and, consequently, the confidence of the prediction is lower. In other intervals that are not well populated with data from the function $f(z)$ with additive noise, the obtained GP model has to extrapolate from the training data and the prior mean function in order to make the prediction. Consequently, the error between the original function and the mean value of the GP model predictions is increasing with the distance from the training points, as is the prediction variance in these intervals due to the selection of the stationary covariance function for modelling. This is fairly obvious in the interval $z < -0.3$. The predicted values smooth the noisy training data when the prediction is drawn for the same value of the input variable as was used for the training.

1.2 Relevance

System identification, hereafter referred to as identification, is composed of methods to build mathematical models of dynamic systems from measured data. It is one of the scientific pillars used for dynamic-systems analysis and control design.

A temporal component is introduced in comparison to the regression problem introduced in the previous section. The identification of a dynamic system means that we are looking for a relationship between past observations and future output values. Identification can be interpreted as the concatenation of a mapping from measured data to a regression vector, followed by a linear or a nonlinear mapping from the regression vector to the output space [18].

For the black-box identification, where in principle no prior knowledge is used, a selected set of delayed samples of inputs and outputs forms the regression vector. Various machine-learning methods and statistical methods are employed to determine the nonlinear mapping from the regression vector to the output space. More details will be presented in Chap. 2.

One of the possible methods for a description of the nonlinear mapping used in identification is GP models. It is straightforward to employ GP models for discrete-time modelling of dynamic systems within the prediction-error framework [15].

What is the rationale for using GP models in system identification?

Many dynamic systems are often considered as complex; however, simplified input–output behaviour representations are sufficient for certain purposes, e.g. feedback control design, prediction models for supervisory control, etc.

GP models have a few properties, already listed at the beginning of this chapter, that make them relevant for modelling of dynamic systems from data.

First, the optimisation of a complex model structure can be avoided due to the nonparametric nature of GP models. The problem of structure determination, so well known in other popular methods like neural networks and fuzzy modelling, is relaxed if not completely avoided.

Second, a Bayesian treatment, more specifically, model selection using marginal likelihood, reduces the risk of overfitting. Such a potential for model overfitting is present with all methods where the number of parameters to be optimised is relatively high and the amount of identification or training data is not large enough with respect to the number of parameters. The potential of overfitting increases with the complexity of the model structure.

Next, limited amounts of data relative to the number of selected regressors, data corrupted with noise and measurement errors and the need for a measure of model-prediction confidence could be the reasons to select identification with the GP model. If there is not enough data or it is heavily corrupted with noise, even the GP model cannot perform well. But in that case the inadequacy of the model and the identification data is indicated through a higher variance of the predictions if the GP model uses a stationary covariance function.

The prediction variance is one of the main differences between the GP model and other black-box models. It can be used effectively in the usefulness validation, where a lack of confidence in the model prediction can serve as the grounds to reject the model as not being useful. The prediction variance can also be used in a falseness validation, whether via specific performance measures or through the observation of confidence limits around the mean values of the model's predictions. Chapter 2 will provide more details on model validation.

It is important to be aware of the fact that a model is always a model and GP models represent yet another step in the approximation of reality.

Another fact to be aware of is that GP models are not to be considered as an alternative to other black-box system-identification methods, but rather as a complementary method, where the context of the problem at hand requires or suits its utility.

Numerous papers, most of which have been published since 2000, describe the use of GP models for modelling of dynamic systems. These publications have explored the use of GP models for various applications:

- dynamic systems modelling, e.g., [19–22]
- time-series prediction, e.g., [23–26]
- dynamic systems control, e.g., [27–33]
- fault detection, e.g., [34–36],
- state estimation, e.g., [25, 37],
- smoothing, e.g., [37, 38].

The ability to provide information about the confidence of the model's prediction made GP models attractive for modelling case studies in various domains like: chemical engineering [39] and process control [40], biomedical engineering [41], biological systems [42], environmental systems [24], power systems [43] and engineering [44], motion recognition [22], etc., to list just a selection. It is worth noting that the utility of GP modelling could also be interesting for use in other domains and applications.

1.3 Outline of the Book

This book consists of six chapters.

The first chapter introduces GP models and provides a simple, illustrative example of modelling of a static mapping function. Next, a brief historical overview of developments in the field of GP models for dynamic systems identification is presented. The chapter continues with a discussion about the rationale and the relevance of using GP modelling for system identification and control design.

The next chapter talks about system identification with GP models. After outlining the complete procedure for system identification, the chapter is focused on issues that are specific to modelling based on GPs. The issues emphasised here are the setting-up of the model, model selection and validation of the identified model. The system identification is illustrated on the bioreactor benchmark model.

The GP modelling framework enables incorporation of prior knowledge of various kinds. Chapter 3 shows the application of GP models in block-oriented nonlinear models, how the local linear dynamic models can be incorporated into GP models and how GP models can be used in the context of the paradigm of linear models with varying parameters.

Chapter 4 describes a range of approaches to GP model-based control system design. The described methods are only those that have been published in the literature. The selection contains control based on an inverse dynamics model, optimal and model-predictive control and various adaptive control design methods.

The trends, challenges and research opportunities related to GP model-based control-systems design are indicated in Chap. 5.

Three case studies of practical applications are given in Chap. 6. The case studies are as follows:

- A gas–liquid separator is used to demonstrate dynamic system identification and model-predictive control design.
- An urban-traffic case study shows time-series modelling and prediction for signal modelling and reconstruction.
- Forecasting the ozone concentration in the air demonstrates the application of online modelling.

References

1. Krige, D.G.: A statistical approach to some basic mine valuation problems on the Witwatersrand. J. Chem., Metal. Min. Soc. S. Afr. **52**(6), 119–139 (1951)
2. O'Hagan, A.: On curve fitting and optimal design for regression (with discussion). J. R. Stat. Soc. Ser. B (Methodol.) **40**(1), 1–42 (1978)
3. Neal, R.M.: Bayesian Learning for Neural Networks. Lecture Notes in Statistics. Springer, New York, NY (1996)
4. Rasmussen, C.E.: Evaluation of Gaussian processes and other methods for nonlinear regression. Ph.D. thesis, University of Toronto, Toronto (1996)
5. Gibbs, M.N.: Bayesian Gaussian processes for regression and classification. Ph.D. thesis, Cambridge University, Cambridge (1997)
6. Williams, C.K.I.: Prediction with Gaussian processes: from linear regression and beyond. In: Jordan, M. (ed.) Learning in graphical models, Nato Science Series D, vol. 89, pp. 599–621. Springer, Berlin (1998)
7. Rasmussen, C.E., Williams, C.K.I.: Gaussian Processes for Machine Learning. MIT Press, Cambridge, MA (2006)
8. Shi, J.Q., Choi, T.: Gaussian process regression analysis for functional data. Chapman and Hall/CRC, Taylor & Francis group, Boca Raton, FL (2011)

9. MAC Multi-Agent Control: Probabilistic reasoning, optimal coordination, stability analysis and controller design for intelligent hybrid systems (2000–2004). Research Training Network, 5th EU framework
10. Johnson, R.A., Miller, I., Freund, J.E.: Probability and Statistics for Engineers, 8th edn. Pearson, Boston, MA (2011)
11. Rohatgi, V.K.: An Introduction to Probability Theory and Mathematical Statistics. Wiley, New York, NY (1976)
12. Bishop, C.M.: Pattern recognition and machine learning. Springer Science + Business Media (2006)
13. Suykens, J.A.K., Gestel, T.V., Brabanteer, J.D., Moor, B.D., Vandewalle, J.: Least Squares Support Vector Machines. World Scientific, Singapore (2002)
14. Peterka, V.: Bayesian approach to system identification (chapter). Trends and Progress in System Identification, pp. 239–304. Pergamon Press, Oxford (1981)
15. Ljung, L.: System identification—Theory for the User, 2nd edn. Prentice Hall, Upper Saddle River, NJ (1999)
16. Deisenroth, M.P.: Efficient reinforcement learning using Gaussian processes. Ph.D. thesis, Karlsruhe Institute of Technology, Karlsruhe (2010)
17. Thompson, K.: Implementation of Gaussian process models for non-linear system identification. Ph.D. thesis, University of Glasgow, Glasgow (2009)
18. Gevers, M.: A personal view of the development of system identification, a 30-year journey through an exciting field. IEEE Control Syst. Mag. **26**(6), 93–105 (2006)
19. Girard, A., Rasmussen, C., Murray-Smith, R.: Gaussian process priors with uncertain inputs: Multiple-step ahead prediction. Tech. Rep. DCS TR-2002-119, University of Glasgow, Glasgow (2002)
20. Gregorčič, G., Lightbody, G.: Gaussian processes for modelling of dynamic non-linear systems. In: Proceedings of the Irish Signals and Systems Conference, pp. 141–147. Cork (2002)
21. Kocijan, J., Girard, A., Banko, B., Murray-Smith, R.: Dynamic systems identification with Gaussian processes. In: Troch, I., Breitenecker, F. (eds.) Proceedings of 4th IMACS Symposium on Mathematical Modelling (MathMod), pp. 776–784. Vienna (2003)
22. Wang, J., Fleet, D., Hertzmann, A.: Gaussian process dynamical models. Adv. Neural Inf. Process. Syst. **18**, 1441–1448 (2005)
23. Babovic, V., Keijzer, M.: A Gaussian process model applied to prediction of the water levels in Venice lagoon. In: Proceedings Of The XXIX Congress Of International Association For Hydraulic Research, pp. 509–513 (2001)
24. Grašič, B., Mlakar, P., Božnar, M.: Ozone prediction based on neural networks and Gaussian processes. Nuovo cimento Soc. ital. fis., C Geophys. space phys. **29**(6), 651–661 (2006)
25. Ko, J., Fox, D.: GP-Bayesfilters: Bayesian filtering using Gaussian process prediction and observation models. Auton. Robot. **27**(1), 75–90 (2009)
26. Deisenroth, M.P., Huber, M.F., Hannebeck, U.D.: Analytic moment-based Gaussian process filtering. In: Proceedings of the 26th Annual International Conference on Machine Learning, pp. 225–232. Montreal (2009)
27. Gregorčič, G., Lightbody, G.: Gaussian processes for internal model control. In: Rakar, A. (ed.) Proceedings of 3rd International PhD Workshop on Advances in Supervision and Control Systems, A Young Generation Viewpoint, pp. 39–46. Strunjan (2002)
28. Kocijan, J.: Gaussian process model based predictive control. Tech. Rep. DP-8710, Jozef Stefan Institute, Ljubljana (2002)
29. Murray-Smith, R., Sbarbaro, D.: Nonlinear adaptive control using nonparametric Gaussian process prior models. In: Proceedings of IFAC 15th World Congress. Barcelona (2002)
30. Kocijan, J., Murray-Smith, R.: Nonlinear predictive control with a Gaussian process model. In: Murray-Smith, R., Shorten, R. (eds.) Switching and Learning in Feedback Systems, pp. 185–200. Springer, Berlin (2005). Lecture Notes in Computer Science
31. Ko, J., Klein, D.J., Fox, D., Haehnel, D.: Gaussian processes and reinforcement learning for identification and control of an autonomous blimp. In: Proceedings of the International Conference on Robotics and Automation, pp. 742–747. Rome (2007)

32. Deisenroth, M.P., Rasmussen, C.E.: PILCO: A model-based and data-efficient approach to policy search. In: Proceedings of the 28th International Conference on Machine Learning (ICML 2011). Bellevue, WA (2011)
33. Deisenroth, M.P., Rasmussen, C.E., Fox, D.: Learning to control a low-cost manipulator using data-efficient reinforcement learning. In: Proceedings of the International Conference on Robotics: Science & Systems (R:SS 2011). Los Angeles, CA (2011)
34. Juričić, D., Kocijan, J.: Fault detection based on Gaussian process model. In: Troch, I., Breitenecker, F. (eds.) Proceedings of the 5th Vienna Symposium on Mathematical Modeling (MathMod). Vienna (2006)
35. Serradilla, J., Shi, J.Q., Morris, J.A.: Fault detection based on Gaussian process latent variable models. Chem. Intel. Lab. Syst. **109**(1), 9–21 (2011)
36. Osborne, M.A., Garnett, R., Swersky, K., de Freitas, N.: Prediction and fault detection of environmental signals with uncharacterised faults. In: 26th AAAI Conference on Artificial Intelligence (AAAI-12). Toronto (2012)
37. Deisenroth, M.P., Turner, R.D., Huber, M.F., Hanebeck, U.D., Rasmussen, C.E.: Robust filtering and smoothing with Gaussian processes. IEEE Trans. Autom. Control **57**(7), 1865–1871 (2012). doi:10.1109/TAC.2011.2179426
38. Deisenroth, M.P., Mohamed, S.: Expectation propagation in Gaussian process dynamical systems. Adv. Neural Info. Process. Sys. **25**, 2618–2626 (2012)
39. Kocijan, J., Likar, B.: Gas-liquid separator modelling and simulation with Gaussian-process models. Simul. Model. Pract. Theory **16**(8), 910–922 (2008)
40. Likar, B., Kocijan, J.: Predictive control of a gas-liquid separation plant based on a Gaussian process model. Comput. Chem. Eng. **31**(3), 142–152 (2007)
41. Faul, S., Gregorčič, G., Boylan, G., Marnane, W., Lightbody, G., Connolly, S.: Gaussian process modeling of EEG for the detection of neonatal seizures. IEEE Trans. Biomed. Eng. **54**(12), 2151–2162 (2007)
42. Ažman, K., Kocijan, J.: Application of Gaussian processes for black-box modelling of biosystems. ISA Trans. **46**, 443–457 (2007)
43. Leith, D., Heidl, M., Ringwood, J.: Gaussian process prior models for electrical load forecasting. In: International Conference on Probabilistic Methods Applied to Power Systems, pp. 112–117 (2004)
44. Leithead, W., Zhang, Y., Neo, K.: Wind turbine rotor acceleration: identification using Gaussian regression. In: Proceedings of International conference on informatics in control automation and robotics (ICINCO). Barcelona (2005)

Chapter 2
System Identification with GP Models

In this chapter, the framework for system identification with GP models is explained. After the description of the identification problem, the explanation follows the system identification framework that consists of roughly six stages:

1. defining the purpose of the model,
2. selection of the set of models,
3. design of the experiment,
4. realisation of the experiment and the data processing,
5. training of the model and
6. validation of the model.

The model identification is always an iterative process. Returning to some previous procedure step is possible at any step in the identification process, and this is usually necessary.

The listed stages are given in the sections describing the model's purpose (Sect. 2.1), the experiment design and data processing (Sect. 2.2), the model setup (Sect. 2.3), the model selection (Sect. 2.4), the model validation (Sect. 2.6), the dynamic model simulation (Sect. 2.7) and ends with an illustrative example of nonlinear system identification with a GP model (Sect. 2.8).

The *identification problem* [1, 2] is as follows: For a given set of past observations, i.e. delayed samples of input and output signals that form a regression vector, we would like to find a relationship with future output values. As already mentioned in the previous chapter, this relation can be presented as the concatenation of a mapping from the measured data to a regression vector, followed by a nonlinear mapping from the regression vector to the output space

$$y(k) = f(\mathbf{z}(k), \boldsymbol{\theta}) + \nu(k), \tag{2.1}$$

J. Kocijan, *Modelling and Control of Dynamic Systems Using Gaussian Process Models*, Advances in Industrial Control,
DOI 10.1007/978-3-319-21021-6_2

where

$$\mathbf{z}(k) = \varphi(y(k-1), y(k-2), \ldots, u(k-1), u(k-2), \ldots), \tag{2.2}$$

k is the sampling instant,
f is a nonlinear mapping from the regression vector $\mathbf{z}$ to the output space,
$\boldsymbol{\theta}$ is the finite-dimensional parameter vector,
$\nu(k)$ represents the noise and accounts for the fact that the next output value $y(k)$ will not be an exact function of past data,
φ is a nonlinear mapping from the finite-dimensional vector of the measurements to the regression vector $\mathbf{z}$, and its components are referred to as regressors,
$y(k-i), i = 1, 2, \ldots, k-1$ are the delayed samples of the measured output signal and
$u(k-i), i = 1, 2, \ldots, k-1$ are the delayed samples of the measured input signal.

The temporal or time component is inevitably present when dealing with dynamic systems. Instead of considering time as an extra input variable to the model, the time is embedded into regressors in the form of delayed samples of input and output signals. In our notation the time, usually denoted with t, has been substituted for k, where k represents the kth subsequent time-equidistant instant sampled with the sample period T_s.

The identification problem has thus to be decomposed into two tasks: (a) the selection of the regression vector $\mathbf{z}(k)$ and (b) the selection of the mapping f from the space of the regressors to the output space.

When the mapping f is presumed linear, we talk about the identification of linear dynamic systems. The more general case is when the mapping is nonlinear. While there are numerous methods for the identification of linear dynamic systems from measured data, the nonlinear systems identification requires more sophisticated approaches.

In general, the identification methods for nonlinear systems can be grouped into those for parametric and those for nonparametric system identification. While *parametric system identification* deals with the estimation of parameters for a known structure of the mathematical model, nonparametric system identification identifies the model of an unknown system without structural information.

Nonparametric system identification can be divided further [3]. The first group of methods is that where the system is approximated by a linear or nonlinear combination of some basis functions ϕ_i with l coefficients $\mathbf{w} = [w_1, w_2, \ldots, w_l]^\mathrm{T}$ to be optimised

$$f(\mathbf{z}, \mathbf{w}) = F(\phi_i(\mathbf{z}), \mathbf{w}), \tag{2.3}$$

where F is a function representing the nonlinear combination of basis functions. The most commonly seen choices in identification practice include artificial neural networks (ANNs), fuzzy models and Volterra-series models, which can be seen as universal approximators.

Let us briefly discuss the use of methods where the nonlinear system is approximated by the combination of basis functions. The problem of nonparametric system identification is translated into the problem of a suitable basis function selection and of the coefficients' estimation, which can be considered further as a parameter estimation problem. The approach is sensitive to the choice of basis function. Depending on the nonlinearity, a fixed basis function approach could need a relatively large number of terms to approximate the unknown nonlinear system. The increase in the number of terms with the increase in the unknown system's complexity is called the '*curse of dimensionality*' [4]—the exponential growth of the modelled volume with the input space dimension [5]—leading to (a) a model with a large number of basis functions with corresponding parameters and (b) a lot of data needed for a system description. An example of such models is an ANN. The local model network (LMN) [4], a form of the fuzzy model, which we address in Sect. 3.3, reduces this problem, but has problems with a description of the off-equilibrium regions of the dynamic system [4, 6].

The second possibility is to estimate the unknown nonlinear system locally, point by point. Representatives of these methods are kernel methods like least-square support vector machines [7]. These methods circumvent the curse of dimensionality in the sense that they do not contain basis functions. These methods are considered local because any of them is actually a weighted average based on measurements in the neighbourhood of the point where the system is estimated [3].

Nevertheless, both possibilities, the one with basis functions and the one that is point-by-point, are based on the amount of measured data in the neighbourhood of the point where the system is identified. If the dimension of the problem is high, the amount of data necessary for training increases.

The curse of dimensionality is not an issue for linear and parametric nonlinear system identification. In such cases, it is not important whether the measured data used for the identification is distributed locally or far away from the point where the unknown system is identified.

For the nonparametric identification methods that estimate the unknown system locally, only local data is useful. For kernel methods, this depends on the kernel selection. Data points that are far away provide little value for these methods [3]. This means that local approaches would not perform well when modelling large regions with only a limited amount of data available.

As an alternative to methods for the identification of nonlinear dynamic systems that are strongly affected by the curse of dimensionality, the GP model was proposed in [6]. In this context, the unknown system to be identified at a given point and the data obtained at other points are assumed to be a joint GP with a mean and a covariance matrix that has some hyperparameters.

The idea of using GP models for system identification differs from both mentioned possibilities of nonlinear system identification that can be described as the local average approaches [3] because it is a probabilistic method. GP models provide a posteriori distribution. Using this distribution, a probabilistic estimate at a point of interest can be made based on the training data that can be close or far away from this point. This prediction is presumed to be Gaussian, characterised by a predictive mean and a predictive variance. As we have seen in the previous chapter, the predictive variance can be interpreted as a level of confidence in the predictive mean. This is important, especially in the case when the predictive mean is not sufficiently close to the ground truth.

In other words, due to its probabilistic nature the GP model provides information about its estimate over the entire space defined by the regressors. The GP model is, therefore, not constrained to the space where the measured data is available.

Furthermore, the GP approach to modelling alleviates any model bias by not focusing on a single dynamics model, but by using a *probabilistic dynamics model*, a distribution over all plausible dynamics models that could have generated the observed experience. The probabilistic model is used to faithfully express and *represents the uncertainty* about the learned dynamics. We use a probabilistic model for the deterministic system. The probabilistic model does not necessarily imply that we assume a stochastic system. In the case of modelling, the deterministic system the probabilistic model is solely used to describe the uncertainty about the model itself. In the extreme case of a test input data $\mathbf{z}^*$ at the exact location $\mathbf{z}_i$ of a training input data, the prediction of the probabilistic model will be absolutely certain about the corresponding function value $p(f(\mathbf{z}^*)) = \delta(f(\mathbf{z}_i))$.

When using GP models for identification, it is important that we are also aware of the disadvantages [3] in this context.

The first one is the computational complexity due to the inverse of a high-dimensional covariance matrix during the training. The computational complexity measured with the number of computer operations rises with the third power of the number of identification points N and is denoted with $\mathcal{O}(N^3)$, while the number of computer operations for the prediction mean is $\mathcal{O}(N)$ and for the prediction variance $\mathcal{O}(N^2)$. The issue of overcoming the computational complexity is addressed in Sect. 2.5.

The second is that the noise that corrupts the measurement used for the system identification does not always have a Gaussian distribution. The hyperparameters of the GP model are frequently optimised to maximise the marginal likelihood conditioned on the measured data where the assumption is that the marginal likelihood is Gaussian. In the case of non-Gaussian noise, this assumption is not correct and it is not known whether the maximum likelihood is achieved. However, this is exactly the same issue with any other known parametric or nonparametric method for system identification. In the case of non-Gaussian noise, the data can be transformed in the form that will be better modelled by the GPs. More details about transformations, called also GP warping (Sect. 2.3.3), can be found in [8, 9].

The third disadvantage is that the performance of GP models for system identification depends on the selection of training data and the covariance function with hyperparameters, which is system dependent. Again, this disadvantage can, in one form or another, be met with any method for system identification where the selection of the basis function or the system structure and the corresponding parameter estimation is system dependent.

2.1 The Model Purpose

The model purpose specifies the intended use of the model and has a major impact on the level of detail of the model. The decision for the use of a specific model derives, besides the model purpose, also from the limitations met during the identification process.

In general, dynamic system models can be used for [10]

- prediction,
- simulation,
- optimisation,
- analysis,
- control and
- fault detection.

Prediction means that on the basis of previous samples of the process input signal $u(k-i)$ and the process output signal $y(k-i)$ the model predicts one or several steps into the future. There are two possibilities: the model is built to directly predict h steps into the future or the same model is used to predict a further step ahead by replacing the data at instant k with the data at instant $k+1$ and using the prediction $\hat{y}(k)$ from the previous prediction step instead of the measured $y(k)$. This is then repeated indefinitely. The latter possibility is equivalent to *simulation*. Simulation therefore means that only on the basis of previous samples of the process input signal $u(k-i)$, and initial conditions for a certain number of samples of output signals, the model simulates future output values. Name prediction will, in our case, mostly mean a one-step-ahead prediction.

Both of these sorts of models can be used for the optimisation of systems, systems analysis, control and fault detection [10]. When a model is used for *optimisation*, the issues of optimisation duration and the disturbance of the process's normal operation, which frequently occur when an optimisation is carried out in the real world, are circumvented.

The procedure of identification can also be considered as a form of system *analysis*. Based on the model's form some process properties, like input–output behaviour, stationary properties, etc., can be inferred from the identified model.

Control design relies on a model of the process to be controlled. The form of the model depends on the control design method, but it has to be kept in mind that it is the closed-loop system performance containing the designed controller that is evaluated on the performance and not the process model itself. Nevertheless, there are control algorithms for which the closed-loop performance is strongly dependent on the model quality. Usually, the models used for control design have to be accurate in terms of some particular properties, e.g. range around the crossover frequency of the closed-loop system.

The idea of *fault detection* is to compare the process behaviour for the time being with its nominal behaviour, which is commonly represented with a process model.

Fault detection is a model-based strategy and the focus of the fault detection also determines the model properties.

Later in the text, some of these modelling applications will be highlighted in the context of GP applications.

2.2 Obtaining Data—Design of the Experiment, the Experiment Itself and Data Processing

The data describing an unknown system is very important in any black-box identification. The data can be obtained from an experiment and is processed afterwards. It is usually collected with measurements on the physical system of interest or, in special cases, from computer simulations. An example of such a special case is when identification is used for complexity reduction of a theoretically obtained model.

Unless the system to be identified does not allow controlled experimentation, the experiment needs to be carefully designed. In the case when experimentation is not possible, the safety of the system or its environment might be jeopardised by the experimentation and, consequently, the data needs to be collected from the daily operation.

The design of the experiment and the experiment itself are important parts of the identification procedure. The quality of the model depends on the system information contained in the measurement data, regardless of the identification method. The design of the experiment for the nonlinear system identification is described in more details in [10–13]. Only the main issues are highlighted in this section.

As already mentioned, the Gaussian process modelling approach relies on the relations among the input–output data and not on an approximation with basis functions. Consequently, this means that the distribution of the identification data within the process operating region is crucial for the quality of the model. Model predictions can only be highly confident if the input data to the model lies in the regions where the training data is available. The GP model provides good predictions when used for interpolation, but these are not necessarily good enough when used for extrapolation, which is indicated by the large variances of the model predictions.

Consequently, the data for the model training should be chosen reasonably, which can be obstructed by the nature of the process, e.g. limitations in the experimental design in industrial processes and the physical limitations of the system.

The experiment has to be pursued in such a way that the experiments provide sets of data that describe how the system behaves over its entire range of operation.

Here we list a number of issues that need to be addressed when the experiment is designed for the acquisition of data and subsequent data processing. The list is not exhaustive. Instead, it is given as a reminder of the important issues that engineers need to address, but the reader is, again, referred to exploit the details in [10–12].

Let us highlight the most important points.

Nonlinearity test. A nonlinearity test should be pursued to see whether a linear or nonlinear system is to be modelled. This is important to know, not only for GP models identification, but also for other methods as well. This can be done by a test of the superposition and homogeneity [12] or by checking the frequency response. The test of the superposition can be done with step changes of the input signal in different operating regions and checking the responses, while the frequency response test can be done by checking the output frequency response for sub-harmonic components when a harmonic signal excites the input.

Sampling frequency. For a good description of the process, the influential variables and a suitable sample time must be chosen. A rule of thumb is that the sampling frequency should be high enough to capture all the interesting dynamics of the system to be modelled. However, in the case that it is too high, problems with numerical ill-conditioning occur in the process of identification. The sampling frequency is therefore a compromise that is usually achieved iteratively.

Selection of excitation signals. The main issues that need to be addressed with the selection of the excitation signals are as follows.

Purpose. The selection of the excitation signals [10, 13] needs to be done based on the purpose of the modelling. For example, if the purpose of the model identification is control design, then good data information content is needed around the regions determined with the envisaged closed-loop set-point signals.

Size of the training set. The maximum amount of training data should be carefully selected due to the trade-off between the complexity of the computation and the information content.

Range of input signals. The input signals should be selected to excite the nonlinear dynamic system across the entire operating region. It is important to realise that it is the range of the input space that is important, i.e. the space determined with the input regressors, and not just the range of the input and output signals and their rates.

Data distribution. The input signals should be selected to populate the region with data homogenously and with sufficient density. A uniform data distribution over the region of interest is the ultimate goal.

Prior knowledge. The 'design of experiment' requires some prior knowledge about the process. An iterative procedure is necessary, in which the design of experiment and model identification are interlaced.

Examples of signals. Signals that can be used for the system identification are the amplitude-modulated PRBS (Pseudo-random Binary Sequence) signals. There are different ways to obtain these kinds of signals [10, 12], but it is important to homogenously cover the input space with samples of these signals.

If the system is unstable in any way or poorly damped, it might be necessary to conduct the experiment in a closed loop [12].

Data preprocessing. Some of the procedures that are important for the quality of the model before the data is used for the modelling are as follows.

Filtering. Aliasing, the effect that results when the signal reconstructed from the samples is different from the original continuous signal, which can be avoided with analogue pre-filtering and digital filtering, can be used to remove some low-frequency disturbances [12].

Removing data redundancy. A large amount of data in certain regions, most frequently the regions around the equilibria of a nonlinear system, dominates the data in other regions and may lead to the poor performance of the model in these regions. A possible solution to this problem is addressed in Sect. 2.5.2.

Removing outliers. Data outliers are not very problematic with GP models due to the smoothing nature of GP models [14].

Scaling. To cancel the influence of different measuring scales, the preprocessing of the measured data can be pursued, e.g. centering and scaling, here referred to as *normalisation*. Normalisation of the input and output signals helps with the convergence of the parameter optimisation that is part of the identification procedure.

2.3 Model Setup

In our case the selection of the GP model is presumed. This approach can be beneficial when the information about the system exists in the form of input–output data, when the data is corrupted by noise and measurement errors, when some information about the confidence in what we take as the model prediction is required and when there is a relatively small amount of data with respect to the selected number of regressors.

After the type of model is selected, the model has to be set up. In the case of the GP model, this means selecting the model regressors, the mean function and the covariance function.

2.3.1 Model Structure

This and the following subsection deal with the choice of suitable regressors. In the first subsection different model structures are discussed, while in the second, methods for the selection of regressors are discussed. The selection of the covariance function is described in the third subsection.

It is mainly nonlinear models that are discussed in the context of GP modelling in general. There are various model structures for nonlinear, black-box systems that can also be used with GP models. An overview of the structures for nonlinear models is given in [1, 2]. The true noise properties that cause uncertainties in the identified model are usually unknown in the black-box identification and therefore different model structures should be seen as reasonable candidate models and not as model structures reflecting the true noise descriptions.

The subsequently used nomenclature for nonlinear models is taken from [1]. The single-input single-output case is described in the continuation, but models can be easily extended to the multiple-input multiple-output case. The nonlinear, black-box models are divided into a number of different groups depending on the choice of regressors. Based on this choice we can divide the models into input–output models and state-space models. The prefix N for nonlinear is added to the names for different model structures. Input–output models that can be utilised for GP models are as follows:

NFIR (nonlinear finite impulse response) models, which use only the input values $u(k-i)$ as the regressors and are usually considered as deterministic input values in GP models. Since the regressors are only input values, the NFIR model is always stable, which is particularly important in the nonlinear case where the stability issue is a complex one. NFIR models are well suited for applications like [15] control, dynamic system identification, noise cancellation, nonstationary time-series modelling, adaptive equalisation of a communication channel and other signal processing applications. An example of the GP-NFIR model structure can be found in [16]. A block diagram of the GP-NFIR model is shown in Fig. 2.1.

NARX (nonlinear autoregressive model with exogenous input) models, which use the input values $u(k-i)$ and the measured output values $y(k-i)$ as the regressors and are usually considered as deterministic input values in GP models. The NARX model, also known as the equation-error or series-parallel model, is a prediction model.

$$\begin{aligned}\hat{y}(k) = & f(y(k-1), y(k-2), \ldots, y(k-n), u(k-1), \\ & u(k-2), \ldots, u(k-m)) + \nu,\end{aligned} \tag{2.4}$$

where n is the maximum lag in the output values, m is the maximum lag in the input values and ν is the white Gaussian noise. The NAR model is the special case of NARX model without the exogenous input and uses only the measured output values $y(k-i)$. The GP-NARX model was introduced in [17] and it is schematically shown in Fig. 2.2.

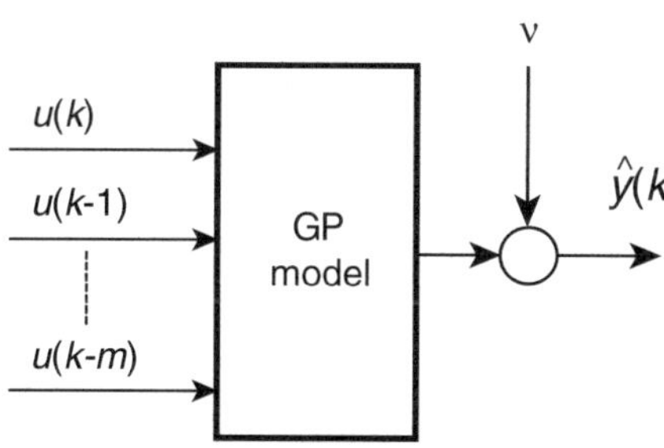

Fig. 2.1 GP-NFIR model, where the output predictions are functions of previous measurements of input signals

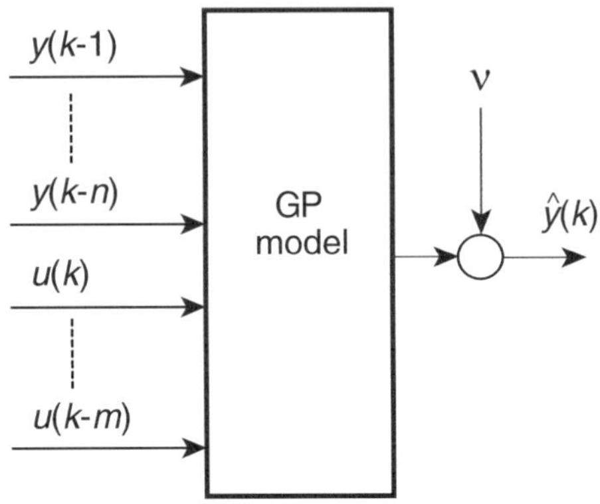

Fig. 2.2 GP series-parallel or equation-error or NARX model, where the output predictions are functions of previous measurements of the input and output signals

NOE (nonlinear output error) models, which use the input values $u(k-i)$ and the output estimates $\hat{y}(k-i)$ as the regressors. The NOE model, also known as the parallel model, is a simulation model.

$$\hat{y}(k) = f(\hat{y}(k-1), \hat{y}(k-2), \ldots, \hat{y}(k-n), u(k-1), \\ u(k-2), \ldots, u(k-m)) + \nu. \tag{2.5}$$

In the case of GP-NOE models $\hat{y}(k)$ is a normal distribution. When the normal distribution and its delayed versions are used as the regressors, the output of a nonlinear model is not a normal distribution anymore, and therefore output predictions are only approximations. The GP-NOE model is discussed in [18] and it is schematically shown in Fig. 2.3.

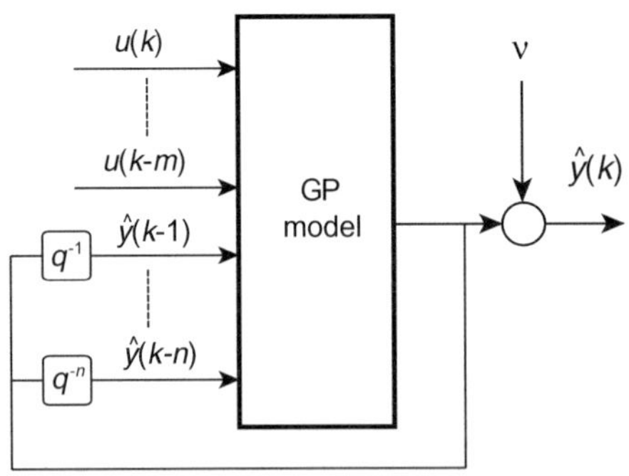

Fig. 2.3 GP parallel or output-error or NOE model, where the output predictions are functions of previous measurements of the input signal only and delayed predictive distributions $\hat{y}$, or their approximations, are fed back to the input. q^{-1} denotes the backshift operator. The time shift operator q influences the instant in the following way: $q^{-1}y(k) = y(k-1)$

In general nonlinear system identification, there are other input–output model structures that have not yet been explored within the context of GP models like **NARMAX** (nonlinear autoregressive and moving average model with exogenous input) and **NBJ** (nonlinear Box–Jenkins) [1].

State-space models: State-space models are frequently used in dynamic systems modelling. Their main feature is the vector of internal variables called the states, which are regressors for these kinds of models. State-space regressors are less restricted in their internal structure. This implies that in general it might be possible to obtain a more efficient model with a smaller number of regressors using a state-space model [1].

The following model, described by a state-space equation, is considered:

$$\mathbf{x}(k+1) = f(\mathbf{x}(k), u(k)) + \boldsymbol{\nu}_1(k) \tag{2.6}$$

$$y(k) = g(\mathbf{x}(k), u(k)) + \nu_2(k) \tag{2.7}$$

where $\mathbf{x} \in \mathbb{R}^n$ is a vector of states, $y \in \mathbb{R}$ is a measurement output, $u \in \mathbb{R}$ is an input, $\boldsymbol{\nu}_1 \in \mathbb{R}^n$ and $\nu_2 \in \mathbb{R}$ are some white Gaussian noise sequences. The noise enters the system at two places. $\boldsymbol{\nu}_1$ is called the process noise and ν_2 is called the measurement noise, f is the transition or system function and g is called the measurement function. In our case both functions can be modelled with GP models, so $f \sim \mathcal{GP}_f$ and $g \sim \mathcal{GP}_g$. Figure 2.4 shows a state-space model based on GP models.

The system identification task for the GP state-space model is concerned with f in particular and can be described as finding the state-transition probability conditioned on the observed input and output values [19]

$$p(\mathbf{x}(k+1)|\mathbf{x}(k), \mathbf{u}(k), \mathbf{y}(k)); \mathbf{u}(k) = [u(k), \ldots, u(1)]^{\mathrm{T}}, \mathbf{y}(k) = [y(k), \ldots, y(1)]^{\mathrm{T}}. \tag{2.8}$$

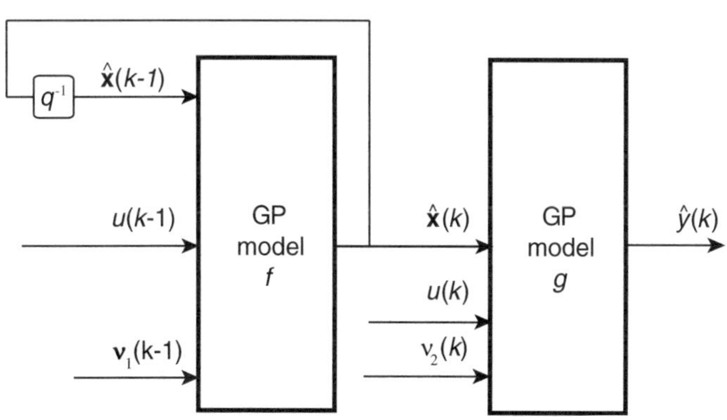

Fig. 2.4 State-space model. q^{-1} denotes the backshift operator

The system function g can often be assumed to be known. Such an example is the situation where g corresponds to a sensor model, where we know which of the states the sensor is measuring. This then simplifies the identification task. It is mentioned in [20] that using models that are too flexible for both f and g can result in problems of non-identifiability.

While using the state-space structure is common, the modelling with GP state-space models [21] is still a field of active research, and so the state-of-the-art is briefly reviewed here. At present, a state-space model of a dynamic system that involves GP models appears in the literature mainly in the context of an unknown state estimation from noisy measurements, i.e. filtering and smoothing. More results, also in the context of modelling for prediction, are to be expected in the future.

In the context of filtering and smoothing, the states $\hat{\mathbf{x}}(k)$ are estimated from the measurements $\mathbf{y}(k)$. More precisely, the posterior distribution of the states $p(\mathbf{x}(k)|\mathbf{y}(k))$ is to be found from measurements of $\mathbf{y}(k)$. If the used data history is up to and including $y(k)$, i.e. $\mathbf{y}(k) = [y(k), \ldots, y(1)]^{\mathrm{T}}$, this is called *filtering*. If we process the entire set of measurement data, then *smoothing* is taking place. Such an inference is usually made with Bayesian filters, like a Kalman filter [22] in the case of linear functions f and g or its nonlinear versions when at least one of the functions f and g is nonlinear.

GP models are used in the context of Bayesian filters, i.e. unscented and extended Kalman filters in [23] and in the context of an assumed density filter in [24]. A GP model with a state-space structure and without external input u appears in [24, 25], where GP models of the functions f and g are obtained from presumed available observations of the states as the identification or training data. These two references describe how the inference of states can be made for dynamic systems after the functions f and g are identified.

We have to keep in mind that variables with a normal distribution at the input do not keep a normal distribution at the output of a nonlinear function. In the case of nonlinear functions, we deal with approximations of posterior distributions. Nevertheless, good results for the inference of posterior distributions $p(\mathbf{x}(k)|\mathbf{y}(k))$ are reported in [24, 25].

Other methods for the inference of the states $\mathbf{x}(k)$ using GP models have been reported, like particle filters, e.g. [26–29], or variational inference, e.g. [30].

Authors of [21] learn a state-space model without input u with GPs by finding a posterior estimate of the latent variables and hyperparameters. The model in [21] is used for motion recognition.

The other situation is when the observations of states are not available for training. Filtering and smoothing in this case is described in [31, 32]. To learn the GP models for the functions f and g in such a case, they are parametrised by *pseudo training sets*, which are similar to the pseudo training sets used in so-called sparse GP approximations [33], which are described in Sect. 2.5.2. The system identification determines the appropriate hyperparameters for both GP models, such that the target time series $\mathbf{y}(k)$ can be explained. The expectation–maximisation algorithm used to determine the parameters is iterated between two steps. In the first a posterior distribution $p(\mathbf{Z_x}|\mathbf{y}(k), \boldsymbol{\theta})$ on the hidden states $\mathbf{Z_x} = [\mathbf{x}(k), \ldots, \mathbf{x}(1)]$ for a fixed parameter setting $\boldsymbol{\theta}$ is determined, and in the second the parameters $\boldsymbol{\theta}^*$ of the GP state-space model that maximise the expected log likelihood $E(\ln p(\mathbf{Z_x}|\mathbf{y}(k), \boldsymbol{\theta}))$ are searched, where the expectation is taken with respect to $p(\mathbf{Z_x}|\mathbf{y}(k), \boldsymbol{\theta})$ from the first step. The log likelihood is decomposed into [32]

$$
\begin{aligned}
E(\ln p(\mathbf{Z_x}|\mathbf{y}(k), \boldsymbol{\theta})) &= E(\ln p(\mathbf{x}(1)|\boldsymbol{\theta}) \\
&\quad + \sum_{k=2}^{N} \underbrace{\ln p(\mathbf{x}(k)|\mathbf{x}(k-1), \boldsymbol{\theta})}_{\text{Transition}} + \sum_{k=1}^{N} \underbrace{\ln p(y(k)|\mathbf{x}(k), \boldsymbol{\theta}))}_{\text{Measurement}}
\end{aligned}
\tag{2.9}
$$

In contrast to the listed methods that tend to model the GP using a finite set of identification data points and identifying functions f and g, the method reported in [20] marginalises the transition function f using sampling with the particle Markov chain Monte Carlo (MCMC) method. Consequently, the identification of f is avoided for the states' estimation.

State-space models utilising GP models can also be found for continuous-time models. References [34–37] describe methods of the states' inference in the continuous-time domain. They deal mainly with linear transition functions to retain the normal distributions on states or combine them in hybrid models with first-principle models. The identification of a discrete-time linear GP state-space model is described in [38]. The signal processing aspect of GP modelling in the spatio-temporal domain is reviewed in [37].

The interested reader who wants more information about states estimation using GP models will find it in the references listed in this section.

2.3.2 Selection of Regressors

A very important step in the model setup is the selection of the model order. The problem of order determination in the case of input–output models is equivalent to the selection of the relevant input variables for the mapping function $f(\cdot)$ in Eq. (2.1). Therefore, the problem is actually the problem of selecting the regressors or input variables in statistics and system theory terminology, or features in the machine-learning terminology. As pointed out in [10], it is important to understand that in the case of the identification, the previous input values $u(k-i)$ and the previous output values $y(k-i)$ are considered as separate regressors. Nevertheless, subsequent input values, e.g. $y(k-1)$ and $y(k-2)$, are typically correlated. This correlation indicates redundancy, but these subsequent input values may both be relevant, so we cannot dismiss any of them. This complicates the order selection problem.

Most of the methods for system identification start with the assumption that the regressors or the input variables in general are known in advance. Nevertheless, it is crucial to obtain a good model that the appropriate regressors are selected. Here we review the selection of regressors from the viewpoint of selection-independent and relevant input variables to the mapping function.

From the identification point of view, the regressors should be independent of each other and must carry as much information as possible for the output value prediction. If the input data is not independent, the information about the output data is doubled, which results in larger model dimensions and a more difficult search for the optimal model. The same is true if the regressors do not carry information about the output data and so represent redundant regressors.

Our goal is to select only as many regressors as are really necessary. Every additional regressor increases the complexity of the model and makes the model optimisation more demanding. Nevertheless, optimal input selection is often an intractable task, and an efficient input selection algorithm is always an important element in many modelling applications.

A quick look at the literature reveals plenty of methods and algorithms for regressor selection. A thorough overview of these methods and algorithms would take up too much space, so only a general overview is presented and a survey of the literature is listed.

Various authors divide the methods differently. We adopt the division of the regressors' selection into three major groups [39–42]: wrappers or *wrapper methods*, *embedded methods* and *filter methods*.

Wrapper methods are the so-called brute-force methods for regressor selection. The basic idea behind these methods is that they form a kind of wrapper around the system model, which is considered as a black box. No prior knowledge or internal state of the model is considered. The search for the optimal vector of regressors is initiated from some basic set of regressors. After the model optimisation and cross-validation, the regressors are added to or taken from the model. Successful models, according to selected performance criteria, are kept, while poorly performing models are rejected.

The wrapper methods are very general and easy to implement. Nevertheless, in the case of a large number of regressor candidates the methods require lots of computational effort, so various search or optimisation methods are used. Some of these methods or groups of methods are [42] forward selection, backward elimination, nested subset, exhaustive global search, heuristic global search, single-variable ranking and other ranking methods. The wrapper methods are also known by the names validation-based regressor selection or exhaustive search for best regressors [43].

Embedded methods have the regressor selection built into the model optimisation procedure. For example, if a certain sort of model has a property that the values of model's parameters correspond to the importance of the used regressors, then properly selected regressors with lower importance can be eliminated. This property is called automatic relevance determination—ARD [44]. GP models possess this property for certain covariance functions, e.g. squared exponential covariance with hyperparameters describing the scales for each regressor. The ARD property assumes that the global minimum of the parameter optimisation cost function is achieved. Some other embedded methods are coupled with model optimisation, e.g. the direct optimisation method, or are weight-based, e.g. stepwise regression, recursive feature elimination.

Filter methods do not rely on the model structure we identify like the other two groups of methods. The measure of relevance for the regressors or combinations of regressors is extracted directly from the identification data. The relevant regressors are selected based on this measure. The relevance measure is usually computed based on the statistical properties of identification data, e.g. correlations for linear systems, conditional probabilities, or based on measures from information theory, e.g. information entropy, mutual information, or based on other properties, e.g. mapping function smoothness. These methods are attractive because they are computationally efficient in comparison with the wrapper and embedded methods. Computational efficiency comes from the fact that multiple optimisation runs are not necessary and from the relatively straightforward computation of the filter method measures.

In the case when the signals from which regressors are to be selected are known, the problem is downsized to the problem of lag or model-order selection. If we eliminate the dead time from the responses, it is often assumed that regressors that have a smaller lag are more relevant for the prediction. This can be a substantial aid for the regressor selection algorithm. The assumption about the smoothness of the regressors of the mapping function is also frequently met in filter methods. Filter methods are the method of He and Asada [12, 45], and the false nearest neighbours method [46].

When it comes to regressor selection, it is also important to touch on embedding theory [47], which describes the theory of mapping continuous-time dynamic systems to discrete-time dynamic models. In the case that prior knowledge about the order of the continuous-time dynamic model is available, or presumed, the non-minimal realisation of the discrete-time dynamic model might be required to capture the dynamic of the nonlinear system. The non-minimal realisation means the order of the discrete-time model is higher than the order known by prior. This is in accordance with Taken's embedding theorem [48], which determines the necessary order of the model obtained from sampled input–output data.

2.3.3 Covariance Functions

The choice of kernel function, which in the context of GP modelling is called the covariance function, is of fundamental importance for successful modelling with kernels. The covariance function reflects the correlations between different training data observations. The parameters of the covariance functions must be determined in order to obtain accurate model predictions. More information on the topic of covariance functions' selection and its use in GP models can be found in [49, 50]. Here we provide an overview of this topic.

The prior over mapping functions that is the Gaussian process is specified by its mean and covariance matrix and the covariance function is used to generate this covariance matrix. The covariance matrix reflects the relationship between the data and the prior knowledge or the assumptions of the system to be modelled. The elements of the covariance matrix are obtained with covariance functions, i.e. kernel functions. The hyperparameters of the covariance functions must be sought to match the behaviour of the model with the original system. However, before the hyperparameters are determined, a suitable covariance function must be selected. When selecting the covariance function, we need to be aware of the inherent assumptions relating to the regression problem that we discuss [50]. The first one is that the collected training data must represent the characteristics of the function to be modelled. The second one is that measurements of a real system contain noise, and therefore a noise model needs to be incorporated into the model. The third assumption is that two data points close together in the input space are likely to have a greater correlation than two points that are distant. We assume that similar input values are likely to result in similar target values and, consequently, the training points near to a test point should be informative about the desired prediction [50]. It is the covariance function that defines the nearness of the individual data points.

We are not so much interested in the covariance between the input and output values or the covariance between pairs of different input values. We are interested in the covariance of pairs of the output values, which is presumed through the relationship between pairs of the input values, as described with Eq. (1.10). Two data points that are close together in the input space are to be informative about each other's respective targets and expose a high covariance between these targets. Consequently, two data points that are distant have a low covariance between two corresponding targets.

The covariance function and its hyperparameters can also be selected to reflect the prior knowledge about the lengthscale property. This means that we can select the covariance function, where the similarity between nearby input data decays more quickly or slowly with their distance.

The covariance function must generate a positive, semi-definite, covariance matrix [49], which limits the choice of functions to be covariance functions. A positive-definite covariance function will ensure a positive-definite covariance matrix, which guarantees the existence of a valid Gaussian process.

A number of valid covariance functions have been defined in the literature, see [49] for an overview. Most commonly, they are divided into stationary and nonstationary covariance functions. The stationary covariance functions are those that are functions of the distance between the input data and therefore invariant to translations in the input space.

Stationary covariance functions are more commonly used for implementation and interpretability reasons. Nevertheless, there are some cases, e.g. the system changes its behaviour during operation, when a nonstationary covariance function might be a better choice.

Another property that is of importance only for stationary covariance functions is a smoothness property of the Gaussian process prior determined by the covariance function. The selection of the covariance function influences the mean-square continuity [49] and differentiability [49] of the functions describing the system to be modelled.

For systems modelling, the covariance function is usually composed of two main parts:

$$C(\mathbf{z}_i, \mathbf{z}_j) = C_f(\mathbf{z}_i, \mathbf{z}_j) + C_n(\mathbf{z}_i, \mathbf{z}_j), \tag{2.10}$$

where C_f represents the functional part and describes the unknown system we are modelling and C_n represents the noise part and describes the model of the noise.

It is often assumed that the noise ν has an independent, identically distributed Gaussian distribution with zero mean and variance σ_n^2; $\nu \sim \mathcal{N}(0, \sigma_n^2)$. This kind of noise is called white Gaussian noise. This means that there is no cross-covariance between the noise and the system input data and it affects only the diagonal elements of the covariance matrix

$$C_n(\mathbf{z}_i, \mathbf{z}_j) = \sigma_n^2 \delta_{ij} \tag{2.11}$$

and $\delta_{ij} = 1$ if $i = j$ and 0 otherwise, which essentially encodes that the measurement noise is independent.

The functions $C_f(\mathbf{z}_i, \mathbf{z}_j)$ and $C_n(\mathbf{z}_i, \mathbf{z}_j)$ can be selected separately, because the sum of two nonnegative definite functions gives a nonnegative definite function. An alternative for noise is that it is encoded in the likelihood [51], as we will see later in the text (Eq. (2.26)).

When the noise is not modelled as white noise, it can be modelled differently, e.g. as an input-dependent noise model like ARMA noise [52, 53].

Let us list the most common stationary and nonstationary covariance functions that can be used with GP modelling.

Stationary Covariance Functions

Constant covariance function

The simplest covariance function is the one that has the same value over the whole domain. This is the constant covariance function given by

$$C_f(\mathbf{z}_i, \mathbf{z}_j) = \sigma_f^2. \tag{2.12}$$

The only hyperparameter σ_f^2 represents the scaling factor of the possible variations of the function. The function is illustrated in Fig. 2.5. Due to the simplicity of the constant covariance function, it is normally used in combination with other covariance functions.

Squared exponential covariance function

This is one of the most commonly used covariance functions in Gaussian process modelling when the function to be modelled is assumed to exhibit smooth and continuous behaviour with a high correlation between the output data and the input data in close proximity.

A squared exponential covariance function is also called a Gaussian covariance function. It is defined by

$$C_f(\mathbf{z}_i, \mathbf{z}_j) = \sigma_f^2 \exp\left(-\frac{r^2}{2l^2}\right). \tag{2.13}$$

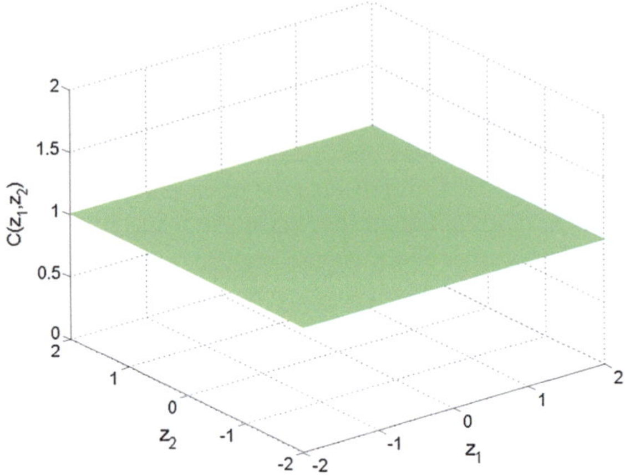

Fig. 2.5 Constant covariance function of two one-dimensional variables with the hyperparameter $\sigma_f = 1$

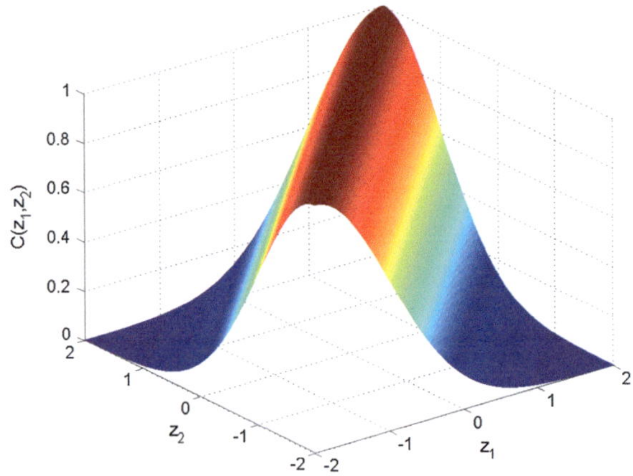

Fig. 2.6 Isotropic squared exponential covariance function of two one-dimensional variables with the hyperparameters $\sigma_f = 1$ and $l = 1$

The hyperparameter σ_f^2 represents the scaling factor of the possible variations of the function or the vertical scaling factor and the hyperparameter l is called the horizontal scaling factor and determines the relative weight on distance for the input variable $\mathbf{z}$. The variable r is the input distance measure and is $r = |\mathbf{z}_i - \mathbf{z}_j|$. The function is illustrated in Fig. 2.6.

Samples of functions with different hyperparameters are given in Fig. 2.7.

The covariance function, as represented with Eq. (2.13), decays monotonically with r. These kinds of functions are called isotropic covariance functions [49]. Anisotropic versions of covariance functions can be created by setting $r^2(\mathbf{z}_i, \mathbf{z}_j) = (\mathbf{z}_i - \mathbf{z}_j)^{\mathrm{T}}\boldsymbol{\Lambda}^{-1}(\mathbf{z}_i - \mathbf{z}_j)$ for some positive, semi-definite matrix $\boldsymbol{\Lambda}^{-1}$:

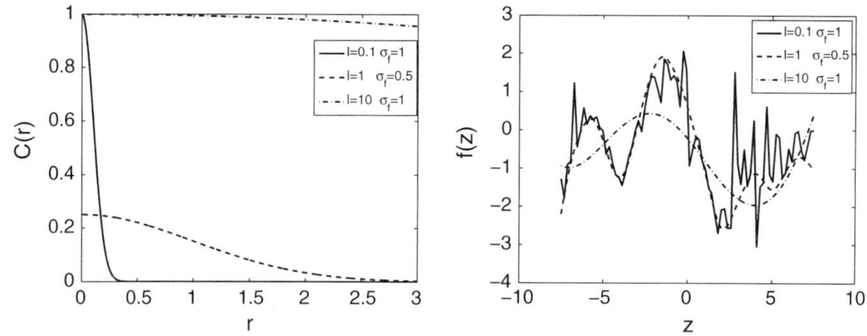

Fig. 2.7 Squared exponential covariance function with different hyperparameters (*left figure*) and random sample functions from the GP model with squared exponential covariance function (*right figure*)

$$C_f(\mathbf{z}_i, \mathbf{z}_j) = \sigma_f^2 \exp\left[-\frac{1}{2}(\mathbf{z}_i - \mathbf{z}_j)^{\mathrm{T}}\mathbf{\Lambda}^{-1}(\mathbf{z}_i - \mathbf{z}_j)\right]$$
$$= \sigma_f^2 \exp\left[-\frac{1}{2}\sum_{d=1}^{D} w_d(z_{di} - z_{dj})^2\right], \qquad (2.14)$$

where $w_d = \frac{1}{l_d^2}$; $d = 1, \ldots, D$.

If $\mathbf{\Lambda}^{-1}$ is diagonal, $\mathbf{\Lambda}^{-1} = \mathrm{diag}([l_1^{-2}, \ldots, l_D^{-2}])$ this implements the use of different length scales on different regressors and can be used to assess the relative importance of the contributions made by each regressor through comparison of their lengthscale hyperparameters. This is a property called *Automatic Relevance Determination* (ARD).

The ARD property was first introduced in [54, 55] in the context of a Bayesian neural network implementation. The ARD property can be used to optimise the structure, i.e. the regressor selection, of the GP model.

See [49] for a discussion on the use of the non-diagonal matrix $\mathbf{\Lambda}^{-1}$.

It must be kept in mind that in the case of dynamic systems identification the input dimension is high and this makes setting the assumptions on the mapping function to be modelled somehow difficult. A squared exponential covariance function is therefore frequently used because smooth and continuous input–output characteristics are expected, commonly from lots of dynamic systems, even though such assumptions are sometimes unrealistic, as argued in [56].

It is shown in [24, 51] that a GP model with a squared exponential covariance function corresponds to a universal function approximator.

Exponential covariance function

An exponential covariance function is used when the function to be modelled is assumed to be continuous, but not smooth and non-differentiable in the mean-square sense [49].

It is defined by

$$C_f(\mathbf{z}_i, \mathbf{z}_j) = \sigma_f^2 \exp\left(-(\frac{r}{l})^d\right) \quad \text{for } 0 < d \leq 2. \tag{2.15}$$

The hyperparameter σ_f^2 represents the scaling factor of the possible variations of the function or the vertical scaling factor, the hyperparameter l or the horizontal scaling factor determines the relative weight on distance for the input variable $\mathbf{z}$ and the hyperparameter d determines the exponent. The variable r is the input distance measure and is $r = |\mathbf{z}_i - \mathbf{z}_j|$. The function is illustrated in Fig. 2.8. Samples of functions with different hyperparameters are given in Fig. 2.9.

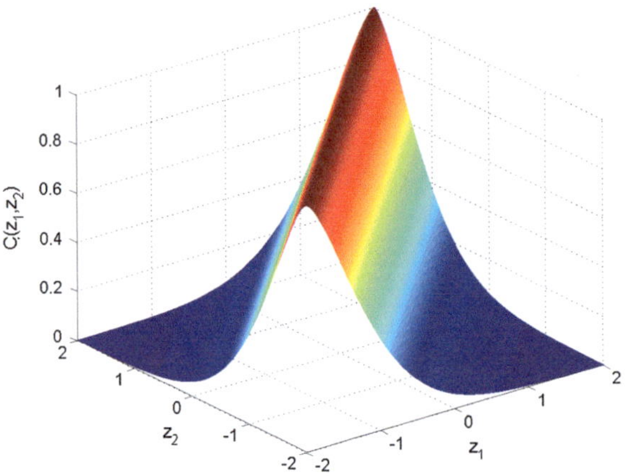

Fig. 2.8 Exponential covariance function of two one-dimensional variables with the hyperparameters $\sigma_f^2 = 1, l = 1$ and $d = 1$

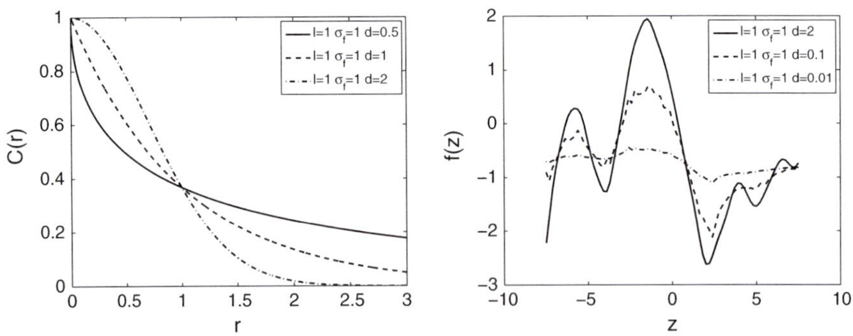

Fig. 2.9 Exponential covariance function with different hyperparameters (*left figure*) and random sample functions from the GP model with an exponential covariance function (*right figure*)

Rational quadratic covariance function

The rational quadratic covariance function is used when the function to be modelled is assumed to be continuous and differentiable in the mean-square sense [49]. Therefore, it is not the appropriate choice when the function to be modelled contains a discontinuity or is discontinuous in its first few derivatives. The rational quadratic covariance function can be seen as a scale mixture or an infinite sum of squared exponential covariance functions with different characteristic length scales.

It is defined by

$$C_f(\mathbf{z}_i, \mathbf{z}_j) = \sigma_f^2 \left(1 + \frac{r^2}{2\alpha l^2}\right)^{-\alpha} . \tag{2.16}$$

The hyperparameter σ_f^2 represents the scaling factor of the possible variations of the function or the vertical scaling factor, the hyperparameter l or the horizontal scaling factor determines the relative weight on distance for the input variable $\mathbf{z}$ and α is a positive hyperparameter. The variable r is the input distance measure and is $r = |\mathbf{z}_i - \mathbf{z}_j|$. The function is illustrated in Fig. 2.10. Samples of functions with different hyperparameters are given in Fig. 2.11.

The ARD version of the rational quadratic covariance function is

$$C_f(\mathbf{z}_i, \mathbf{z}_j) = \sigma_f^2 \left(1 + \frac{1}{2}(\mathbf{z}_i - \mathbf{z}_j)^{\mathrm{T}} \mathbf{\Lambda}^{-1} (\mathbf{z}_i - \mathbf{z}_j)\right)^{-\alpha} . \tag{2.17}$$

Matérn covariance functions

The Matérn covariance function is used when assumptions about the function to be modelled are less stringent regarding the smoothness or differentiability in the mean-square sense.

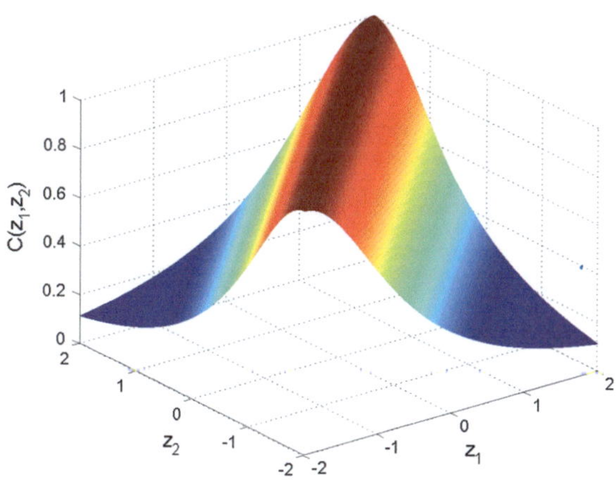

Fig. 2.10 Isotropic rational quadratic covariance function of two one-dimensional variables with the hyperparameters $\sigma_f = 1, l = 1$ and $\alpha = 1$

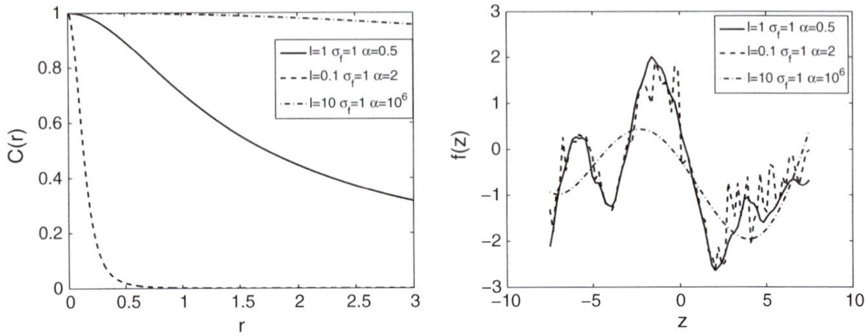

Fig. 2.11 Rational quadratic covariance function with different hyperparameters (*left figure*) and random sample functions from the GP model with rational quadratic covariance function (*right figure*)

It is defined by

$$C_f(\mathbf{z}_i, \mathbf{z}_j) = \sigma_f^2 \left(\frac{2^{1-d}}{\Gamma(d)} \right) \left(\frac{\sqrt{2d}r}{l} \right)^d K_d \left(\frac{\sqrt{2d}r}{l} \right). \tag{2.18}$$

The hyperparameter σ_f^2 represents the scaling factor of the possible variations of the function or the vertical scaling factor, the hyperparameter l or the horizontal scaling factor determines the relative weight on distance for the input variable $\mathbf{z}$, K_d is a modified Bessel function and the hyperparameter d can be seen to control the differentiability of the modelled mapping function. The variable r is the input distance measure and is $r = |\mathbf{z}_i - \mathbf{z}_j|$. Often, d is fixed to be $d = \frac{3}{2}$ or $d = \frac{5}{2}$. Increasing the value of d makes the sample function smoother. In [49] it is stated that in the cases where $d > \frac{5}{2}$ it is probably difficult to distinguish between the properties of the sample functions in the case of noisy training data.

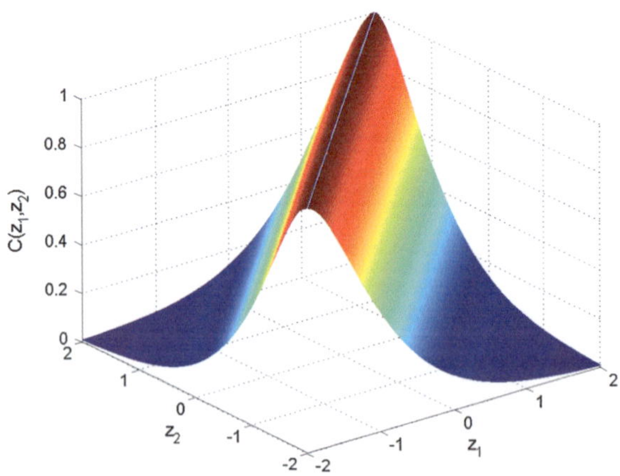

Fig. 2.12 Matérn covariance function of two one-dimensional variables with the hyperparameters $\sigma_f = 1$, $l = 1$ and $d = 3/2$

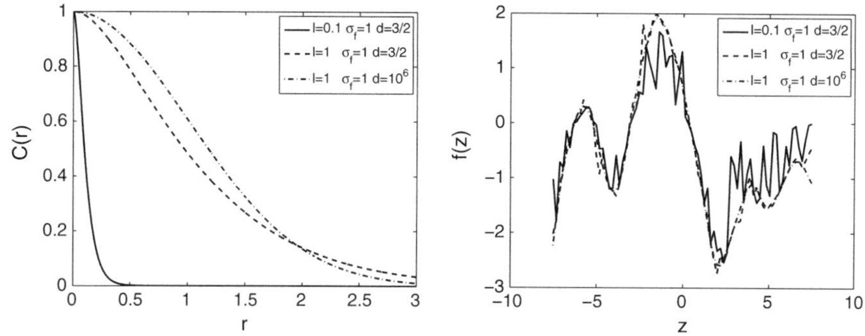

Fig. 2.13 Matérn covariance function with different hyperparameters (*left figure*) and random sample functions from the GP model with the Matérn covariance function (*right figure*)

The function is illustrated in Fig. 2.12. Samples of functions with different hyperparameters are given in Fig. 2.13.

Periodic covariance functions

A periodic covariance function does not have a large value only between two data points that are close together, but also between data points that are on a fixed distance, i.e. period. There exist many periodic covariance functions. A representative one is defined by

$$C_f(\mathbf{z}_i, \mathbf{z}_j) = \sigma_f^2 \exp\left(-\frac{2}{l^2}\sin^2\left(\frac{\pi}{T_p}r\right)\right) \tag{2.19}$$

It can be used for the modelling of functions that repeat themselves exactly. The hyperparameter σ_f^2 represents the scaling factor of the possible variations of the function or the vertical scaling factor, the hyperparameter l or the horizontal scaling factor determines the relative weight on distance for the input variable $\mathbf{z}$ and the hyperparameter T_p defines the period. The variable r is the input distance measure and is $r = |\mathbf{z}_i - \mathbf{z}_j|$.

The function is illustrated in Fig. 2.14. Samples of functions with different hyperparameters are given in Fig. 2.15.

Nonstationary Covariance Function

Linear Covariance Function

A linear covariance function is used when the function to be modelled is assumed to be linear.

It is defined by

$$C_f(\mathbf{z}_i, \mathbf{z}_j) = \sigma_f^2(\mathbf{z}_i \cdot \mathbf{z}_j + 1), \tag{2.20}$$

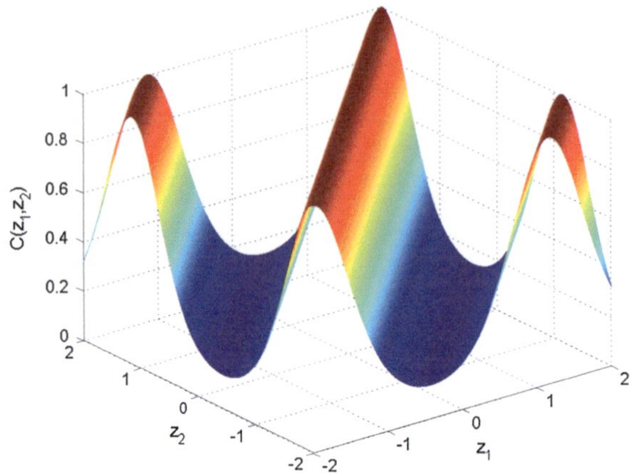

Fig. 2.14 Periodic covariance function of two one-dimensional variables with the hyperparameters $\sigma_f = 1$, $l = 1$ and $T_p = \pi$

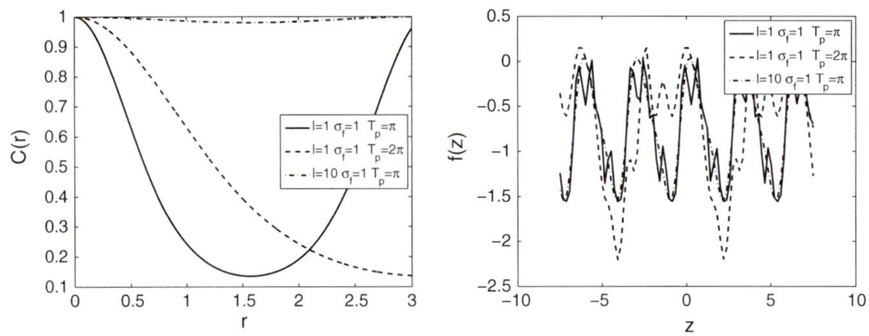

Fig. 2.15 Periodic covariance function with different hyperparameters (*left figure*) and random sample functions from the GP model with a periodic covariance function (*right figure*)

or without a bias term by

$$C_f(\mathbf{z}_i, \mathbf{z}_j) = \sigma_f^2(\mathbf{z}_i \cdot \mathbf{z}_j), \tag{2.21}$$

The hyperparameter σ_f^2 represents the scaling factor of the possible variations of the function or the vertical scaling factor. The function in Eq. (2.20) is illustrated in Fig. 2.16 together with samples of functions with different hyperparameters. The linear covariance function without a bias term can be generalised to

$$C_f(\mathbf{z}_i, \mathbf{z}_j) = \mathbf{z}_i^{\mathrm{T}} \boldsymbol{\Lambda}^{-1} \mathbf{z}_j. \tag{2.22}$$

where $\boldsymbol{\Lambda}^{-1}$ is a general positive semi-definite matrix used on the components of $\mathbf{z}$. If $\boldsymbol{\Lambda}^{-1}$ is diagonal, we obtain a linear covariance function with the ARD property.

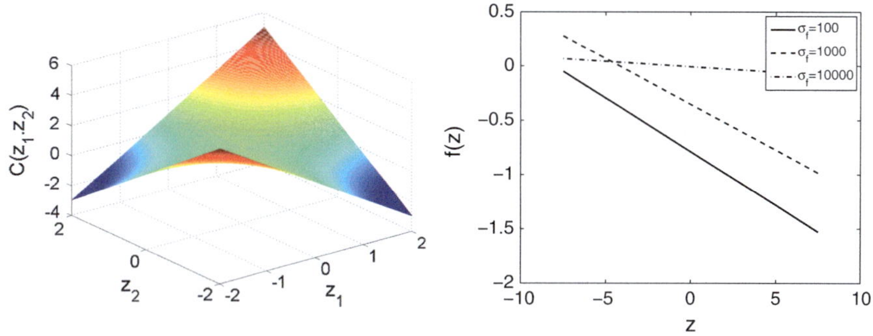

Fig. 2.16 Linear covariance function of two one-dimensional variables with the hyperparameters $\sigma_f^2 = 1$ (*left figure*) and random sample functions from the Gaussian process with a linear covariance function (*right figure*)

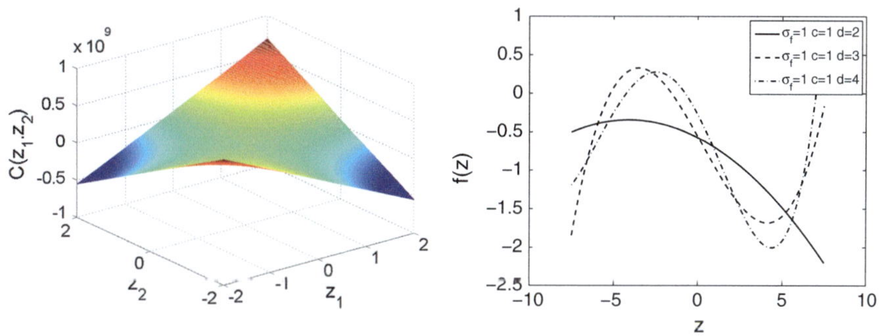

Fig. 2.17 Polynomial covariance function of two one-dimensional variables with the hyperparameters $\sigma_f = 1$, $c = 1$ and $d = 10$ (*left figure*) and random sample functions from the GP model with a polynomial covariance function (*right figure*)

Polynomial covariance function

The polynomial covariance function is defined by

$$C_f(\mathbf{z}_i, \mathbf{z}_j) = (\sigma_f^2 \mathbf{z}_i \cdot \mathbf{z}_j + c)^d. \tag{2.23}$$

The hyperparameter σ_f^2 represents the scaling factor of the possible variations of the function or the vertical scaling factor, the hyperparameter c determines the vertical bias and the hyperparameter d determines the exponent.

The polynomial covariance function is not thought to be very useful for regression problems, as the prior variance will become very large with $|\mathbf{z}|$ as $|\mathbf{z}| > 1$.

The function is illustrated in Fig. 2.17 together with samples of functions with different hyperparameters.

Neural network covariance function

The neural network covariance function [49] is defined by

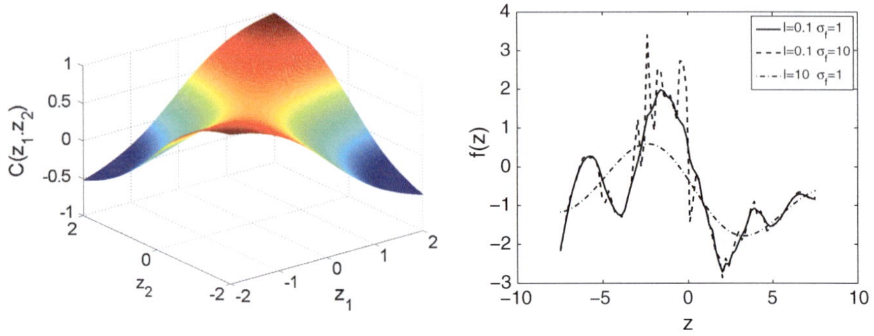

Fig. 2.18 Neural network covariance function of two one-dimensional variables with the hyperparameter $\sigma_f = 1$ (*left figure*) and random sample functions from the GP model with a neural network covariance function

$$C_f(\mathbf{z}_i, \mathbf{z}_j) = \sigma_f^2 \frac{2}{\pi} \sin^{-1}\left(\frac{2\tilde{\mathbf{z}}_i^{\mathrm{T}} \mathbf{\Lambda}^{-1} \tilde{\mathbf{z}}_j}{\sqrt{1 + 2\tilde{\mathbf{z}}_i^{\mathrm{T}} \mathbf{\Lambda}^{-1} \tilde{\mathbf{z}}_i}\sqrt{1 + 2\tilde{\mathbf{z}}_j^{\mathrm{T}} \mathbf{\Lambda}^{-1} \tilde{\mathbf{z}}_j}}\right), \tag{2.24}$$

where $\tilde{\mathbf{z}}_i = [1, \mathbf{z}_i] = [1, z_1, \ldots, z_D]^{\mathrm{T}}$, $\mathbf{\Lambda}^{-1}$ is the covariance matrix on components of $\mathbf{z}$ and is often set as a unity matrix multiplied by l^{-2}.
For neural networks, a more common sigmoid function, e.g. $\tanh(\cdot)$, is not positive definite so it cannot be used as a covariance function [49].
The covariance function in Eq. (2.24) has been successfully applied to modelling the input–output mapping function of the step form in [49].
This function is illustrated in Fig. 2.18 together with samples of functions with different values of the hyperparameters.

The covariance functions can be manipulated in different ways to form a new composite covariance function:

Sum of covariance functions
A sum of two kernels is also a kernel. Therefore, a sum of two covariance functions $C_1(\mathbf{z}_i, \mathbf{z}_j)$ and $C_2(\mathbf{z}_i, \mathbf{z}_j)$ is a nonnegative definite function $C(\mathbf{z}_i, \mathbf{z}_j) = C_1(\mathbf{z}_i, \mathbf{z}_j) + C_2(\mathbf{z}_i, \mathbf{z}_j)$ and consequently a covariance function. This property is, for example, useful where a number of different characteristic length scales can be observed. If you sum together two kernels, then the resulting kernel will have a high value if either of the two summed kernels has a high value. An example is the summation of a linear and a periodic covariance function for obtaining a kernel that can be used for modelling functions that are periodic with an increasing mean, as we move away from the origin.

Product of covariance functions
Similarly, a product of two kernels is also a kernel. Therefore, a product of two covariance functions $C_1(\mathbf{z}_i, \mathbf{z}_j)$ and $C_2(\mathbf{z}_i, \mathbf{z}_j)$ is a nonnegative definite function $C(\mathbf{z}_i, \mathbf{z}_j) = C_1(\mathbf{z}_i, \mathbf{z}_j) \cdot C_2(\mathbf{z}_i, \mathbf{z}_j)$ and consequently a covariance function. If you

multiply together two kernels, then the resulting kernel will have a high value only if both of the two used kernels have a high value. An example is the multiplication of a periodic and a squared exponential covariance function for obtaining a kernel that can be used for modelling functions that do not repeat themselves exactly.

Vertical rescaling

This is the operation where a stationary covariance function is transformed into a nonstationary one. Let it be $g(\mathbf{z}_i) = a(\mathbf{z}_i)f(\mathbf{z}_i)$, where $a(\mathbf{z}_i)$ is a deterministic function, $f(\mathbf{z}_i)$ is a random process and $C_1(\mathbf{z}_i, \mathbf{z}_j) = \text{cov}(f(\mathbf{z}_i), f(\mathbf{z}_j))$. The new covariance function is $C(\mathbf{z}_i, \mathbf{z}_j) = \text{cov}(g(\mathbf{z}_i), g(\mathbf{z}_j)) = a(\mathbf{z}_i)C_1(\mathbf{z}_i, \mathbf{z}_j)a(\mathbf{z}_j)$. This method can be used to normalise kernels.

Convolution of covariance functions

A new covariance function can also be obtained with convolution. This operation also converts a stationary covariance function into a nonstationary one. If $C_2(\mathbf{z}_i, \mathbf{y}_i)$ is an arbitrary fixed covariance function, $C_1(\mathbf{y}_i, \mathbf{y}_j) = \text{cov}(f(\mathbf{y}_i), f(\mathbf{y}_j))$ and the transformation is $g(z) = \int C_2(\mathbf{z}_i, \mathbf{y}_i)f(\mathbf{y}_i)d\mathbf{y}_i$ then the transformed new covariance function is $C(\mathbf{z}_i, \mathbf{z}_j) = \text{cov}(g(\mathbf{z}_i), g(\mathbf{z}_j)) = \int C_2(\mathbf{z}_i, \mathbf{y}_i)C_1(\mathbf{y}_i, \mathbf{y}_j)$ $C_2(\mathbf{z}_j, \mathbf{y}_j)d\mathbf{y}_j d\mathbf{y}_i$.

Warping—nonlinear mapping of covariance function

Another possibility is to employ an arbitrary nonlinear mapping, also known as warping of the input to handle the nonstationary nonlinearity of the function in tandem with a stationary covariance function. More on warping using the parametric functions can be found in [8] and using the nonparametric functions in [9]. This method can be also used to transform data when the noise that corrupts measurement data does not have Gaussian distribution.

Composite covariance functions that are suited to the problem at hand can be developed using listed operations. An overview [57] gives further directions for the selection and combination of covariance functions. Examples of customary covariance functions can be found in [58–62].

2.4 GP Model Selection

This section explains how we get from the Bayesian model selection to the optimisation of the hyperparameters for the selected covariance function explained in Sect. 2.3 so that the GP model is a model of an unknown system.

Due to the probabilistic nature of the GP model, the model optimisation approach where the model parameters and the structure are optimised by the minimisation of a loss function defined in terms of model error only is not really applicable. Since the GP is a Bayesian probabilistic method, a probabilistic approach to the model selection is appropriate. The probabilistic approach to model selection is described in the following subsection. According to [49], the term model selection covers the selection of a covariance function for a particular model, the selection of the mean function, the choice of the hyperparameters and a comparison across different families of models.

Several possibilities for hyperparameter determination exist. A very rare possibility is that the hyperparameters are known in advance as prior knowledge. Almost always, however, they must be determined from the training data.

In the following subsections, the background for hyperparameter selection is described. First, the evidence or marginal likelihood maximisation is described. The next subsection deals with the mathematical and computational implementations, where the methods can be divided into direct and approximate implementations.

The alternative to the case where the set of identification data is available in advance is when the identification data is collected online. This alternative is practical when the unknown system is modelled as a time-varying model or as a method to circumvent the computational burden due to a large amount of data being used for the identification. In both cases, only the data with sufficient information content is selected for the training process.

2.4.1 Bayesian Model Inference

Bayesian inference is a process in which a prior on the unknown quantity, i.e. an input–output mapping function f, has to be specified. Next, the identification data is observed. Afterwards, a posterior distribution over f is computed that refines the prior by incorporating the observations, i.e. the identification data.

For example, the identification data for the GP-NARX model is

$$
\begin{aligned}
\mathcal{D} &= \{(\mathbf{Z}, \mathbf{y})\}, \\
\mathbf{y} &= \begin{bmatrix} y(k) \\ \vdots \\ y(k+i) \\ \vdots \\ y(k+N) \end{bmatrix}, \mathbf{Z} = \begin{bmatrix} \mathbf{z}_1, \ \ldots, \ \mathbf{z}_i, \ \ldots, \mathbf{z}_N \end{bmatrix}, \\
\mathbf{z}_i &= [y(k+i-1, \ldots, y(k+1-n), u(k+i-1), \ldots, u(k+i-m)]^{\mathrm{T}}.
\end{aligned}
$$

We set a GP prior for a function f, i.e. a presumption on the distribution of the model candidates. We specify the prior mean function and the prior covariance function. The prior mean function is often set as $m_f = 0$.

Inference of the GP posterior [51] can be pursued in more levels based on the unknowns to be inferred. There are at least two levels above the data $\mathcal{D} = \{(\mathbf{Z}, \mathbf{y})\}$: one for inferring the distribution over function f and the second, on the top of it, to determine the hyperparameters $\boldsymbol{\theta}$ that specify the distribution over the function values f. A further model inference level can be the level of different covariance functions.

We start with the GP posterior of the function presuming that the data and the hyperparameters are given:

$$p(f|\mathbf{Z}, \mathbf{y}, \boldsymbol{\theta}) = \frac{p(\mathbf{y}|f, \mathbf{Z}, \boldsymbol{\theta})p(f|\boldsymbol{\theta})}{p(\mathbf{y}|\mathbf{Z}, \boldsymbol{\theta})}, \tag{2.25}$$

where $p(\mathbf{y}|f, \mathbf{Z}, \boldsymbol{\theta})$ is the likelihood of the function f and $p(f|\boldsymbol{\theta})$ is the GP prior on f.

The likelihood of the function f, with the assumption that the observations, i.e. measurements, y_i are conditionally independent given $\mathbf{Z}$, is [51]

$$\begin{aligned} p(\mathbf{y}|f, \mathbf{Z}, \boldsymbol{\theta}) &= p(y_1, y_2, \ldots, y_N|f(\mathbf{Z}), \mathbf{Z}, \boldsymbol{\theta}) = \prod_{i=1}^{N} p(y_i|f(\mathbf{z}_i), \boldsymbol{\theta}) \\ &= \prod_{i=1}^{N} \mathcal{N}(y_i|f(\mathbf{z}_i), \sigma_n^2) = \mathcal{N}(\mathbf{y}|f(\mathbf{Z}), \sigma_n^2\mathbf{I}), \end{aligned} \tag{2.26}$$

where σ_n^2 is a variance of additive noise on the output identification data. The likelihood in Eq. (2.26) encodes the assumed noise model.

The normalising constant in Eq. (2.25)

$$p(\mathbf{y}|\mathbf{Z}, \boldsymbol{\theta}) = \int p(\mathbf{y}|\mathbf{Z}, f, \boldsymbol{\theta})p(f|\boldsymbol{\theta})df \tag{2.27}$$

is the evidence or marginal likelihood. The evidence is the likelihood of the hyperparameters given the data after having marginalised out the function f.

Because we do not have the hyperparameters given, we have to infer them. The next level is, therefore, to infer a posterior probability distribution over the hyperparameters $\boldsymbol{\theta}$ that is conditional on the data $\mathcal{D}$:

$$p(\boldsymbol{\theta}|\mathbf{Z}, \mathbf{y}) = \frac{p(\mathbf{y}|\mathbf{Z}, \boldsymbol{\theta})p(\boldsymbol{\theta})}{p(\mathbf{y}|\mathbf{Z})}, \tag{2.28}$$

where $p(\boldsymbol{\theta})$ is the prior on the hyperparameters.

The evidence of Eq. (2.28) is

$$p(\mathbf{y}|\mathbf{Z}) = \int p(\mathbf{y}|\mathbf{Z}, \boldsymbol{\theta})p(\boldsymbol{\theta})d\boldsymbol{\theta} \tag{2.29}$$

where we marginalise out the hyperparameters $\boldsymbol{\theta}$.

Note that in Bayesian inference there is no model overfitting to the identification data common to other methods mentioned at the beginning of this chapter. This is because in essence there is no data fitting in the Bayesian inference.

However, the integral in Eq. (2.29) is analytically intractable in most interesting cases. A possible solution is to use numerical approximation methods, such as the Markov chain Monte Carlo method, to obtain the posterior. Unfortunately, significant computational efforts may be required to achieve a sufficiently accurate approximation.

In addition to the numerical approximation methods, another standard and general practice for estimating the hyperparameters is the evidence maximisation with regards to the hyperparameters. This approximation means that instead of a posterior distribution over the hyperparameters $p(\boldsymbol{\theta}|\mathbf{Z}, \mathbf{y})$ we look for a point estimate $\hat{\boldsymbol{\theta}}$. This method has several names, among others it is called empirical Bayes or type-2 maximum likelihood [44]. Nevertheless, with this approximation we are not completely in line with the coherent Bayesian inference anymore.

Hyperparameters estimation via evidence maximisation is described in the next section.

2.4.2 Marginal Likelihood—Evidence Maximisation

Using evidence maximisation is based on the presumption that the most likely values of the hyperparameters are noticeably more likely than other values. This means that the posterior distribution of hyperparameters is unimodal and is described with a narrow Gaussian function. This is usually valid for a larger number of identification datapoints relative to the number of hyperparameters. For the smaller number of datapoints relative to the number of hyperparameters, the posterior distribution is usually not unimodal.

The values of the hyperparameters depend on the data-at-hand and it is difficult to select their prior distribution. If a uniform prior distribution is selected, which means that any values for the hyperparameters are equally possible a priori, then the hyperparameters' posterior distribution is proportional to the marginal likelihood in Eq. (2.25), that is,

$$p(\boldsymbol{\theta}|\mathbf{Z}, \mathbf{y}) \propto p(\mathbf{y}|\mathbf{Z}, \boldsymbol{\theta}). \tag{2.30}$$

This means that the maximum a posteriori (MAP) estimate of the hyperparameters $\boldsymbol{\theta}$ equals the maximum marginal likelihood estimate of Eq. (2.27) [51].

The hyperparameters $\boldsymbol{\theta}$ are therefore obtained with the maximisation of evidence or marginal likelihood in Eq. (2.25) on the first level of inference with respect to the hyperparameters:

$$p(\mathbf{y}|\mathbf{Z}, \boldsymbol{\theta}) = \frac{1}{(2\pi)^{\frac{N}{2}}|\mathbf{K}|^{\frac{1}{2}}} e^{-\frac{1}{2}\mathbf{y}^{\mathrm{T}}\mathbf{K}^{-1}\mathbf{y}}. \tag{2.31}$$

with the $N \times N$ covariance matrix $\mathbf{K}$ of the identification or training data.

Due to the mathematical properties of the logarithm function for the numerical scaling purposes, the logarithm of the evidence is used as the objective function for the optimisation:

$$\ln p(\mathbf{y}|\mathbf{Z}, \boldsymbol{\theta}) = \ell(\boldsymbol{\theta}) = -\overbrace{\frac{1}{2}\ln(|\mathbf{K}|)}^{\text{complexity term}} - \overbrace{\frac{1}{2}\mathbf{y}^{\mathrm{T}}\mathbf{K}^{-1}\mathbf{y}}^{\text{data-fit term}} - \overbrace{\frac{N}{2}\ln(2\pi)}^{\text{normalisation const.}}, \tag{2.32}$$

where ℓ is the value of the logarithm of the evidence or marginal likelihood. Equation (2.32) also shows interpretations of its three terms.

The posterior model we obtain with evidence maximisation trades off data fit and model complexity. Hence, it avoids overfitting [49, 51] by implementing *Occam's razor*, which tells us to use the simplest model that explains the data.

Maximising the evidence using Eq. (2.32) is a nonlinear, non-convex *optimisation* problem. Nevertheless, even though a global optimum of this optimisation problem has not necessarily been found, the GP model can always explain the data [51].

The fact that the posterior GP model is obtained with optimisation can be, on one hand, considered as a disadvantage, because Bayesian inference is not pursued strictly all the way to the end. On the other hand, however, a lot of people involved in system identification are very familiar with optimisation-based methods for experimental modelling, which brings GP modelling closer to their attention. This might not be the case with full Bayesian inference.

The following sections describe some possibilities for the optimisation implementation.

2.4.2.1 Deterministic Optimisation Methods

Deterministic optimisation methods are well known and are widely used in system identification. Here, we are not going into a detailed description. An overview of these methods can be found in many references, e.g. [10].

In the case of using one of the gradient optimisation methods, the computation of the partial derivatives of marginal likelihood with respect to each of the hyperparameters is required:

$$\nabla(\ell(\boldsymbol{\theta})) = \frac{\partial \ln p(\mathbf{y}|\mathbf{Z}, \boldsymbol{\theta})}{\partial \theta_i} = -\frac{1}{2}\text{trace}\left(\mathbf{K}^{-1}\frac{\partial \mathbf{K}}{\partial \theta_i}\right) + \frac{1}{2}\mathbf{y}^{\mathrm{T}}\mathbf{K}^{-1}\frac{\partial \mathbf{K}}{\partial \theta_i}\mathbf{K}^{-1}\mathbf{y}. \quad (2.33)$$

The computation of the partial derivatives involves the computation of the inverse of the $N \times N$ covariance matrix $\mathbf{K}$ during every iteration. This means that the computational complexity is $\mathcal{O}(N^3)$. However, there are alternative approaches, and some of them will be touched upon in Sect. 2.5.2.

A frequently used method for optimising the cost function is a conjugate gradient method—a Polack–Ribiere version utilised in tandem with the Wolfe-Powell stopping conditions [49]. Compared with numerical approximation methods of the Monte Carlo type used for obtaining the hyperparameters' posterior, this conjugate gradient approach can find a reasonable approximation to a local maximum after a relatively acceptable number of evaluations.

The trust-region optimisation method is proposed as an alternative by [63], where the Hessian matrix is simplified and then the trust-region algorithm is used for the GP hyperparameters' optimisation.

A property of deterministic optimisation methods is that their result heavily depends on the initial values of the hyperparameters, especially for complex multidimensional systems, where the objective function has many local optima. Therefore, whatever optimisation method is used it should be run repeatedly with various initial values of the hyperparameters. While the space of possible values is huge, the initial values are often chosen randomly. Therefore, stochastic optimisation methods can be considered as an alternative approach. Three stochastic algorithms are described in the next subsection: genetic algorithms, differential evolution and particle swarm optimisation.

2.4.2.2 Stochastic Optimisation Methods

Stochastic optimisation methods are used as an alternative to deterministic optimisation methods when local extremes are expected in the optimisation of hyperparameters. The following description is adopted from [64].

Evolutionary algorithms (EAs) are generic, population-based, stochastic optimisation algorithms inspired by biological evolution. They use mechanisms similar to those known from the evolution of species: reproduction, mutation, recombination and selection. Candidate solutions to the optimisation problem play the role of individuals in a population, and the objective function determines the environment within which the solutions 'live'. Simulated evolution of the population then takes place after the repeated application of the above operators.

Evolutionary algorithms [65] perform well at approximating solutions to diverse types of problems because they make no assumptions about the underlying fitness landscape; this generality is shown by their success in fields as diverse as science, engineering, economics, social sciences and art.

In most real-world applications of evolutionary algorithms, computational complexity is a prohibiting factor. In fact, this computational complicity is due to a objective function evaluation. In our case, as was shown in the previous section, an evaluation of the objective function contains an inversion of the covariance matrix, of which the computational time rises with the third power of the amount of data. However, this 'inconvenience' is unfortunately inescapable without an approximation.

The most commonly used evolutionary algorithms for numerical optimisation, such as maximisation of the logarithmic marginal likelihood, are genetic algorithm with real numbers representation, differential evolution and particle swarm optimisation.

Genetic algorithm (GA) is a flexible search technique [65] used in computing to find exact or approximate solutions to optimisation and search problems in many areas. Traditionally, the solutions are represented as binary strings of 0s and 1s, but other encodings are also possible. The simulated evolution usually starts from a population of randomly generated individuals and proceeds in generations. In each generation, the fitness of every individual in the population is evaluated, and multiple individuals are stochastically selected from the current population (based on their fitness) and modified (recombined and possibly randomly mutated) to form a new population. The new population is then used in the next generation of the algorithm. Commonly, the algorithm terminates when either a maximum number of generations have been iterated, or a satisfactory fitness level has been reached for the population. If the algorithm has terminated due to a maximum number of generations, a satisfactory solution may or may not have been reached. Examples of GP model optimisation with a GA can be found in [66, 67].

Differential evolution (DE) is a method for numerical optimisation without explicit knowledge of the gradients. It was presented by Storn and Price [68] and works on multidimensional real-valued functions that are not necessarily continuous or differentiable. DE searches for a solution to a problem by maintaining a population of candidate solutions and creating new candidate solutions by combining existing ones according to its simple formulae of vector crossover and mutation, and then keeping whichever candidate solution has the best score or fitness on the optimisation problem at hand. In this way, the optimisation problem is treated as a black box that merely provides a measure of quality given a candidate solution and the gradient is therefore not needed. More details about DE can be found in [69].

Example of GP model optimisation with DE can be found in [64].

Particle swarm optimisation (PSO) is a method proposed by Kennedy and Eberhart [70] that is motivated by the social behaviour of organisms such as bird flocking and fish schooling. Like DE it is used for numerical optimisation without explicit knowledge of the gradients. PSO provides a population-based search procedure in which individuals called particles change their position (state) with time. In a PSO system, particles 'fly' around in a multidimensional search space. During flight, each particle adjusts its position according to its own experience and the experience of a neighbouring particle, making use of the best position encountered by itself and its neighbour. Thus, a PSO system combines local search with global search, attempting to balance exploration and exploitation. Further details about PSO can be found in [71].

Examples of GP model optimisation with a PSO can be found in [64, 72].

Example 2.1 (*CO_2 concentration modelling*) This example compares different stochastic optimisation methods and is adopted from [64].

To assess the potential of evolutionary algorithms in the optimisation of GP model hyperparameters, a problem concerning the concentration of CO_2 in the atmosphere from [49] was chosen. The data consists of monthly average atmospheric CO_2 concentrations derived from in situ air samples collected at the Mauna Loa Observatory, Hawaii, between 1959 and 2009 (with some missing data).[1] The goal is to model the CO_2 concentration as a function of time.

Although the data is one-dimensional, and therefore easy to visualise, a complex covariance function is used. It is derived by combining several kinds of simple covariance functions. First, a squared exponential covariance function in Eq. (2.13) is used to model the long-term smooth rising trend. With the product of a periodic

[1]The data is available from http://cdiac.esd.ornl.gov/ftp/trends/co2/maunaloa.co2.

of Eq. (2.19) and the squared exponential covariance function of Eq. (2.13), a seasonal component is modelled. To model the medium-term irregularities, a rational quadratic covariance function in Eq. (2.16) is used. Finally, the noise is modelled as the sum of a squared exponential and a constant covariance function of Eq. (2.12).

$$\begin{aligned}
C(\mathbf{z}_i, \mathbf{z}_j) &= C_1(\mathbf{z}_i, \mathbf{z}_j) + C_2(\mathbf{z}_i, \mathbf{z}_j) + C_3(\mathbf{z}_i, \mathbf{z}_j) + C_4(\mathbf{z}_i, \mathbf{z}_j),\\
C_1(\mathbf{z}_i, \mathbf{z}_j) &= \theta_{12}^2 \exp\left(-\frac{r^2}{2\theta_{11}^2}\right),\\
C_2(\mathbf{z}_i, \mathbf{z}_j) &= \theta_{22}^2 \exp\left(-\frac{r^2}{2\theta_{21}^2}\right)\theta_{24}^2 \exp\left(-\frac{2}{\theta_{23}^2}\sin^2\left(\frac{\pi}{\theta_{25}}r\right)\right),\\
C_3(\mathbf{z}_i, \mathbf{z}_j) &= \theta_{32}^2 \exp\left(1 + \frac{r^2}{2\theta_{33}\theta_{31}^2}\right)^{-\theta_{33}},\\
C_4(\mathbf{z}_i, \mathbf{z}_j) &= \theta_{42}^2 \exp\left(-\frac{r^2}{2\theta_{41}^2}\right) + \theta_{43}^2\delta_{ij}.
\end{aligned} \tag{2.34}$$

This complex covariance function involves 13 hyperparameters. Note that in [49] the covariance function involves only 11 hyperparameters due to the fixed period of the periodic covariance function to one year.

For differential evolution an implementation from [69], for particle swarm optimisation an implementation from [73], and for genetic algorithms an implementation from [74] were used. These methods were compared to a commonly used deterministic conjugate gradient method (CG) as a frequently used deterministic optimisation method for GP modelling.

All the stochastic algorithms had the same population size, number of generations and number of solution evaluations. The population size was set to 50 individuals, the number of generations to 2000 and the number of iterations to 10. Although the tested algorithms have various properties, the fairness of experimental evaluation was tried to be guaranteed. The parameters of these algorithms were tuned in preliminary experiments. For a comparison of the tested stochastic methods with the conjugate gradient method, the same number of evaluations was used with it as well. Thus the conjugate gradient method was executed 10 times with 100,000 solution evaluations available. That means, in one iteration, the conjugate gradient methods are repeatedly executed with random initial values and possibly restarted until all the available evaluations are spent.

For each algorithm the following statistics were calculated: minimum, maximum, average and standard deviation. They are given in Table 2.1.

Figure 2.19 shows the performance traces of each algorithm averaged over 10 runs. At first sight, it appears that the DE and PSO perform similarly and a lot better than the GA and CG. The DE and PSO reached a very similar maximum value that was very close to the result from [49]. However, the PSO reached a lower minimum value, which means it has a larger variance than the DE. In other words, using the DE will more probably find an optimal value or at least a value near to it.

Table 2.1 Statistical values—maximum, minimum, average and standard deviation—of 10 iterations for each algorithm: CG, GA, DE, PSO

	CG	GA	DE	PSO
Maximum	−598.8627	−639.0114	−142.2658	−142.4529
Minimum	−638.9760	−703.9570	−176.2658	−216.9546
Average	−630.9088	−645.8582	−154.5382	−177.6854
Standard deviation	12.5953	20.4178	11.3083	26.9605

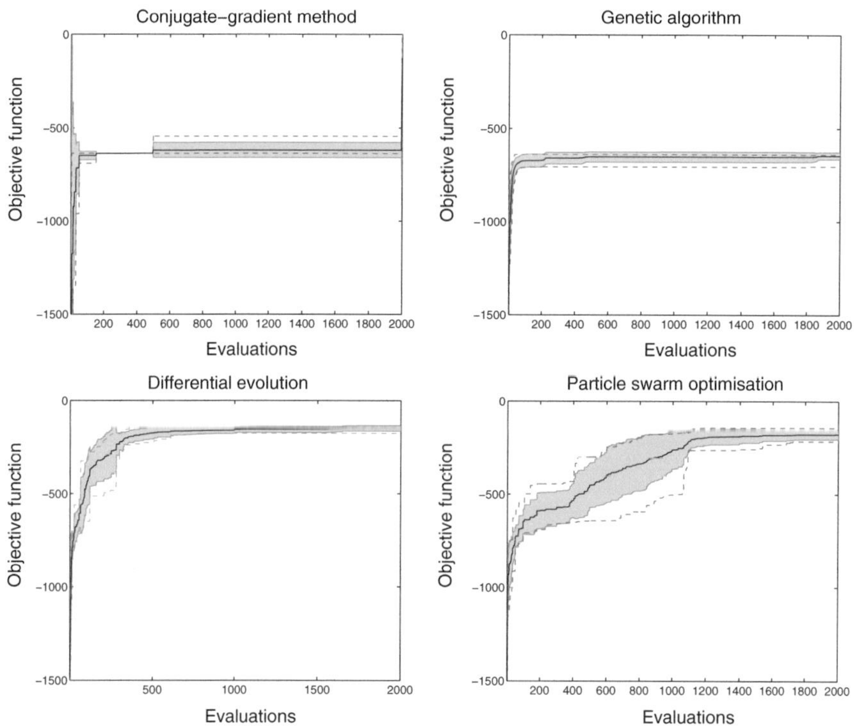

Fig. 2.19 Performance traces for the tested algorithms. *Solid lines* denote mean values, *dashed lines* denote maximum and minimum values, *grey areas* present standard deviations

Unsatisfactory results obtained by the CG imply the difficulty of the chosen problem for this traditional method.

However, Fig. 2.20, which shows the predictions of CO_2 for the next 20 years based on the model obtained from the best hyperparameter values found with DE and shown in Table 2.2, confirms the superiority of the DE. While the best hyperparameter values obtained by the PSO differ from the DE's at most by 10^{-5}, consequently the

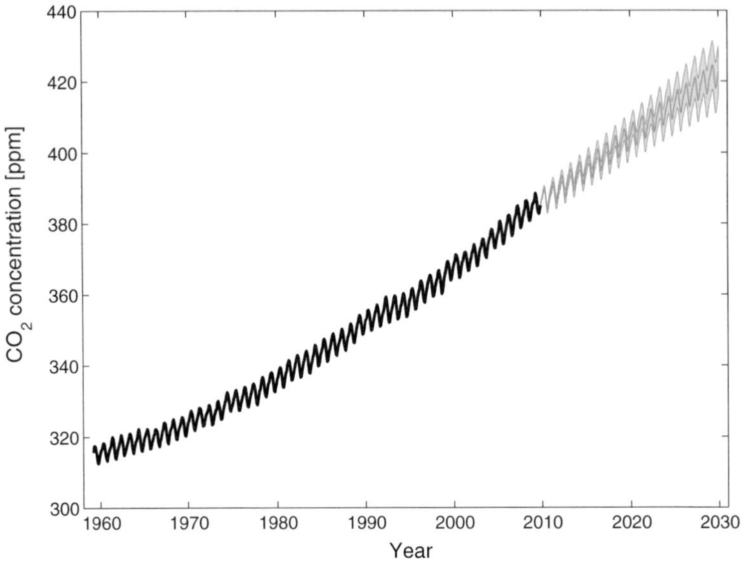

Fig. 2.20 Concentration of CO_2 in the atmosphere and its predictions based on a GP model with hyperparameters optimised using differential evolution

Table 2.2 Best hyperparameters of our experiment, which were obtained with the DE

i	θ_{i1}	θ_{i2}	θ_{i3}	θ_{i4}	θ_{i5}
1	5.0000	5.0000			
2	5.0000	−1.5668	0.2406	2.1831	2.4843
3	1.3454	−0.7789	5.0000		
4	0.4580	−1.7163	−1.6161		

log-marginal likelihood and the predictions of the PSO are very similar to those of the DE.

Please note that our predictions are almost identical to the originals from [49], but the log-marginal likelihood is smaller, due to the wider range of measurements for training and the wider range of predictions used in our experiment.

2.4.3 Estimation and Model Structure

Up until now, we have not been taking into account the different model structures, as listed in Sect. 2.3.1.

Let us explore the differences between the NARX and NOE general kinds of models, not only GP models in particular, with an emphasis on modelling for simulation, and therefore training in a parallel configuration as described in [18]. This is of particular interest in dynamic systems identification. The NFIR model can in this context be taken as a special case of the NARX model.

Consider again the system with added white Gaussian noise

$$y(k) = f(\mathbf{z}(k)) + \nu \tag{2.35}$$

where $\mathbf{z}$ is a vector of regressors at the time instant k and ν is the measurement noise, also the observation noise, at output y at the time instant k. If there is no noise in the output measurements, then the NARX and NOE models are the same and the differences between them do not matter. Nevertheless, noise always exists in real-life measurements. In the NARX model, these measurements are contained within the regressors that are delayed samples of the output signal. This means that when the NARX model is identified the *errors-in-variables* problem occurs. There are structures with other noise models that are more complex.

The problem of learning GP models with noisy input variables is investigated in [75]. Another possible solution to this problem is the automated filtering of input and output signals. This solution is equivalent to the optimisation of a noise filter, i.e. a filter that filters white noise to the one that corresponds to the measurement noise. In system identification this method is known as the *instrumental variables* method [2]. A similar principle in the context of GP models is described in [19].

If, on the other hand, the model is trained without using delayed measurements as the regressors, i.e. an output-error model, then it is assumed that the noise is entering the output signal after the system.

For a dynamic system of order n, the one-step-ahead prediction is calculated with the previous process output values as described with Eq. (2.4)

$$\begin{aligned}\hat{y}(k) = f(&y(k-1), y(k-2), \ldots, y(k-n), u(k-1),\\ &u(k-2), \ldots, u(k-m)) + \nu,\end{aligned}$$

while the simulation is evaluated with the previous model's output estimates as described with Eq. (2.5)

$$\begin{aligned}\hat{y}(k) = f(&\hat{y}(k-1), \hat{y}(k-2), \ldots, \hat{y}(k-n), u(k-1),\\ &u(k-2), \ldots, u(k-m)) + \nu.\end{aligned}$$

In the first case, we have a feedforward prediction, while in the second case we have a recurrent one and, in the case of nonlinear system's modelling with GP models, an approximation of the prediction distribution. A simulation is required whenever the process output cannot be measured during the operation, which is always when the system is simulated independently from the real system. This is frequently the case for the design of fault detection and control systems.

Nevertheless, the NARX model is by far the most applied model for dynamic systems, not because it is realistic, but because it is easier to train than the NOE model.

In the NARX case, the model is trained based on loss functions that are dependent on the prediction error, while in the NOE case the model is trained based on loss functions that are dependent on the simulation error. The hyperplane of the loss function is much more complicated in the second case. However, the model that is used in the parallel configuration does not necessarily have to be trained in the parallel configuration as well, if the noise assumptions are relaxed, which is often the case in engineering practice. The disadvantage of using NARX models for the simulation is the error accumulation, which does not happen with the prediction. In extreme situations, the NARX model used for the simulation can become unstable, regardless of a satisfactory one-step prediction performance.

The trade-offs between the advantages and disadvantages of the NARX and NOE models need to be evaluated when the model is developed for a model simulation.

The case when GP models are used is specific. Since the output of the GP model $\hat{y}(k)$ is a normal distribution in the case of the NOE model, this output represents only an approximation of the distribution that should appear on the output of a nonlinear model. The level of the approximation can be very different, as will be discussed in Sect. 2.6. It can range from the simplest case, where only expectations of the outputs are fed back, to more informative analytical or numerical approximations.

General GP optimisation algorithms for parallel-series or the GP-NARX model described with Eq. (2.4) and for parallel or GP-NOE model in Eq. (2.5) are as follows.

Optimisation algorithm for GP-NARX model:

Algorithm: OPTIMISENARXMODEL(*Inputs*)

---

set input data, target data, covariance function, initial hyperparameters
repeat
  calculate $-\ell(\boldsymbol{\theta})$ and its derivative based on input data
  $\{y(k-1), y(k-2), \ldots, y(k-n), u(k-1), u(k-2), \ldots, u(k-m)\}$
  change hyperparameters
until $-\ell(\boldsymbol{\theta})$ is minimal

Optimisation algorithm for GP-NOE model:

Algorithm: OPTIMISENOEMODEL(*Inputs*)

---

set input data, target data, covariance function, initial hyperparameters
repeat
  calculate the model simulation response
  calculate $-\ell(\boldsymbol{\theta})$ and its derivative based on input data
  $\{\hat{y}(k-1), \hat{y}(k-2), \ldots, \hat{y}(k-n), u(k-1), u(k-2), \ldots, u(k-m)\}$
  change hyperparameters
until $-\ell(\boldsymbol{\theta})$ is minimal

The difference between considering the prediction and simulation error has a major impact on the complexity of the optimisation. This difference is illustrated in the following example.

Example 2.2 (*Difference in GP-NARX and GP-NOE optimisation*) The difference in the optimisation for the GP-NARX and GP-NOE models is illustrated with an example where two loss functions are calculated. The input and output data is obtained from the selected dynamic GP model of the first order. The regressors are delayed samples of the output and input signals, $y(k-i)$ and $u(k-i)$, respectively, in the case of the GP-NARX model; and delayed samples of expectations of the output predictions $E(\hat{y}(k))$ and delayed samples of the input signals $u(k-i)$ in the case of the GP-NOE model. This means that the simplest possible approximation has been used in this example where no prediction distribution is propagated through the model. Even though this is the simplest approximation, the main differences between the model structures are noticeable.

Since there are two regressors and the GP model uses squared exponential and constant covariance functions, there are four hyperparameters in this process.

We fix the two hyperparameters that are parameters that control the variances σ_f^2 and σ_n^2 and calculate the surfaces of the loss functions for the GP-NARX and GP-NOE models, when the two remaining hyperparameters controlling the two regressors are varied. The loss function is a negative logarithm of the marginal likelihood in both cases, but in the case of the GP-NARX model the output observations are used for the calculation of the loss function and in the case of the GP-NOE model the mean values of the output predictions are used for the calculation of the loss function. The results can be seen in Fig. 2.21.

Figure 2.21 shows that not only the loss functions surfaces, but also the optimal hyperparameters for the GP-NARX and GP-NOE models are significantly different. It is apparent that the optimisation of the model parameters for the prediction is a much easier task than the optimisation of the model parameters for simulation.

2.4.4 Selection of Mean Function

We have mentioned in several places that the zero mean function is often presumed as the prior mean function of the GP model. This is quite common with system identification where the identification and validation data is preprocessed to ensure, among others, a zero mean. Nevertheless, this is not necessary and in some cases an explicit modelling of the mean function is required. The mean function can be specified using explicit basis functions. The following text is summarised from [49], but the topic is also elaborated in [76].

There are three main ways that the mean function can be used together with GP models:

1. The mean function is fixed and deterministic, which is usually the case when it is determined before the identification. An example would be when the data

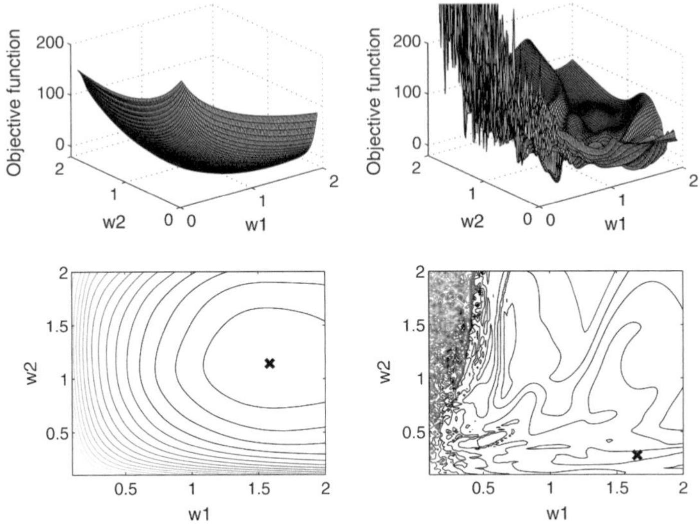

Fig. 2.21 NARX (*left*) and NOE (*right*) model loss functions (*top*) and their contour plots (*bottom*) with *black crosses* at the positions of the minima

is detrended during the preprocessing. In this case, the usual zero mean GP is applied to the difference between the measurements and the fixed mean function. Using

$$f(\mathbf{z}) \sim \mathcal{GP}(m_f(\mathbf{z}), C(\mathbf{z}_i, \mathbf{z}_j)), \tag{2.36}$$

the predictive mean becomes

$$E(y_f^*) = m_f(\mathbf{z}^*) + \mathbf{k}(\mathbf{z}^*)\mathbf{K}^{-1}(\mathbf{y} - m_f(\mathbf{z})), \tag{2.37}$$

where $\mathbf{K} = \mathbf{\Sigma}_f + \sigma_n^2\mathbf{I}$ and the predictive variance is according to Eq. (1.25).

2. The mean function is identified as a deterministic function that is a combination of a few preselected fixed basis functions with the coefficients collected in vector $\boldsymbol{\beta}$. We can write

$$g(\mathbf{z}) = f(\mathbf{z}) + \boldsymbol{\phi}(\mathbf{z})^{\mathrm{T}}\boldsymbol{\beta}, \quad \text{where } f(\mathbf{z}) \sim \mathcal{GP}(0, C(\mathbf{z}_i, \mathbf{z}_j)), \tag{2.38}$$

where $f(\mathbf{z})$ is a zero mean GP, $\boldsymbol{\phi}(\mathbf{z})$ is a vector of fixed basis functions and $\boldsymbol{\beta}$ are the additional coefficients to be identified together with the hyperparameters.

3. The mean function is identified as a stochastic function with a Gaussian prior on $\boldsymbol{\beta}$ so that $p(\boldsymbol{\beta}) = \mathcal{N}(\mathbf{b}, \mathbf{B})$. We get another GP with the covariance function expanded with uncertainty in the parameters of the mean:

$$g(\mathbf{z}) \sim \mathcal{GP}(\boldsymbol{\phi}(\mathbf{z})^{\mathrm{T}}\mathbf{b}, C(\mathbf{z}_i, \mathbf{z}_j) + \boldsymbol{\phi}(\mathbf{z}_i)^{\mathrm{T}}\mathbf{B}\boldsymbol{\phi}(\mathbf{z}_j)^{\mathrm{T}}). \tag{2.39}$$

The objective function that is used for the identification of the hyperparameters and the mean function parameters is expanded from Eq. (2.32) with $\mathbf{B}$ set to 0 in the case of deterministic β:

$$\begin{aligned}\ln p(\mathbf{y}|\mathbf{Z}, \mathbf{b}, \mathbf{B}) \\ &= -\frac{1}{2}\ln(|\mathbf{K} + \mathbf{\Phi}^{\mathrm{T}}\mathbf{B}\mathbf{\Phi}|) - \frac{1}{2}(\mathbf{\Phi}^{\mathrm{T}}\mathbf{b} - \mathbf{y})^{\mathrm{T}}(\mathbf{K} + \mathbf{\Phi}^{\mathrm{T}}\mathbf{B}\mathbf{\Phi})^{-1}(\mathbf{\Phi}^{\mathrm{T}}\mathbf{b} - \mathbf{y}) \\ &- \frac{N}{2}\ln(2\pi),\end{aligned} \tag{2.40}$$

where the matrix $\mathbf{\Phi}$ collects the vectors $\phi(\mathbf{z})$ for all the identification regression vectors.

The predictions are obtained as

$$\begin{aligned}E(\mathbf{y}_g^*) &= E(\mathbf{y}_f^*) + \mathbf{R}^{\mathrm{T}}E(\beta), \\ \mathrm{cov}(\mathbf{y}_g^*) &= \mathrm{cov}(\mathbf{y}_f^*) + \mathbf{R}^{\mathrm{T}}(\mathbf{B}^{-1} + \mathbf{\Phi}(\mathbf{Z})\mathbf{K}^{-1}\mathbf{\Phi}(\mathbf{Z})^{\mathrm{T}})\mathbf{R},\end{aligned} \tag{2.41}$$

where $\mathbf{R} = (\mathbf{\Phi}(\mathbf{Z}^*) - \mathbf{\Phi}(\mathbf{Z})\mathbf{K}^{-1}\mathbf{k}(\mathbf{Z}^*, \mathbf{Z}))$ and $E(\beta) = (\mathbf{B}^{-1} + \mathbf{\Phi}(\mathbf{Z})\mathbf{K}^{-1}$ $\mathbf{\Phi}(\mathbf{Z})^{\mathrm{T}})^{-1}(\mathbf{\Phi}(\mathbf{Z})\mathbf{K}^{-1}\mathbf{y} + \mathbf{B}^{-1}\mathbf{b})$.

In cases when the prior knowledge about $\mathbf{B}$ is vague, which means that $\mathbf{B}^{-1}$ approaches to the matrix of zeros, which is elaborated in [49].

2.4.5 Asymptotic Properties of GP Models

In model identification, it is important that the model is consistent. This means that the posterior distribution concentrates around the true distribution of the parameters as more and more data is observed or the sample size increases. The theory on consistency of GP models with sufficient conditions for the *posterior consistency* of GP regression is explained in [76] and reviewed in [49], with further references describing studies in various general settings in both listed references.

As stated in [76], these sufficient conditions are difficult to validate and may not be intuitive when applied to concrete models. In [77], the alternative concept of the so-called *information consistency* for GP regression models is described. The consistency of the information is checked with the information criterion that is the Césaro average of the sequence of prediction errors. See [76, 77] for more details and the background theory.

2.5 Computational Implementation

A noticeable drawback of the system identification with GP models is the computation time necessary for the modelling. GP regression, on which system identification is based, involves several matrix computations. This increases the number of operations with the third power of the number of input data, i.e. $\mathcal{O}(N^3)$, such as matrix inversion and the calculation of the log determinant of the used covariance matrix. This computational greed restricts the number of training data to at most a few thousand cases on modern workstations.

A common approach to computing the objective function from Eq. (2.32) and its gradient makes use of the Cholesky decomposition [49] of $\mathbf{K}$ to compute $\boldsymbol{\alpha} = \mathbf{K}^{-1}\mathbf{y}$.

The training algorithm in pseudocode is as follows:

Algorithm: GP TRAINING($\mathbf{Z}$, y, C, initial $\boldsymbol{\theta}$)

---

repeat
  change hyperparameters $\boldsymbol{\theta}$
  compute $\mathbf{K}(\boldsymbol{\theta}) = [C(\mathbf{z}_p, \mathbf{z}_q)]_{N\times N}$
  compute Cholesky decomposition $\mathbf{L} = \mathrm{chol}(\mathbf{K})$
  solve $\mathbf{L}\gamma = \mathbf{y}$ for γ and $\mathbf{L}^T\boldsymbol{\alpha} = \gamma$ for $\boldsymbol{\alpha}$ to get $\boldsymbol{\alpha} = \mathbf{K}^{-1}\mathbf{y}$
  compute $\ell(\boldsymbol{\theta})$ and $\nabla\ell(\boldsymbol{\theta})$ using $\boldsymbol{\alpha}$
until $-\ell(\boldsymbol{\theta})$ is minimal

2.5.1 Direct Implementation

One option to deal with the computational implementation is to approach the computation problem from the utilised hardware technology point of view. Since hardware capabilities are increasing every day, this approach might seem inefficient when looking over the longer term, but it is undoubtedly effective in the short term.

Parallel processing [78] is a popular way to deal with a large amount of data and a large number of computations. The authors of [79] envision the future computing systems as being hybrid computers, where the two major types of integrated components will be *multi-core central processing units (CPUs)* and *massively parallel accelerators*. While the 'standard' CPUs will continue to provide users with more and more computing power [80], many computer scientists will migrate towards general-processing graphics processing unit (GPGPU) applications [81], using graphics-card processors as the parallel accelerators for memory-dense, floating-point-intensive applications. The graphics processing unit (GPU) is, therefore, currently a low-cost, high-performance computing alternative to larger, stand-alone, parallel computer systems. An accelerated linear algebra package exploiting the hybrid computation paradigm is currently under development [82].

The concept of a GPGPU processor evolved from the needs of 3D-graphics-intensive applications. This need dictated the design of a processor such that more transistors were dedicated to the data processing than to the control and data caching, as in a regular CPU. Next, the processor was designed to be able to execute a data-parallel algorithm on a stream of data, which is the reason why GPGPU processors are sometimes called 'stream processors'. The currently dominant architectures for GPGPU computing are the nVidia CUDA [83] and the AMD APP (formerly ATI Stream) [84].

Some computation acceleration results for kernel methods are reported in [85, 86], while some preliminary results for GP modelling computation with a GPU are reported in [87].

The intrinsic parallel structure of a GPU allows a significant speed-up in comparison to the single-processor architecture. Although it is relatively easy to set up and perform basic operations, it quickly becomes more complex when dealing with more demanding numerical problems. Additionally, special care must be taken when performing memory operations:

- due to the relatively slow memory transfer, data transfers between the host system and the GPU device shall be as few as possible, and shall be asynchronous if possible;
- improper kernel code design with respect to the operation on different memory types (main GPU memory, shared memory, constant memory, texture memory) and ignoring memory access coalescing on the GPU device can cause a significant performance loss; and
- shared memory is organised into banks and accessing elements not consecutively will cause a bank conflict.

The modern NVIDIA GPUs come with an application programming interface (API), which had to be integrated into the implemented code, for example, Matlab's MEX functions. Additionally, several other programming libraries had to be used in order to implement the time-critical operations, such as the matrix factorisation and the inverse. The code makes use of CUBLAS, which is NVIDIA's implementation of the popular BLAS library [83, 88, 89] and of the CULA premium libraries [90], which provide a CUDA-accelerated implementation of linear algebra routines from the LAPACK library [91], including the time-critical Cholesky decomposition. Additionally, several other matrix operations had to be implemented as custom CUDA kernels.

As hardware capabilities are improving constantly and research on efficient algorithms is on-going, the presented hardware solutions might not be of long-term value. However, they offer a state-of-the-art, affordable, hardware configuration that might help to circumvent the computational issue in an intermediate time frame before more efficient algorithms or better technology arrive.

2.5.2 Indirect Implementation

To overcome the computational limitation issues and make use of the method also for large-scale dataset applications, a lot of authors have suggested various approximations, e.g. [92–94]. However, the research in this field is still active. A unified view of approximations of GP with a comparison of the methods is provided in [95] and some newer developments are compared in [96, 97].

Section 2.5 mentions that the computational demand of the GP regression direct implementation increases with the third power of the size of training set—$\mathcal{O}(n^3)$. This is due to the calculation of the covariance matrix inverse in

$$\alpha = \mathbf{K}^{-1}\mathbf{y} \tag{2.42}$$

or better finding the solution to the linear system of equations

$$\mathbf{K}\alpha = \mathbf{y} \tag{2.43}$$

for α. This becomes an issue when we work on problems that involve large quantities of training data, e.g. more than a couple of thousand for the presently available average computation power. Consequently, the developments of methods designed to reduce the computational demands of the GP modelling approach have received a great deal of attention.

These so-called approximation methods can, in general, be divided as follows:

- fast matrix-vector multiplication (MVM) methods, which use efficient computational methods for solving the system of linear equations. An example would be the use of iterative methods such as conjugate gradients;
- sparse matrix methods, which approximate the covariance matrix. The idea of the sparse matrix methods is to reduce the rank of the covariance matrix, i.e. the number of linearly independent rows, and to keep as much information contained in the training set as possible.

Figure 2.22 shows a schematic representation of an overview of the approximation methods.

Fast Matrix-Vector Multiplication

MVM methods treat all the identification data and build a full covariance matrix $\mathbf{K}$ of size $N \times N$, but use various, more efficient computation methods or their approximations for the calculation of computationally demanding expressions, i.e. the inverse of the covariance matrix and the logarithm of the covariance matrix determinant. An overview of the MVM methods is given in [98].

An example of such a method for solving the problem described with Eq. (2.43) is the conjugate gradient optimisation method, which reduces the computational demand of one iteration down to the order of $\mathcal{O}(N^2)$ [49], but an approximate solution can be obtained if the algorithm is terminated after k iterations, which results in an overall computational demand of $\mathcal{O}(kN^2)$.

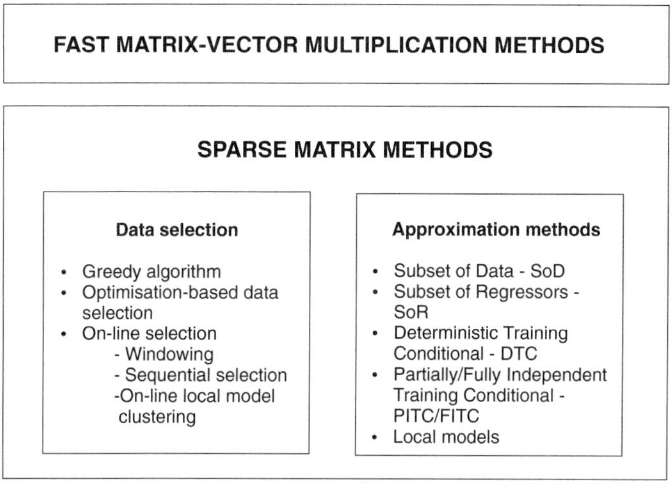

Fig. 2.22 Schematic representation of an overview of the approximation methods for GP modelling

Some other methods that reduce the number of computations of the covariance matrix inverse and the logarithm of the covariance determinant can be found in [99–101].

The MVM methods are the most efficient when the number of data is relatively small. However, these methods are also necessary with large amounts of data in order to decrease the computer memory demand during optimisation.

Sparse Matrix Methods

The fundamental property of the sparse matrix methods is that only a subset of the variables is treated exactly, with the remaining variables given some approximate, but computationally cheaper, approach. This is the most straightforward method for reducing the computational burden.

The first step in constructing a reduced rank or a sparse version of the covariance matrix $\mathbf{K}$ is the selection of a subset of datapoints. The selection of this ‘active’ subset of data, or active dataset, also called the induction variables, is something in common to all sparse matrix methods.

The second step is the approximation or treatment of the used latent stochastic variables that are to be treated exactly, by the GP model framework and of the remaining variables that are to be approximated by a less computationally demanding method. The used subset is of size $M \ll N$, where N is the size of the overall training dataset.

Only the active dataset is used for GP modelling. The rest of the $N - M$ data is treated differently using different approximation methods.

First, we list the methods for the selection of data to be included in the active subset and, second, we list the approximation methods. However, it needs to be emphasised that a number of other sparse matrix methods have appeared in recent years as this is a field of active research.

Methods for Data Selection

As only the active dataset is to be treated fully in the sparse model, the process of determining which data points are to be included is critical to the success of the approximation [50]. One possible strategy is to build the subset of data through the manual selection of data points based on a preliminary knowledge of the system to be modelled. However, this is difficult where the preliminary knowledge about the system to be modelled is limited and when the system contains complex, multidimensional nonlinearities, which is a common situation with systems identification.

It makes sense to search the active dataset based on a selected criterion. The criterion-based evaluation of all the possible combinations of an active dataset of dimension M is not viable, because of the $\frac{N!}{M!(N-M)!}$ combinations. There have been a lot of methods developed based on different criteria to overcome the computational issue.

Greedy algorithm is one of the more popular methods. Apart from the random selection of data points, with limited applicability, *greedy approximation* methods [102] have been shown to have great potential. The idea of these methods is that the active dataset is selected and updated according to some criterion. Such an algorithm would initiate with an empty active set $\mathcal{I}$, with the remaining set $\mathcal{R}$ containing all the indexed training observations. Then, using an iterative method, each indexed training example is added to the active set in turn and the selection criterion is evaluated. If the criterion is met, the training example under review will be included in the active set.
Various authors, e.g. [102–104], have suggested different criteria for the selection of data.
This method can be used with most of the approximation methods. The question that arises is what kind of selection criteria should be used to determine the active subset of data. Some of the selection criteria are those used in the following methods: informative vector machine [102], informative gain [105], sparse spectral sampling [106], iterative sparse [103] and matching pursuit [104].

Optimisation-based data selection Greedy methods select the active dataset out of the available identification data. A method is proposed in [33] called *Sparse Pseudo-input Gaussian Processes (SPGP)*, which does not select the dataset $\mathcal{I}$, but optimises it. This means that the active dataset $\mathcal{I}$ can at the end of the optimisation contain arbitrary data. Optimisation of this arbitrary data, called *pseudo-inputs*, is pursued simultaneously with the optimisation of the hyperparameters based on the log-marginal likelihood loss function. This is perceived as an advantage of this method.
Based on the same idea, the authors of [107] have proposed a new sparse matrix method called *Sparse Spectrum Gaussian Process Regression (SSGP)*, which is based on a spectral representation of GP.

Because of the increased number of optimisation parameters, both methods are susceptible to overfitting, particularly when there is no available information on the initial values of the hyperparameters and the pseudo-inputs. To overcome this issue, a variational method is proposed in [108]. This method maximises the lower boundary of the log-marginal likelihood instead of the exact value when it optimises the active dataset $\mathcal{I}$ and the hyperparameters.

Online data selection In the case that the GP model is identified based on the incoming data stream that provides new information about the modelled system online, online identification, or training, methods are in place.

Online data selection can be roughly divided as follows.

Windowing is a method where the most recent m data is used for the identification. The active dataset $\mathcal{I}$ is therefore composed of the most recent m data, which are all weighted equally for the optimisation. The windowing method is called also the time-stamp method. Examples of using windowing in the GP modelling context can be found in [109–111].

The situation is different with the method of *forgetting*, where the significance of the data is decreasing, usually exponentially, with age. The forgetting method is also called the weight-decay method [110]. When the data point significance is below the threshold, the data point is discarded. The forgetting method is still closely related to the pure windowing.

Sequential selection follows the idea that the size of the active dataset $\mathcal{I}$ has to be constrained, which is implemented with a selected criterion for the inclusion of data. This criterion is similar to that used by the greedy method, but in this case each data point is evaluated separately with already-selected data in the active dataset $\mathcal{I}$ and not with those in the remaining dataset $\mathcal{R}$, as is the case with the greedy method. The criterion that has been proposed in [103], on the other hand, is based on a change of the mean value of the posterior distribution when a new data point is included in the model. Following this idea, the authors of [112] proposed the method for online training, which also enables the online optimisation of the hyperparameters.

Online model clustering Slightly different from the other methods for online data selection is the method described in [113]. This method is based on the online clustering of data, where these clusters of data are used for the identification of local GP models. The advantage of this method is its computational speed, because online clustering is computationally much less demanding than online GP model training. Nevertheless, the quality of the prediction could be slightly worse in comparison with using data selection methods.

Methods for Approximation

The idea behind sparse matrix methods is to change the joint prior distribution of the identification and validation data $p(f^*; f)$ from Eq. (2.44) so that the computation of the predictive distribution based on the validation data from Eq. (2.45) is less time consuming.

$$p(f^*; f) = \mathcal{N}\left(\mathbf{0}, \begin{bmatrix} \mathbf{K} & \mathbf{k}(\mathbf{z}^*) \\ \mathbf{k}^{\mathrm{T}}(\mathbf{z}^*) & \kappa(\mathbf{z}^*) \end{bmatrix}\right), \tag{2.44}$$

where * denotes the validation data.

$$p(f^*|y) = \mathcal{N}\left(\mathbf{k}(\mathbf{z}^*)\mathbf{K}^{-1}\mathbf{y}, \kappa(\mathbf{z}^*) - \mathbf{k}^{\mathrm{T}}(\mathbf{z}^*)\mathbf{K}^{-1}\mathbf{k}(\mathbf{z}^*)\right) \quad (2.45)$$

We divide the approximation methods into five groups.

Subset of data (SoD) method appears to be the most straightforward method for a sparse matrix approximation, where the active set of the data M is selected from the whole training data set of size N. The equations for the calculation of the predictions and the optimisation remain unchanged by the method. The calculation method for such a method depends only on the dataset of size M and is therefore $\mathcal{O}(M^3)$, where $M < N$. The efficiency of this method depends very much on the selection of the subset of data, while the remaining dataset is discarded rather than approximated. Nevertheless, with carefully selected data the method may still provide a good approximation to the GP model with complete dataset and is consequently frequently used in comparison with the more sophisticated sparse methods.

Subset of regressors (SoR) method takes advantage of the equivalence between the GP model's mean predictor and that of a finite-dimensional, generalised, linear regression model. Consequently, the SoR model is a finite, linear-in-the-parameters model with a particular prior knowledge put on the weights. In contrast to the SoD method, the SoR method is to employ all N datapoints of the training set in the approximation. The major disadvantage of the SoR method is that, due to its basis upon a linear-in-the-parameters model, the GP model becomes degenerate and restricts the variety of possible functions that will be plausible under the posterior [96]. The main disadvantage of this degeneracy is that the predictive distributions of the GP model can become unreasonable. The computation demand of the SoR method is $\mathcal{O}(M^2N)$ for the initial matrix computations, and $\mathcal{O}(M)$ and $\mathcal{O}(M^2)$ for the calculation of the predictive mean and variance, respectively.

Deterministic training conditional (DTC) method applies the model of the entire N size dataset and therefore does not use a degenerated model like the SoR method does. The obtained GP model represents only $M < N$ function values, but uses the entire dataset so that it projects M input values to N dimensions. The computational method for the DTC method is, like in the case of SoR method, $\mathcal{O}(M^2N)$, for the initial matrix computations, and $\mathcal{O}(M)$ and $\mathcal{O}(M^2)$ for the calculation of the predictive mean and variance, respectively. The method was initially named projected latent variables [105], but is also known as projected processes [49].
The predictive mean value when using the DTC method is identical to that of using the SoR method. The predictive variance is never less than the variance when using the SoR method. The computation demand of the DTC method is the same as that of the SoR method.

Partially and fully independent training conditional (PITC, FITC) The previously described methods use exclusively identification data for the input–output mapping function modelling, so the covariance matrix for the training conditional distribution $p(f|\mathbf{u})$ is zero. Another possibility is that the covariance matrix has a block-diagonal form, with data obtained from the covariance matrix of the complete dataset. This means that we assume conditional independence, which is assumed in previously described methods for only a part of the modelled function values. Such a method is called *Partially Independent Training Conditional (PITC)*.

The PITC method's computational demand depends on the used block-diagonal form. In the case of $l = \frac{N}{M}$ blocks of size $M \times M$, as in [114] the computational demand is $\mathcal{O}(NM^2)$.

A special form of the PITC method is the approximation method called *Fully Independent Training Conditional (FITC)*, which was proposed in [33], where it was named the SPGP method. The name FITC comes from the fact that the training set function observations are presumed to be completely independent. The FITC method is almost identical to the PITC method, except for the fact that the covariances on a diagonal of the covariance matrix are exact. This means that instead of approximated prior variances the exact prior variances are on the covariance matrix diagonal.

The predictive distribution obtained with the FITC method is identical to the predictive distribution obtained with the PITC method. The computational demand is equal to those of the SoR, DTC and PITC methods.

Local models An alternative approach to the listed methods for the approximation of the covariance matrix or the model likelihood is the method that replaces one GP model with *local models*. The obtained model is known as a *mixture of GP models* or *mixture of GP experts*. This method does not match exactly to the listed ones, because the local models replace the entire GP model. Some research about using local models for an approximation in the context of dynamic systems modelling can be found in [113, 115, 116].

The idea of the method is based on the *divide-and-conquer* principle. The method divides the dataset of N identification data into $l = \frac{N}{M}$ subsets or *clusters* of size M. The data in these clusters is used as the identification data for separate, locally valid GP models. The various local-model-based methods differ, among others, in the way in which the various existing clustering methods are pursued.

The hyperparameters can be optimised for each of the local models separately or for all the local GP models together. The latter is an interesting option only when M is small enough. Combination of posterior distributions of local models into one distribution that represents the model prediction is another difference between the various methods.

All the listed approximation methods have comparable computational demands. The overview in [96] shows that the obtained predictive mean values are very similar, while the predictive variances are quite different. Therefore, it is sensible to decide before selecting an approximation method, whether the information about the exact model's predictive variance is important, or we are focused more on the model's predictive mean value.

The SoR method is appropriate in the case that the predictive variance is not very important, because the predictive variance approximation is computationally the simplest one.

In the case, however, that the accurate information about the predictive variance is important, it is sensible to use one of the more advanced approximation methods, i.e. DTC, PITC or FITC.

In the case of computationally simpler identification problems, which are those with the smaller number of identification data, it is sensible to use the SoD method. It competes well with the more advanced approximation methods.

Another aspect interesting for the approximation-method selection is the implementation of the method, the computational demand for the model identification and the demand necessary for the obtained model prediction.

It is very much worth considering a combination of methods, like the one proposed in [97]. This approach combines the SoD method for the model training and the FITC method for the model prediction. It uses the selected active dataset $\mathcal{I}$ and the hyperparameters obtained with the SoD method to save computational time for the hyperparameters' optimisation, which for the SoD method is $\mathcal{O}(M^3)$, instead of $\mathcal{O}(NM^2)$ for the FITC method.

2.5.3 *Evolving GP Models*

In this section, an approach to the online training of GP models is described. Such an approach is needed when the dynamic system to be identified is represented as a time-varying one or when the training data is not available for the whole region of interest, and so not all the dynamics of the system can be trained at once. In these cases, the model needs to be constantly adapted in accordance with the new operating region and/or the changing dynamics. Such an approach can be used in environments that are constantly changing. For this purpose, a method for the online adaptation of GP models is proposed in [117], and the models obtained with this method are named *Evolving GP models*.

Evolving systems [118] are self-developing systems inspired by the idea of system model evolution in a dynamically changing and evolving environment. They differ from other traditional adaptive systems known from control theory [119, 120] in that they online adapt both the structure and parameter values of the model using the incoming data [118].

The term evolving is used in the sense of the iterative adaptation of the structure of the GP model and the hyperparameter values. This term was introduced in the

1990s for neural networks [121], in 2001 for fuzzy rule-based systems [122] and in 2002 for neuro-fuzzy systems [123].

The GP models depend on the data and the covariance function. More precisely, the data is defined with various regressor variables, in short regressors, and corresponding observations grouped in the so-called basis vectors. The covariance function is defined with the type and hyperparameter values.

Therefore, there are at least four parts that can evolve

- the regressors,
- the basis vectors,
- the type of covariance function and
- the hyperparameter values.

The ideas of the online adaptation of GP models can be found in literature implemented mainly as online data selection, e.g. [111, 124], online data selection and hyperparameters determination, e.g. [125], and active learning for control, e.g. [117, 126]. A method that does not select an active dataset for GP model identification, but only iteratively corrects the identified model's predictive mean and variance with new data, is proposed in, e.g. [127, 128]. The criterion for the correction is based on the prediction error.

As already stated in Sect. 2.5, the training of GP models for a large amount of data is very time consuming. The condition under which the evolving GP model does not 'decay in time' with in-streaming data is the so-called *persistent excitation* condition, which means that the in-streaming signal has enough information content. This can be achieved so that only a subset of the most informative data, the so-called **basis vectors** set, is used. With a type or a combination of various types of covariance functions a prior knowledge of the system is included in the model. Nevertheless, by optimising **the hyperparameter values** the model response evolves closer to the response of the real system. However, in dynamic, nonlinear systems, where the nonlinear mapping between the input and output data cannot be easily formulated, the squared exponential covariance function is frequently used, assuming the smoothness and stationarity of the system. This means that the covariance function may be kept fixed and does not need to evolve. Nevertheless, the structure-discovery algorithm described in [129] can be used for the online **covariance function** selection. Furthermore, the squared exponential covariance function can be used with ARD, which is able to find the influential **regressors** [49]. With the optimisation of the hyperparameter values, the noninfluential regressors have smaller values and, as a consequence, they have a smaller influence on the result. Therefore, all the available regressors can be used and, as a result, only the set of basis vectors and the hyperparameter values remain to be evolved.

The general concept of evolving GP models, presented in [117], with a fixed covariance function and regression vector, contains the following steps:

Algorithm: MODELEVOLVING($\mathbf{z}^*, y^*$)

comment: Add new input data $\mathbf{z}^*$ to the informative dataset $\mathcal{I}$

1. $\mathcal{I} \leftarrow$ ADD($\mathcal{I}, \mathbf{z}^*$)

if LENGTH($\mathcal{I}$) $>$ *maxLength*

then
- **comment:** Calculate the information gain for each data point in $\mathcal{I}$
- 2. *iGains* $\leftarrow$ INFORMATIONGAIN($\mathcal{I}$)
- **comment:** Remove worst data from $\mathcal{I}$
- 3. $\mathcal{I} \leftarrow$ REMOVEWORST($\mathcal{I}$, *iGains*)

comment: Calculate hyperparameter values $\boldsymbol{\theta}$

4. $\boldsymbol{\theta} \leftarrow$ CALCHYPVAL($\mathcal{I}, \boldsymbol{\theta}$)

comment: Update covariance matrix

5. $\mathbf{K} \leftarrow$ UPDATECOVMAT($\mathcal{I}, \boldsymbol{\theta}$)

These basic steps are repeated for every incoming sample of data until there is no more available data or until a requirement to stop the process is received.

To keep the subset of the most informative data small enough to process all the steps before new data arrives, the maximum length of the subset should be set with the parameter *maxLength*. This means that the parameter *maxLength* is a design parameter.

Operations in the pseudocode can be implemented in various ways. There are two critical operations: the calculation of the information gain and the calculation of the hyperparameter values. Both of these operations can be implemented using various well-known methods. For the information gain calculation, any data selection criterion from online learning methods for GP models can be used, e.g. [103, 105], etc. Also, for the hyperparameter value calculation any suitable optimisation method or even some heuristic method can be used. Our implementation of the operations in the concept is described below.

First, the basic elements and some operations will be described. The core of the concept is a set of the most informative data. Actually, it is a subset of the data that was already considered. It is denoted as $\mathcal{I}$ and defined as

$$\mathcal{I} \subset \mathcal{D}. \tag{2.46}$$

To operate with the set $\mathcal{I}$ two basic operations are needed: adding elements and removing elements. Both operations are very straightforward. Adding the new element ζ^+ to the set $\mathcal{I}$ is defined as

$$\mathcal{I}^+ = \left\{\mathcal{I}, \zeta^+\right\}, \tag{2.47}$$

where $\mathcal{I}^+$ is a new, extended set of the most informative data. Removing the ith element ζ_i from the set $\mathcal{I}$ is defined as

$$\mathcal{I}^{-} = \{\zeta_1, \ldots, \zeta_{i-1}, \zeta_{i+1}, \ldots, \zeta_M\}, \tag{2.48}$$

where M is the number of elements in the set $\mathcal{I}$.

The main operations of the algorithm are implemented as follows:

1. ADD($\mathcal{I}, \mathbf{z}^*$): Adds new data to the set of the most informative data $\mathcal{I}$. It is implemented in such a way that it adds new data $\mathbf{z}^*$ to $\mathcal{I}$ only when the data contributes new information to the current model. This improves the robustness of the algorithm. The contribution of the data, according to the current model, is scored on the prediction distribution for $\mathbf{z}^*$. If the absolute difference between the prediction mean value $\mu(\mathbf{z}^*)$ and the measured value $y(\mathbf{z}^*)$ is greater than the pre-set threshold, this means the current model cannot predict the prediction based on $\mathbf{z}^*$ accurately enough. Therefore, $\mathbf{z}^*$ should be added to $\mathcal{I}$.
If the absolute difference $|y(\mathbf{z}^*) - \mu(\mathbf{z}^*)|$ is small enough, the prediction variance is also taken into consideration. If the prediction variance $\sigma^2(\mathbf{z}^*)$ is smaller than the pre-set threshold, it can be considered that the model is also confident enough in its prediction; therefore, there is no need to include the new data. If the prediction variance is high, the model is not confident in the prediction. This means that the model does not have enough information in that region; therefore, $\mathbf{z}^*$ should be added to $\mathcal{I}$.
To summarise, $\mathbf{z}^*$ is added only when the absolute difference between the prediction mean value $\mu(\mathbf{z}^*)$ and the measured value $y(\mathbf{z}^*)$ or the prediction variance $\sigma^2(\mathbf{z}^*)$ is greater than the pre-set thresholds for the mean value and variance, respectively.
Thresholds can be set heuristically and are, therefore, design parameters. If the condition is fulfilled, $\mathbf{z}^*$ is added to $\mathcal{I}$ using Eq. (2.47). This operation as such might not be necessary, but it avoids any unnecessary updates of $\mathcal{I}$, which improves the computational efficiency of the algorithm.

Procedure 1: ADD($\mathcal{I}, \mathbf{z}^*, y(\mathbf{z}^*)$)

---

comment: Prediction of the GP model is calculated

if $|y(\mathbf{z}^*) - \mu(\mathbf{z}^*)| >$ threshold$_\mu$ AND $\sigma^2(\mathbf{z}^*) >$ threshold$_{\sigma^2}$

then $\begin{cases} \textbf{comment: } \text{Adding } \mathbf{z}^* \text{ to } \mathcal{I} \text{ using Eq. (2.47)} \\ \mathcal{I} \leftarrow \{\mathcal{I}, \mathbf{z}^*\} \end{cases}$

return ($\mathcal{I}$)

---

2. INFORMATIONGAIN($\mathcal{I}$): Calculates the information gain for each element in $\mathcal{I}$. Actually, it calculates the log-marginal likelihood for each subset $\mathcal{I}^-$ of size $M-1$, where M is the number of elements in $\mathcal{I}$. It should be noted that this operation is performed only when the subset $\mathcal{I}$ has exceeded the pre-set size L. A higher log-marginal likelihood means a higher information gain for the processed element

in $\mathcal{I}$. As calculated for the subset without the data for which the information gain is to be calculated, it should be inverted. Therefore, it is multiplied by -1, so that a higher log-marginal likelihood means a lower information gain.
This principle is in its essence the Bayesian information criteria (BIC) [44]:

$$\begin{aligned} \mathrm{BIC} &= -2 \ln p(\mathcal{I}^- | \boldsymbol{\theta}) + (D+2) \ln(M) \\ &= -2\tilde{\ell}(\mathcal{I}^-, \boldsymbol{\theta}) + (D+2) \ln(M), \end{aligned} \tag{2.49}$$

where $D+2$ is the number of hyperparameters to be estimated.
In the case of a time series, forgetting is also used. It is implemented as exponential forgetting

$$\tilde{\ell}(\mathcal{I}^-, \boldsymbol{\theta}) = \lambda^{i-c} \cdot \ell(\mathcal{I}^-, \boldsymbol{\theta}) \tag{2.50}$$

where $\lambda \in [0, 1]$ is the forgetting factor, i is the current sequence number, c is the number of the sequence when the currently considered data was added to $\mathcal{I}$, $\tilde{\ell}$ is the log-marginal likelihood considering the exponential forgetting and $\mathcal{I}^-$ is the subset of $\mathcal{I}$ of size $M-1$. The forgetting can be easily turned off by setting $\lambda = 1$.

Procedure 2: INFORMATIONGAIN($\mathcal{I}$)

---

for $i \leftarrow 1$ **to** LENGTH($\mathcal{I}$)
do
- **comment:** removing ith element from $\mathcal{I}$ using Eq. (2.48)
- $\mathcal{I}^- \leftarrow \{\zeta_1, \ldots, \zeta_{i-1}, \zeta_{i+1}, \ldots, \zeta_M\}$
- **comment:** calculating information gain using Eq. (2.50)
- $iGains[i] \leftarrow -\tilde{\ell}(\mathcal{I}^-, \boldsymbol{\theta})$

return ($iGains$)

---

3. REMOVEWORST($\mathcal{I}$, *iGains*): Removes element with the worst information gain from the set $\mathcal{I}$. It is performed simply using Eq. (2.48).

Procedure 3: REMOVEWORST($\mathcal{I}$, *iGains*)

$ind \leftarrow -1$
$min \leftarrow \infty$
for $i \leftarrow 1$ **to** LENGTH($iGains$)
do
- **if** $iGains[i] <$ min
  then
  - $ind \leftarrow i$
  - $min \leftarrow iGains[i]$

comment: removing *ind*th element from $\mathcal{I}$ using Eq. (2.48)
$\mathcal{I} \leftarrow \{\zeta_1, \ldots, \zeta_{ind-1}, \zeta_{ind+1}, \ldots, \zeta_M\}$
return ($\mathcal{I}$)

---

4. CALCHYPVAL($\mathcal{I}$, $\boldsymbol{\theta}$): Hyperparameter values are calculated by maximising the marginal log likelihood. This can be done with any suitable optimisation method off-line or with iterative calculations online.
In the off-line case, the evolving method downgrades to the so-called online modelling with a fixed set of input data, examples can be found in [7, 76].
5. UPDATECOVMAT($\mathcal{I}$, $\boldsymbol{\theta}$): Updates the covariance matrix when $\mathcal{I}$ or $\boldsymbol{\theta}$ have changed.
Updates the covariance matrix and its inversion. If the hyperparameter values have changed, both the covariance matrix $\mathbf{K}$ and its inversion $\mathbf{K}^{-1}$ must be fully recalculated. However, in cases when only $\mathcal{I}$ has changed, the covariance matrix and its inversion can be updated more efficiently. The covariance matrix is updated by appending $\mathbf{k}(\mathbf{z}^*)$ and $\mathbf{k}^{\mathrm{T}}(\mathbf{z}^*)$ as presented in Eq. (2.51) and removing the ith row and column if the ith data was removed from $\mathcal{I}$, as shown in Eq. (2.52).

$$\mathbf{K}^{+} = \left[\begin{array}{ccc|c} \mathbf{K}_{1,1} & \cdots & \mathbf{K}_{1,M} & \mathbf{k}_1 \\ \vdots & \ddots & \vdots & \vdots \\ \mathbf{K}_{M,1} & \cdots & \mathbf{K}_{M,M} & \mathbf{k}_M \\ \hline \mathbf{k}_1 & \cdots & \mathbf{k}_M & \kappa \end{array}\right], \tag{2.51}$$

where $\mathbf{k}$ is the vector of covariances between the data in the active dataset $\mathcal{I}$ and the new data $\mathbf{z}^{+}$ and κ is the autocovariance of $\mathbf{z}^{+}$.

$$\mathbf{K}^{-} = \left[\begin{array}{ccc|ccc} \mathbf{K}_{1,1} & \cdots & \mathbf{K}_{1,i-1} & \mathbf{K}_{1,i+1} & \cdots & \mathbf{K}_{1,M} \\ \vdots & \ddots & \vdots & \vdots & \ddots & \vdots \\ \mathbf{K}_{i-1,1} & \cdots & \mathbf{K}_{i-1,i-1} & \mathbf{K}_{i-1,i+1} & \cdots & \mathbf{K}_{i-1,M} \\ \hline \mathbf{K}_{1+1,1} & \cdots & \mathbf{K}_{i+1,i-1} & \mathbf{K}_{i+1,i+1} & \cdots & \mathbf{K}_{i+1,M} \\ \vdots & \ddots & \vdots & \vdots & \ddots & \vdots \\ \mathbf{K}_{M,1} & \cdots & \mathbf{K}_{M,i-1} & \mathbf{K}_{M,i+1} & \cdots & \mathbf{K}_{M,M} \end{array}\right] \tag{2.52}$$

As the Cholesky decomposition is used for the calculation of the covariance matrix inversion, it can be updated in a similar way as updating the covariance matrix using the rank-1 update and downdate operations [130]. First, the Cholesky decomposition is updated by $\mathbf{k}(\mathbf{z}^*)$ and later downdated by $\mathbf{k}(\mathbf{z}_i)$ if the ith data was removed from $\mathcal{I}$, which is relatively efficient ($\mathcal{O}(M^2)$).

2.6 Validation

Validation concerns the level of agreement between the mathematical model and the system under investigation [131] and it is many times underemphasised despite its importance. The model validation in the context of identification is the phase following the model selection, where the regressors, covariance function, mean function and hyperparameters are selected. Validation is eliminated in a full Bayesian approach where these elements are not selected by optimisation but integrated out. Nevertheless, the GP model selection, which is based on evidence or marginal likelihood maximisation, requires this phase too. In validation, we are concerned with evaluating the performance of the model based on datasets different from those used for the modelling.

The data that is used for modelling is called *identification data*, also *estimation data*, which are names common to identification literature. An identification dataset can be, depending on the model used, split into the subset of data used for the parameters' estimation, and the subset of data for the structure identification as well as the model order and regressors. The purpose of this division is to use the other subset of data for monitoring the performance during the identification for model structure reduction. The data for evaluating the performance, which is different from that used for the identification, is called *validation data*.

This division of data subsets is referred to differently in machine-learning literature, though the purpose is the same. The data for parameter estimation is called *training data*, the data for monitoring the performance is called *validation data* and the data used for the validation of the obtained final model is called *test data*. It has to be noted that this division can also be met in the literature describing system identification, e.g. [10].

The concept of splitting empirical data into separate sets for identification and validation is generally known as *cross-validation*. An analysis of importance to use separate sets for validation also in the context of GP classification is done in [132], but can be generalised also for regression.

The quality of the model that is in the focus of the validation can be represented by several features. Their overview can be found in [131, 133]. The most important are the model plausibility, model falseness and model purposiveness, explained as follows.

Model *plausibility* expresses the model's conformity with the prior process knowledge by answering two questions: whether the model 'looks logical' and whether the model 'behaves logical'. The first question addresses the model structure, which in the case of GP models means mainly the plausibility of the hyperparameters. The second one is concerned with the responses at the model's output to typical events on the input, which can be validated with a visual inspection of the responses, as is the case with other black-box models.

Model *falseness* reflects the agreement between the process and the model's output or the process input and the output of the inverse model. The comparison can be made in two ways, both applicable to GP models: qualitatively, i.e. by visual inspection of the differences in the responses between the model and the process, or quantitatively, i.e. through the evaluation of the performance measures, some of them listed later in the section.

Model *purposiveness* or usefulness tells us whether or not the model satisfies its purpose, which means the model is validated when the problem that motivated the modelling exercise can be solved using the obtained model. Here, again, the prediction variance can be used, e.g. when the prediction confidence is too low, the model can be labelled as not purposive.

As already mentioned in Sect. 2.1, various purposes are served with two sorts of models: for prediction, i.e. one-step-ahead prediction, and for simulation. Section 2.3 described different model structures in this context. Nevertheless, it is a very common practice to train the model for prediction and to call out its purpose during the validation, where especially the dynamic system model is often validated for simulation.

The cross-validation concept requires a relatively large amount of data. This is provided by properly designed experiments, which are, commonly, repeated if the amount of data does not correspond to the modelling needs. That is why the experiment design and the experiments themselves are very important parts of system identification.

However, for cases where the amount of empirical data is limited and new experiments cannot be pursued, methods have been suggested that seek to maximise the exploitation of the available data. A lot of research to solve this problem has been done in machine learning. Here, we highlight only some more frequently used methods for the more efficient use of the identification data.

One such method is *k-fold cross-validation* where the available empirical data is partitioned into k data subsets. Each subset is then used in turn as a dataset for the evaluation of the model trained on the other $k - 1$ data subsets. The overall error rate is taken as the average of these k data subset evaluations.

The extreme version of the k-fold cross-validation is when only a single observation, data piece, of the overall data is to be left out and used as an evaluation example. The remaining data is used for training. The method is called *leave-one-out-validation* (LOO). It is a method that can be used for small datasets.

While GP model validation in the context of one-step-ahead prediction is elaborated already in [49], where marginal likelihood is discussed in detail, cross-validation, especially leave-one-out cross-validation (LOO-CV), is described. The expression 'validation' in the listed terms is used here in the machine-learning sense and not in the system identification sense, where the validation data is a separate and fresh dataset.

We have mentioned that model falseness can be evaluated with different performance measures. There exists an abundance of performance measures, each suited for a different purpose with the emphasis on a different property and with different expressiveness. Here we list only a few that serve our purposes.

A commonly used performance measure, especially when the model is identified using a method that is based on a square error, is the mean-squared error (MSE):

$$\mathrm{MSE} = \frac{1}{N}\sum_{i=1}^{N}(y_i - E(\hat{y}_i))^2, \tag{2.53}$$

where y_i and $\hat{y}_i$ are the system's output measurement, i.e. observation and model's output in ith step, respectively. The model's output can be, based on investigation, a posterior probability distribution of a prediction or a simulation.

The measure that normalises the mean-squared error between the mean of the model's output and the measured output of the process by the variance of the output values of the validation dataset is the standardised mean-squared error (SMSE):

$$\mathrm{SMSE} = \frac{1}{N} \frac{\sum_{i=1}^{N} (y_i - E(\hat{y}_i))^2}{\sigma_y^2}, \tag{2.54}$$

where y_i and $\hat{y}_i$ are the system's output measurement, i.e. observation and the model's output in the ith step.

The mean relative square error (MRSE) is calculated by taking the square root of the measure MSE divided by the average of output measurements:

$$\mathrm{MRSE} = \sqrt{\frac{\sum_{i=1}^{N} (y_i - E(\hat{y}_i))^2}{\sum_{i=1}^{N} y_i^2}}. \tag{2.55}$$

Some authors call this performance measure the relative-root-mean-square error (RRMSE).

The performance measures described with Eqs. (2.53)–(2.55) deal with the mean values of outputs and do not take the entire output distribution into account.

The performance measures such as the log predictive density error (LPD) [134, 135] can be used for evaluating GP models, taking into account not only the mean of the model prediction, but also the entire distribution:

$$\mathrm{LPD} = \frac{1}{2} \ln(2\pi) + \frac{1}{2N} \sum_{i=1}^{N} \left(\ln(\sigma_i^2) + \frac{(y_i - E(\hat{y}_i))^2}{\sigma_i^2} \right) \tag{2.56}$$

where σ_i^2 is the model's output variance in the ith step. The performance measure LPD weights the output error $E(\hat{y}_i) - y_i$ more heavily when it is accompanied by a smaller output variance σ_i^2, thus penalising the overconfident model's output values more than the acknowledged bad model's output values, indicated by a higher variance.

The mean standardised log loss (MSLL) [49] is obtained by subtracting the loss of the model that predicts using a Gaussian with the mean $E(\mathbf{y})$ and the variance σ_y^2 of the measured data from the model LPD and taking the mean of the obtained result

$$\mathrm{MSLL} = \frac{1}{2N} \sum_{i=1}^{N} \left[\ln(\sigma_i^2) + \frac{(y_i - E(\hat{y}_i))^2}{\sigma_i^2} \right] - \frac{1}{2N} \sum_{i=1}^{N} \left[\ln(\sigma_y^2) + \frac{(y_i - E(\mathbf{y}))^2}{\sigma_y^2} \right]. \tag{2.57}$$

The MSLL is approximately zero for simple models and negative for better ones.

The smaller the listed measures are, the better the model response is, irrespective of whether it is a prediction or simulation.

The variance of the GP model predictions on a validation signal can be a plausibility measure itself, as it indicates whether the model operates in the region where the identification data were available. Nevertheless, it should be used carefully and in combinations with other validation tools, as predictions with a small variance are not necessarily good, as will be shown in the example at the end of this chapter.

To avoid the computational complexity of cross-validation, alternative methods of evaluating validation errors have been developed. These include the use of various *information criteria* methods, such as the final prediction error information criterion, or Akaike's information criterion [44], where the normalised log likelihood is used as the prediction error criterion

$$\mathrm{AIC} = -2\ln(p(\mathcal{D}|\boldsymbol{\theta}_{\mathrm{ML}})) + 2n, \tag{2.58}$$

where n is a number of adjustable parameters in the model and $\boldsymbol{\theta}_{\mathrm{ML}}$ is the maximum likelihood solution for $\boldsymbol{\theta}$, or Akaike's Bayesian information criterion [44] that uses the marginal likelihood of the observed data $\mathcal{D}$ given the model:

$$\mathrm{BIC} = -2\ln(p(\mathcal{D}|\boldsymbol{\theta}_{\mathrm{MAP}})) + n\ln N, \tag{2.59}$$

where N is the number of data and $\boldsymbol{\theta}_{\mathrm{MAP}}$ is the value of $\boldsymbol{\theta}$ at the mode of the posterior distribution. Statistical hypothesis tests can also be used for the model validation. See [2] or [10] for more information on these validation strategies.

The quality of the obtained model can also be evaluated based on a residual analysis. Residuals are the differences between the most likely model's output values and the measured output values of the system to be modelled. Residual analysis [2] is evaluating statistics of residuals like correlation tests, whiteness tests and analyses of average generalisation errors. It is often used with methods that are concern mainly with the model's posterior mean values.

The authors of [12] discuss methods for the validation of prediction models in the context of neural networks with the residual analysis. They also provide a discussion on the visualisation of predictions, which is also a useful method with the validation of simulation models.

Criteria that are concerned mainly with the model's posterior mean values do not take account the entire posterior distributions as in a fully Bayesian approach [44, 49], which is explained in Sects. 1.1 and 2.4.1.

Since a more than fair portion of dynamic system models is validated for simulation purposes, the next section is devoted to the implementation of a model simulation in the context of GP models.

2.7 Dynamic Model Simulation

The simulation of dynamic system models can be used for the evaluation of the model behaviour or for the model validation. Simulation is a multistep-ahead prediction when the number of steps in the prediction horizon is infinite or at least as large as the time horizon of interest for the foreseen analysis of the model's behaviour.

There are two implementation options for the simulation or multistep-ahead prediction:

- a direct method, where different models are learnt for every perceived horizon h or
- an iterative method, where the one-step-ahead prediction is iteratively repeated.

The problem of the direct method is that the horizon needs to be known and fixed in advance. In the case that the horizon is changed, the model, or better models, has to be learnt again. The second issue with the direct method is that highly nonlinear systems need a large horizon and, consequently, a large amount of learning data [135]. An example of using the direct method for multistep-ahead prediction is given in [66].

The iterative method for Gaussian process models of dynamic systems means that the current output estimate depends on previous model estimations and on the measured inputs.

$$\hat{y}(k) = f(\hat{y}(k-1), \hat{y}(k-2), \ldots, \hat{y}(k-n), u(k-1), u(k-2), \ldots, u(k-m)), \tag{2.60}$$

where the regression vector is composed of the previous model estimations $\hat{y}$ and measured input values u up to a given lag. The model is therefore treated as a model with a NOE structure.

When only the mean values of the model predicted values are fed back, the simulation is named *naive*. However, when we want to obtain a more realistic picture of the dynamic model multistep-ahead prediction, we have to take into account the uncertainty of future predictions, which provide the 'input data' for estimating further means and uncertainties. A partial overview of the results given in [136] is given as follows.

In the case of a multistep-ahead prediction, we wish to make a prediction at $\mathbf{z}^*$, but this time the input vector $\mathbf{z}^*$ contains uncertain input values fed back from the outputs. Within a Gaussian approximation, the input values can be described by the normal distribution $\mathbf{z}^* \sim \mathcal{N}(\boldsymbol{\mu}_{\mathbf{z}^*}, \Sigma_{\mathbf{z}^*})$, where $\boldsymbol{\mu}_{\mathbf{z}^*}$ and $\boldsymbol{\Sigma}_{\mathbf{z}^*}$ are the vector and the matrix of the input mean values and variances, respectively. To obtain a prediction, we need to integrate the predictive distribution $p(y^*|\mathbf{z}^*, \mathcal{D})$ over the input data distribution, that is

$$p(y^*|\boldsymbol{\mu}_{\mathbf{z}^*}, \boldsymbol{\Sigma}_{\mathbf{z}^*}, \mathcal{D}) = \int p(y^*|\mathbf{z}^*, \mathcal{D}) p(\mathbf{z}^*) d\mathbf{z}^*, \tag{2.61}$$

where

$$p(y^*|\mathbf{z}^*, \mathcal{D}) = \frac{1}{\sqrt{2\pi\sigma^2(\mathbf{z}^*)}} \exp\left[-\frac{(y^* - \mu(\mathbf{z}^*))^2}{\sigma^2(\mathbf{z}^*)}\right]. \tag{2.62}$$

Since $p(y^*|\mathbf{z}^*, \mathcal{D})$ is in general a nonlinear function of $\mathbf{z}^*$, the new predictive distribution $p(y^*|(\boldsymbol{\mu}_{y^*}, \boldsymbol{\Sigma}_{y^*}, \mathcal{D}))$ is not Gaussian and this integral cannot be solved without using an approximation. In other words, when the Gaussian distribution is propagated through a nonlinear model, it is not a Gaussian distribution at the output of the model.

Approximations can be roughly divided into numerical methods, for example Monte Carlo methods, and analytical methods.

2.7.1 Numerical Approximation

Eq. (2.61) can be solved by performing a numerical approximation of the integral, using a simple Monte Carlo approach:

$$p(y^*|\boldsymbol{\mu}_{\mathbf{z}^*}, \boldsymbol{\Sigma}_{\mathbf{z}^*}, \mathcal{D}) \approx \frac{1}{S}\sum_{i=1}^{S} p(y^*|\mathbf{z}^{*i}, \mathcal{D}) \tag{2.63}$$

where S is a number of samples and $\mathbf{z}^{*i}$ is a sample from the input data distribution $p(\mathbf{z}^*)$. The output distribution is therefore not a Gaussian, but can be seen as a Gaussian mixture:

$$p(y^*|\boldsymbol{\mu}_{\mathbf{z}^*}, \boldsymbol{\Sigma}_{\mathbf{z}^*}, \mathcal{D}) \approx \frac{1}{S}\sum_{i=1}^{S} \mathcal{N}(\mu(\mathbf{z}^{*i}), \sigma^2(\mathbf{z}^{*i})). \tag{2.64}$$

When applying this approximation in a simulation, it means that in every following time step it can happen that we sample a more complicated Gaussian mixture, so the algorithm has to be implemented efficiently. See [135] for hints on an efficient numerical implementation for multistep-ahead prediction.

Other numerical approximations that have been used for the uncertainty propagation, mainly in the context of state-space models, are sequential Monte Carlo methods, e.g. [20, 26, 28].

2.7.2 Analytical Approximation of Statistical Moments with a Taylor Expansion

In order to achieve computational simplicity, an analytical approximation that consists of computing only the first two moments, namely, the mean and variance of $p(y^*|\mathbf{z}^*, \mathcal{D})$ can be used.

The mean and variance of the predictive distribution which in general is a non-Gaussian predictive distribution, are approximated with a Gaussian approximation, such that

$$p(y^*|\boldsymbol{\mu}_{\mathbf{z}^*}, \boldsymbol{\Sigma}_{\mathbf{z}^*}, \mathcal{D}) \approx \mathcal{N}(\mu^*, \sigma^{2*}). \tag{2.65}$$

The predictive mean and variance at the model's output corresponding to a noisy input value $\mathbf{z}^*$ are obtained by solving [136]

$$\mu^* = E(\mu(\boldsymbol{\mu}_{\mathbf{z}^*})), \tag{2.66}$$

$$\begin{aligned}\sigma^{2*} &= E(\sigma^2(\boldsymbol{\mu}_{\mathbf{z}^*})) + \mathrm{var}(\mu(\boldsymbol{\mu}_{\mathbf{z}^*}))\\ &= E(\sigma^2(\boldsymbol{\mu}_{\mathbf{z}^*})) + E(\mu^2(\boldsymbol{\mu}_{\mathbf{z}^*})) - (E(\mu(\boldsymbol{\mu}_{\mathbf{z}^*})))^2,\end{aligned} \tag{2.67}$$

where $\mu(\boldsymbol{\mu}_{\mathbf{z}^*})$ and $\sigma^2(\boldsymbol{\mu}_{\mathbf{z}^*})$ denote the mean and variance of the Gaussian predictive distribution in the case when there are no uncertain input values, respectively.

Instead of working with the expressions of $\mu(\boldsymbol{\mu}_{\mathbf{z}^*})$ and $\sigma^2(\boldsymbol{\mu}_{\mathbf{z}^*})$, Eqs. (2.66) and (2.67) are solved by approximating directly μ^* and σ^{2*} using their first- and second-order Taylor expansions, respectively, around $\boldsymbol{\mu}_{\mathbf{z}^*}$. The second-order expansion is required in order to get a correction term for the new variance. This is a relatively rough approximation.

Consequently, within a Gaussian approximation and a Taylor expansion μ^* and σ^{2*} around $\mathbf{z}^* = \boldsymbol{\mu}_{\mathbf{z}^*}$, the predictive distribution is again Gaussian with a mean and variance [136]

$$\mu^* = E(\mu(\boldsymbol{\mu}_{\mathbf{z}^*})) \approx \mathbf{k}(\boldsymbol{\mu}_{\mathbf{z}^*})^{\mathrm{T}}\mathbf{K}^{-1}\mathbf{y}, \tag{2.68}$$

$$\begin{aligned}\sigma^{2*} &= E(\sigma^2(\boldsymbol{\mu}_{\mathbf{z}^*})) + \mathrm{var}(\mu(\boldsymbol{\mu}_{\mathbf{z}^*}))\\ &\approx \sigma^2(\boldsymbol{\mu}_{\mathbf{z}^*}) + \frac{1}{2}\mathrm{tr}\left(\left.\frac{\partial^2\sigma^2(\mathbf{z}^*)}{\partial\mathbf{z}^*\partial\mathbf{z}^{*\mathrm{T}}}\right|_{\mathbf{z}^*=\boldsymbol{\mu}_{\mathbf{z}^*}} \boldsymbol{\Sigma}_{\mathbf{z}^*}\right)\\ &\quad + \left.\frac{\partial\mu(\mathbf{z}^*)}{\partial\mathbf{z}^*}\right|_{\mathbf{z}^*=\boldsymbol{\mu}_{\mathbf{z}^*}}^{\mathrm{T}} \boldsymbol{\Sigma}_{\mathbf{z}^*} \left.\frac{\partial\mu(\mathbf{z}^*)}{\partial\mathbf{z}^*}\right|_{\mathbf{z}^*=\boldsymbol{\mu}_{\mathbf{z}^*}}.\end{aligned} \tag{2.69}$$

Equations (2.68) and (2.69) can be applied in a calculation of the multistep-ahead prediction with the propagation of uncertainty. For a more detailed derivation, see [136] and for further details see Appendix B.

2.7.3 Unscented Transformation

The unscented transformation also does not make assumption about the structural nature of the model. It estimates the posterior distribution applying a given nonlinear transformation to a probability distribution that is characterised only in terms of a mean value and covariance.

The unscented transformation takes a finite number of 'sigma points' with the same statistical moments as the input probability distribution, and then maps these sigma points through the mean of the probabilistic dynamics model to obtain the transformed set of points. The mean and covariance are then set to that of the weighted statistics of the transformed dataset to give an unbiased prediction.

More details of applying the unscented transformation and GP models in the context of state-space models and Kalman filtering are in [27].

2.7.4 Analytical Approximation with Exact Matching of Statistical Moments

The alternative approach to approximation is that instead of an approximation of the entire mean and variance, only the integral of Eq. (2.61) is approximated. A simulation with this kind of approximation is named *exact*. In every time step, the model prediction is based on stochastic input data that has a normal distribution and the prediction is a Gaussian mixture, which is approximated with a normal distribution, as depicted in Fig. 2.23 [51] for the case of one input variable for demonstration purposes.

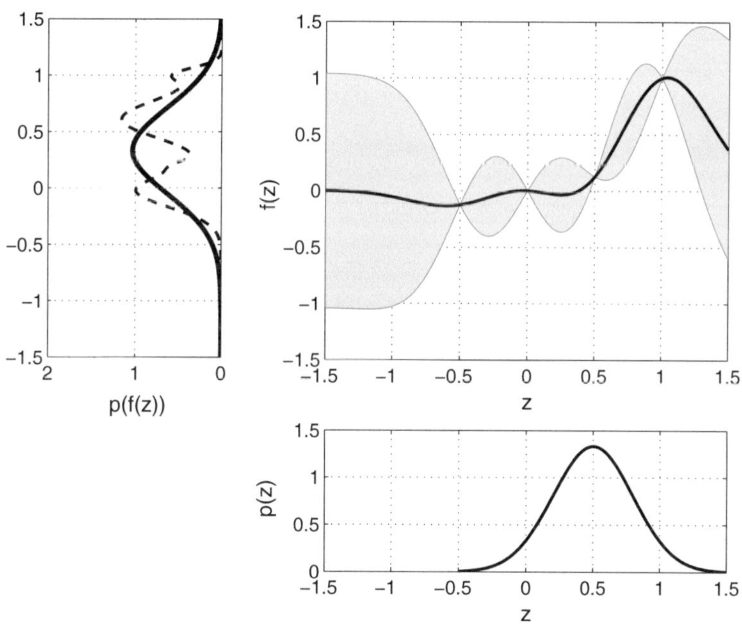

Fig. 2.23 GP prediction at a stochastic input variable. The input distribution $p(z)$ is the figure on the *bottom right*. The figure on the *right* shows the mean function (*full line*) and the 95 % confidence interval (*shaded*) based on the training data points (points with zero confidence interval in the figure). To determine the expected function value, we average over both the input distribution (*bottom right*) and the function distribution (GP model). The *shaded* distribution represents the exact distribution over the function values. The exact predictive distribution (*dashed line* in the *left figure*) is approximated by a Gaussian (*full line* in the *left figure*) that possesses the mean and the covariance of the exact predictive distribution (known as moment matching)

The expressions for mean and variance are Eqs. (2.66) and (2.67):

$$\begin{aligned}
\mu^* &= E(\mu(\boldsymbol{\mu}_{\mathbf{z}^*})), \\
\sigma^{2*} &= E(\sigma^2(\boldsymbol{\mu}_{\mathbf{z}^*})) + \mathrm{var}(\mu(\boldsymbol{\mu}_{\mathbf{z}^*})) \\
&= E(\sigma^2(\boldsymbol{\mu}_{\mathbf{z}^*})) + E(\mu^2(\boldsymbol{\mu}_{\mathbf{z}^*})) - (E(\mu(\boldsymbol{\mu}_{\mathbf{z}^*})))^2.
\end{aligned}$$

which can be derived further using

$$E(\zeta(\mathbf{z}^*)) = \int \zeta(\mathbf{z}^*)p(\mathbf{z}^*)d\mathbf{z}^* \tag{2.70}$$

for each of the components in Eqs. (2.66) and (2.67), with $\zeta(\mathbf{z}^*)$ denoting a particular component of these equations.

$$\begin{aligned}
\mu^* &= \int \mu(\boldsymbol{\mu}_{\mathbf{z}^*})p(\mathbf{z}^*)d\mathbf{z}^* \\
&= \int \mathbf{k}(\mathbf{z}^*)\mathbf{K}^{-1}\mathbf{y}p(\mathbf{z}^*)d\mathbf{z}^*, \qquad (2.71)\\
\sigma^{2*} &= \int (\kappa(\mathbf{z}^*) - \mathbf{k}^{\mathrm{T}}(\mathbf{z}^*)\mathbf{K}^{-1}\mathbf{k}(\mathbf{z}^*))p(\mathbf{z}^*)d\mathbf{z}^* \\
&\quad + \int \mathbf{k}^{\mathrm{T}}(\mathbf{z}^*)\mathbf{K}^{-1}\mathbf{y}\mathbf{y}^{\mathrm{T}}(\mathbf{K}^{-1})^{\mathrm{T}}\mathbf{k}(\mathbf{z}^*)p(\mathbf{z}^*)d\mathbf{z}^* \\
&\quad - (\mu^*)^2. \qquad (2.72)
\end{aligned}$$

The exact derivations for particular covariance functions can be found in [135]. The final results for the case of single and multiple outputs for squared exponential and linear covariance functions can be found in Appendix B.

Predictions for a sparse GP, namely, the FITC, also named SPGP, method in the case of uncertain, i.e. stochastic, input values are introduced in [137] and for the SoR and DTC methods, together with other predictions for stochastic methods in [138].

2.7.5 Propagation of Uncertainty

The iterative, multistep-ahead prediction is made by feeding back the mean of the predictive distribution as well as the variance of the predictive distribution at each time step, thus taking the uncertainty attached to each intermediate prediction into account. In this way, each input variable for which we wish to predict becomes a normally distributed random variable. Nevertheless, this is an approximation of the Gaussian mixture at the output of the model. The illustration of such a dynamic model simulation is given in Fig. 2.24.

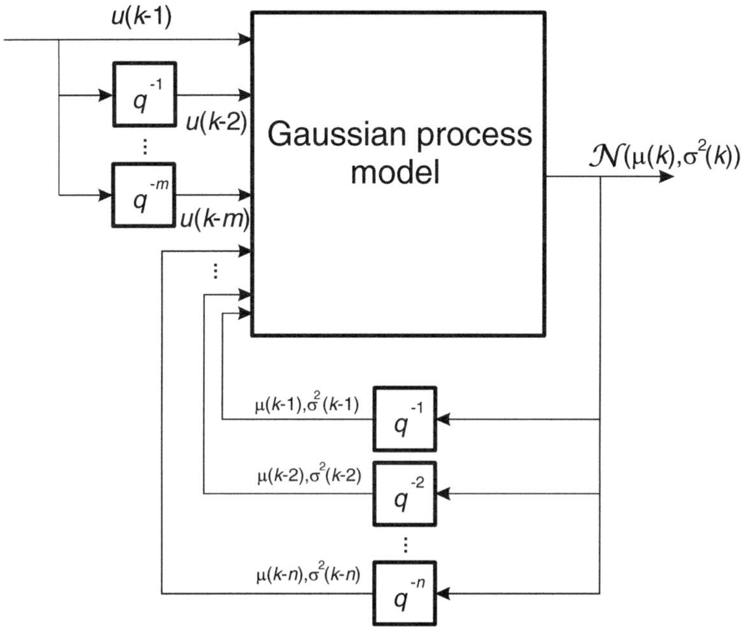

Fig. 2.24 Block scheme of dynamic system simulation with the iterative method where variance is propagated through the model

These results presume the iterative method, where a one-step-ahead prediction is iteratively repeated. Note that when predicting ahead in time and propagating the uncertainty, the exogenous inputs u are usually assumed to be known and are treated like a deterministic approach. This is also the situation shown in Fig. 2.24. However, the following explanation is general and presumes stochastic input variables.

As with [139], in the case of function observations only, we can predict h steps ahead and propagate the uncertainty of the successive predictions by considering each feedback data $y(k+h-i)$ as a Gaussian random variable, resulting in an $D \times 1$; $D = n+m$ input into the model $\mathbf{z}(k+h) = [y(k+h-1), \ldots, y(k+h-n), u(k+h-1), \ldots, u(k+h-m)]^{\mathrm{T}} \sim \mathcal{N}(\boldsymbol{\mu}_{\mathbf{z}}, \boldsymbol{\Sigma}_{\mathbf{z}})$ at each time step with the mean

$$\boldsymbol{\mu}_{\mathbf{z}} = \begin{bmatrix} \mu_y(k+h-1) \\ \vdots \\ \mu_y(k+h-n) \\ \mu_u(k+h-1) \\ \vdots \\ \mu_u(k+h-m) \end{bmatrix} \tag{2.73}$$

and the covariance matrix

$$\boldsymbol{\Sigma}_{\mathbf{z}} = \begin{bmatrix} \mathrm{var}(y(k+h-1)) & \cdots & \mathrm{cov}(u(k+h-m), y(k+h-1)) \\ \vdots & \ddots & \vdots \\ \mathrm{cov}(y(k+h-1), u(k+h-m)) & \cdots & \mathrm{var}(u(k+h-m)) \end{bmatrix}, \tag{2.74}$$

where the mean values and variances for each entry are computed using one of the approximation methods described beforehand with the equations given in Appendix B.

In general, at time sample $k+l$, we have the random input vector $\mathbf{z}(k+l) = [y(k+l-1), \ldots, y(k+l-n), u(k+h-1), \ldots, u(k+h-m)]^{\mathrm{T}} \sim \mathcal{N}(\boldsymbol{\mu}_{\mathbf{z}}, \boldsymbol{\Sigma}_{\mathbf{z}})$ with the vector of means $\boldsymbol{\mu}_{\mathbf{z}}$ formed by the mean of the predictive distribution of the lagged output data and input data $y(k+l-\tau)$, $\tau = 1, \ldots, n$; $u(k+l-\tau)$, $\tau = 1, \ldots, m$ and the diagonal elements of the $D \times D$; $D = n+m$ input covariance matrix $\boldsymbol{\Sigma}_{\mathbf{z}}$ containing the corresponding predictive covariances. The cross-covariance terms $\mathrm{cov}(y(k+l-i), u(k+l-j))$, for $i, j = 1, \ldots, D$ with $i \neq j$, are obtained by computing $\mathrm{cov}(y(k+l), \mathbf{z}(k+l))$, disregarding the last, the oldest element of $\mathbf{z}(k+l)$:

$$\mathrm{cov}(y(k+l), \mathbf{z}(k+l)) = E(y(k+l)\mathbf{z}(k+l)) - E(y(k+l))E(\mathbf{z}(k+l)). \tag{2.75}$$

Again, the equations for calculating the cross-covariances can be found in Appendix B.

2.7.6 *When to Use Uncertainty Propagation?*

The uncertainty propagation seems to be an issue mainly with two applications of GP models. The first one is the simulation of dynamic system and the second one is the inference of state-vector distribution in GP state-space model.

In the case of GP state-space model of nonlinear systems, the uncertainty propagation is inevitable to get usable approximation of state-vector distribution. Therefore, the rest of this discussion is devoted to uncertainty propagation in the case of the dynamic system's simulation.

A simulation is of major importance as the validation tool for systems identification. However, as the uncertainty propagation extension to the GP modelling method adds a considerable level of complexity, it is worth discussing when uncertainty propagation is best employed.

The uncertainty propagation usually has an effect to the shape of the predictive distribution. It mainly affects, usually increases, its variance, because the predictive distribution becomes wider. Nevertheless, it also affects the mean value. The examples presented in [134, 135, 140] show that the differences between the means of naive and non-naive cases are usually not huge. It is difficult to argue for the inclusion of uncertainty propagation purely for the sake of improving the quality of the mean prediction. Nevertheless, the computational load for uncertainty propagation is considerable when a model with a large input dataset is employed.

The trade-off between the computational load and the accuracy of the predictive distribution is certainly of importance for an engineer using GP models for dynamic systems identification. The GP modelling approach results in models with an output estimate in the form of a predictive distribution. Through the variance of this output distribution, the GP model becomes more informative. The question is, however, whether we are interested in a precise quantitative value of the predicted variance or we are more interested in qualitative information that the predicted variance carries.

This issue is closely related to the issue of purposiveness of the dynamic system model. In the case that the accurate model multistep-ahead prediction means and variances are of importance for the accuracy of the final product for which the model is developed, then the computational load needs to be taken into account. In general, implementing uncertainty propagation would increase the robustness of the model's response. This means that the increased variance, and improved mean value, obtained with the uncertainty propagation might have a greater chance of enveloping the real response. Such cases would be some cases of predictive control where the mismatch between the model predictions and the real system response makes a difference in the optimisation of future control input values.

On the other hand, when the dynamic system model's predictive variances are used to determine whether the system's output values are predicted outside the region where the identification dataset was available, the qualitative information about the predicted variance's magnitude already serves the purpose. Therefore, the concept of taking into account the uncertainty of the input values and propagating uncertainty for subsequent predictions would not seem to be sensible for applications where the focus is on predicted mean values.

A possible rule of thumb for uncertainty propagation use with the dynamic system's simulation is that the use is decided upon the importance of the exactness of the variance's magnitude. If the variance is to be actively employed in some manner, such as in the design of control systems, the uncertainty propagation may prove to be an important addition. In general, for a lot of the dynamic system's application for engineering purposes, a naive simulation will do.

2.8 An Example of GP Model Identification

Example 2.3 (*Bioreactor identification*) The purpose of this example, adapted from [141], is to demonstrate the GP model identification procedure with a special emphasis on the utility of the prediction variance and other GP model-specific measures for the model validation. The example illustrates how the model is selected. The selected model is then used to demonstrate the influence of increased noise variance on the system's output, the behaviour of the model prediction in unmodelled regions and the behaviour of the model when a new, unmodelled input is introduced to the system.

A second-order discrete bioreactor model [142, 143] is taken as the system to be identified for demonstration purposes. With the selection of discrete model, the issue of sampling time selection is avoided. In the bioreactor, the microorganisms grow by consuming the substrate.

The bioreactor is given as discrete second-order dynamic system [142] with the sampling time $T_s = 0.5$ s:

$$\begin{aligned} x_1(k+1) &= x_1(k) + 0.5\frac{x_1(k)x_2(k)}{x_1(k)+x_2(k)} - 0.5u(k)x_1(k) \\ x_2(k+1) &= x_2(k) - 0.5\frac{x_1(k)x_2(k)}{x_1(k)+x_2(k)} - 0.5u(k)x_2(k) + 0.05u(k) \\ y(k) &= x_1(k) + \nu(k) \end{aligned} \tag{2.76}$$

where x_1 is the concentration of the microorganisms and x_2 is the concentration of the substrate. The control input u is the output flow rate, which is limited between $0 \leq u(k) \leq 1$. The output of the system y is the concentration of microorganisms, which is corrupted by white Gaussian noise ν with a standard deviation $\sigma_\nu = 0.0005$. Our task is to model this system with the GP model and validate the acquired model. The *purpose of the model* is the simulation of the bioreactor.

In the *experiment design*, two separate identification and one validation input signals are acquired, from which the identification and validation data are sampled. Two separate identification signals are acquired so that one is used for the hyperparameter estimation and the other with structure, regressors and covariance function selection. To acquire the first set of identification data, the system described with Eq. (2.76) is excited with the signal u in the form of 4-seconds-long stairs with random amplitude values between $0 \leq u(k) \leq 0.7$. The second set of identification data is obtained with the same kind of signal, but with stairs that last longer. Note that the upper limit of both input signals is chosen so that a part of the operating region remains unmodelled. Before the identification of the models, the signals are normalised around a constant mean value, so that they had a maximum value of one and a minimum value of minus one. From the normalised signals, 602 training points are composed. Later, when the identification results are presented, the data will be scaled into the original range and a constant value for the *mean function* is added.

The GP-NARX *model structure* is used for the model identification. The ith training point at the sample step k for the nth-order GP model is composed from the input regressors:

$$\mathbf{z}_i = [y(k-1), \ldots, y(k-n),\ u(k-1), \ldots, u(k-m)]^{\mathrm{T}}$$

and the output value $y_i = y(k)$, where u and y are normalised input and output signals. The GP-NOE model structure is going to be used for validation due to the model's purpose.

The squared exponential *covariance function* with the ARD property described with Eq. (2.14) is selected as the initial covariance function that is used for the regressor selection procedure. The ARD property of the selected covariance function ensures that different length scales on different regressors can be used to assess the relative importance of the contributions made by each regressor through a comparison of their lengthscale hyperparameters.

$$C_f(\mathbf{z}_i, \mathbf{z}_j) = \sigma_f^2 \exp\left[-\frac{1}{2}(\mathbf{z}_i - \mathbf{z}_j)^{\mathrm{T}} \mathbf{\Lambda}^{-1} (\mathbf{z}_i - \mathbf{z}_j)\right]$$
$$= \sigma_f^2 \exp\left[-\frac{1}{2}\sum_{d=1}^{D} w_d (z_{di} - z_{dj})^2\right],$$

where $w_d = \frac{1}{l_d^2}$; $d = 1, \ldots, D$. The reason behind the selection of the squared exponential covariance function is that we do not know much about the mapping between the input and the output data. Nevertheless, the prior knowledge about most physical systems, among which is the bioreactor, is that the mapping can be modelled with stationary and smooth covariance functions. As we will see later the validation results confirm this prior knowledge.

The *regressor selection* is done with validation-based regressor selection [144] on the second set of data for identification, i.e. where a low-order model is expected based on prior knowledge of the process. The fourth-, third- and second-order models are evaluated with a simulation to obtain an appropriate set of regressors.

All three initial GP models with the same number of delays in the input and output values used for regressors, i.e. $n = m$, are evaluated with a simulation, where the second identification dataset as well as the validation dataset is obtained by simulating the system described with Eq. (2.76) using a different input signal u than for obtaining the first set of identification data.

The results of the regressor selection procedure can be seen in Table 2.3. The log likelihood of the identified model ℓ_1 described with Eq. (2.32) is used as the objective function for optimisation during the identification and performance measures SMSE, Eq. (2.54) and MSLL, Eq. (2.57), are used for the validation of the simulated model on the second set of identification data and on the validation data. The model has been tested with a naive simulation and with a simulation based on the Monte Carlo numerical approximation. The same conclusions can be drawn from the results of both methods, because the obtained numerical results do not differ significantly. The figures presented in the continuation show the results of the naive simulation.

From the performance measures used on the identification results, shown in the first three rows of Table 2.3, it can be seen that the differences between the identified models in terms of the SMSE and MSLL values evaluating the model simulations on the second set of data are slightly significant. The results on the second set of data which have not been used for the hyperparameters estimation favour a simpler model. The second-order model is also favoured by the principle of Occam's razor [49],

Table 2.3 Values of the validation performance measures and the hyperparameters of different bioreactor GP models with the best measure values in bold

Model	Ident. data			Valid. data		Hyperparameters									Noise‡
Order	ℓ_1	SMSE_2	MSLL_2	SMSE	MSLL	$l_{y_4} = \frac{1}{\sqrt{w_{y_4}}}$	$l_{y_3} = \frac{1}{\sqrt{w_{y_3}}}$	$l_{y_2} = \frac{1}{\sqrt{w_{y_4}}}$	$l_{y_1} = \frac{1}{\sqrt{w_{y_1}}}$	$l_{u_4} = \frac{1}{\sqrt{w_{u_4}}}$	$l_{u_3} = \frac{1}{\sqrt{w_{u_3}}}$	$l_{u_2} = \frac{1}{\sqrt{w_{u_2}}}$	$l_{u_1} = \frac{1}{\sqrt{w_{u_1}}}$	σ_f	σ_n
4	1628	3.1×10^{-3}	−3.31	5.1×10^{-3}	−3.22	9.2	6.3	4.8	977	406	7.2	7.0	8.9	2.5	5.0×10^{-4}
3	1621	2.3×10^{-3}	−3.38	3.8×10^{-3}	−3.32	×	20.0	3.90	49.3	×	17.4	17.3	15.6	4.2	5.1×10^{-4}
2	1612	1.2×10^{-3}	−3.55	1.9×10^{-3}	**−3.50**	×	×	5.8	29.8	×	×	13.3	14.7	5.3	5.3×10^{-4}
2♠	1603	**7.8×10^{-4}**	**−3.57**	**1.2×10^{-3}**	−3.49	×	×	5.2	×	×	×	13.1	14.4	5.6	5.4×10^{-4}
2♣	785	3.5×10^{-3}	−2.36	3.1×10^{-4}	−2.37	×	×	4.07	×	×	×	10.7	8.7	3.3	2.1×10^{-3}

Notes
The index $_1$ denotes the first set of data for identification
The index $_2$ denotes the second set of data for identification
‡the identified standard deviation of noise σ_n
♠reduced number of regressors by masking the regressor $y(k-1)$
♣identified on the output signal with the increased standard deviation of noise, $\sigma_\nu = 0.002$

Table 2.4 Values of log-marginal likelihood ℓ_1 for the prediction results on the first set of identification data and simulation performance measures for the second set of identification data for different covariance functions with the best measure values in bold

2nd order model	Ident. data		
Covariance function	ℓ_1	$SMSE_2$	$MSLL_2$
Squared exponential + ARD	**1603**	7.8×10^{-4}	**−3.57**
Matérn + ARD $d = \frac{3}{2}$	1591	$\mathbf{7.2 \times 10^{-4}}$	−3.54
Rational quadratic + ARD	**1603**	4.1×10^{-3}	−3.31
Linear + ARD	1281	1.7×10^{-2}	−1.46
Matérn $d = \frac{3}{2}$	1587	1.2×10^{-3}	−2.03
Matérn $d = \frac{5}{2}$	1597	1.7×10^{-3}	−1.58
Neural network	1596	5.2×10^{-3}	−1.00
Squared exponential	1600	1.6×10^{-3}	−1.35

stating that the most simple explanation for the given problem should be used. The results obtained on the validation data confirm these results.

The hyperparameters l_{z_i} reflect the relative importance of the regressors $z(k-i)$ and in all the model structures the regressor $y(k-1)$ exhibits a lower importance due to the large value of the associated hyperparameter l_{y_1}. The removal of this regressor from the selected second-order model results in even better results. Note that the variance of modelled noise is likely to be a small amount greater than the ground truth. This effect is related to errors-in-variables problem.

This regressor selection procedure led us to the GP model with the following regressors: $y(k-2)$, $u(k-1)$, $u(k-2)$.

Covariance function selection is illustrated with Table 2.4, where the same performance measures as for the regressors' selection are gathered for different covariance functions with and without the ARD property for the second-order model.

The squared exponential covariance function and Matérn covariance function with the ARD property have the best results from the tested covariance functions. Consequently, the squared exponential covariance function with the ARD property is kept for the model.

The *model validation* is pursued with the dynamic model simulation according to the purpose of the model. The naive simulation is selected in our case, because no special accuracy needs are expressed for the posterior variance in the model purpose and the obtained accuracy for mean values is acceptable. In this way, we are avoiding computational complexity due to the propagation of uncertainty through the model, at the cost of only a slightly lower accuracy for this case.

The simulation results on the validation data for the selected second-order model can be seen in Fig. 2.25, where the model's output and the noise-free target are depicted. It can be seen that most of the time the value of the predicted standard deviation σ_n is around 5×10^{-4}, corresponding to the noise level present at the

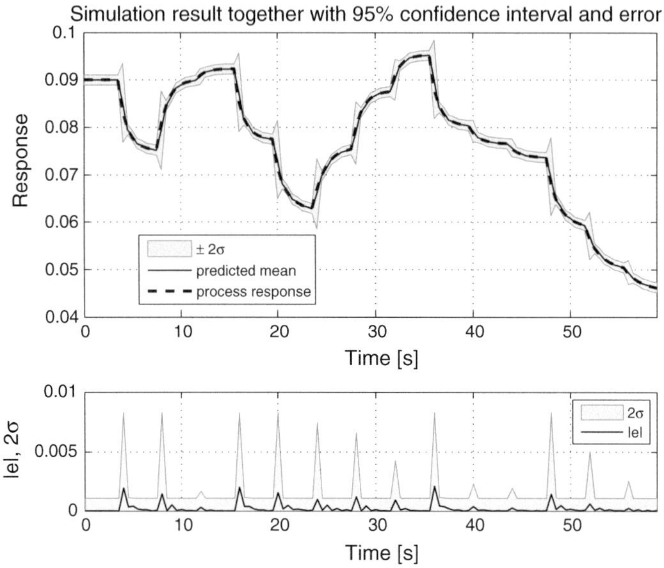

Fig. 2.25 Validation with simulation for the bioreactor GP model, $|e|$ is the absolute value of the error between the predicted mean and the true values

system's output. The prediction variance increases at the steps where the input variable u changes its value due to the small number of training points describing those regions. Note that the error of the model's predicted mean values remains inside the 95 % confidence limits, defined within $\pm 2\sigma$, indicating the level of trust that can be put in the prediction.

These model validation results will serve as the reference for the observation of how different conditions can influence the model prediction and validation.

First, it will be shown how the model prediction changes when the model reaches the unmodelled region of the system. As there is no identification data available nearby, the model must extrapolate from the data describing the neighbouring regions and the prior mean function in order to make the predictions. This worsens the prediction mean, but is also accompanied by an increase of the prediction variance, thus widening the model's confidence limits. This effect can be observed in Fig. 2.26, where the values of the control input were increased above the $u(k) > 0.7$ at time $t > 12$ s.

Second, we would like to show how the increase in the system's output noise variance reflects in the identified model. For this purpose, the standard deviation of the system's output noise was increased to $\sigma_\nu = 2 \times 10^{-3}$. The set of control input signals, used for generating the identification and validation data, is the same as in the reference example. The second-order GP model is identified. The values of the GP model's hyperparameters can be seen in Table 2.3.

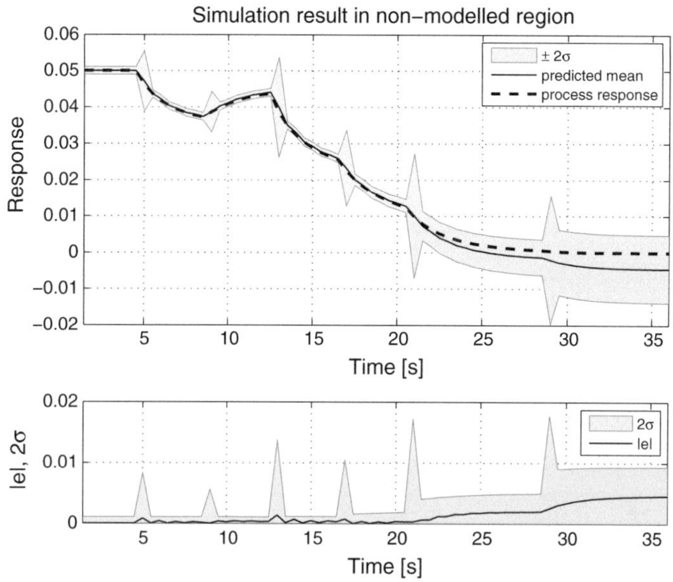

Fig. 2.26 GP model prediction in the unmodelled region

The mean model prediction is satisfactory and the prediction variance is increased at the expense of a higher output noise variance being predicted, as can be observed from the simulation results on the validation data in Fig. 2.27. The estimation of the system's output noise is satisfactorily close to the real value, i.e. $\sigma_n = 2.1 \times 10^{-3}$, which also shows that the value of the hyperparameter σ_n^2 tends to the value of the system's output noise when enough identification data is used. The value of SMSE is slightly worse than in the reference example, as this model is identified with more noise present in the identification, i.e. training, data. Also, the value of MSLL is slightly worse as, despite the increased variance, the influence of the prediction error prevails.

Finally, we would like to show how the unmodelled regressor influences the model. For this purpose, an additional control input signal υ, not correlated to the input signal u, was added to the system. The effect of this unmeasured input variable is the same as the effect of the control input signal u and could represent an additional outlet or leak of the system described with Eq. (2.76), which changes the description to

$$
\begin{aligned}
x_1(k+1) &= x_1(k) + 0.5\frac{x_1(k)x_2(k)}{x_1(k)+x_2(k)} - 0.5u(k)x_1(k) - 0.5\upsilon(k)x_1(k) \\
x_2(k+1) &= x_2(k) - 0.5\frac{x_1(k)x_2(k)}{x_1(k)+x_2(k)} - 0.5u(k)x_2(k) \\
&\quad - 0.5\upsilon(k)x_2(k) + 0.05u(k) + 0.05\upsilon(k) \\
y(k) &= x_1(k) + \nu(k)
\end{aligned}
\tag{2.77}
$$

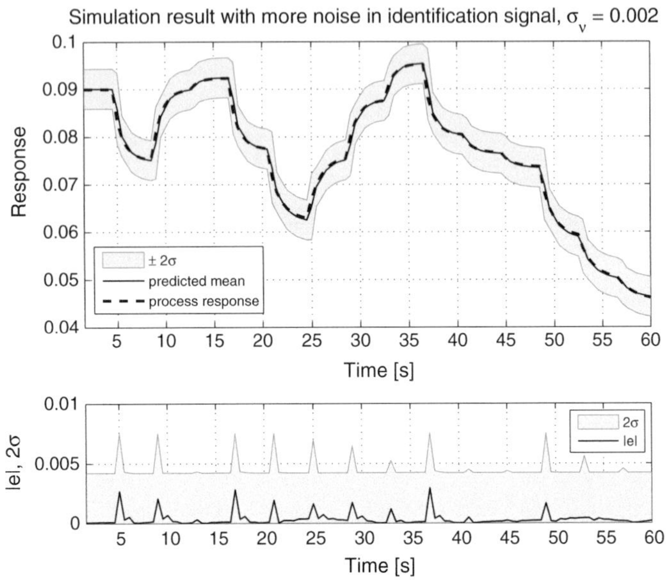

Fig. 2.27 Influence of the increased system's output noise variance on the GP model

The reference GP model is used for the simulation, where the input signal v is not present and therefore neglected for the identification of the model. The control input signal in the form of a step $v = 0.05$ is introduced into the system at the simulation time 30 s. The (non)influence of the unmodelled input signal on the prediction variance, when the model operates in the region with sufficient identification data, can be seen in Fig. 2.28. The model's simulation response from time $t = 30$ s worsens, but the 95 % confidence interval remains tight. This example shows that the variance, obtained with the model simulation, cannot be informative with respect to the unaccounted influences on the system in the identification data. Note that the results are different if the model prediction is pursued instead of the model simulation.

With the bioreactor example, the following properties of the GP model have been illustrated:

1. The hyperparameters' ARD property can be effectively used to reduce the number of regressors of the identified model.
2. There are two possible causes for the increase of the prediction variance:
    - the particular region of the system, where the model makes predictions, is described with insufficient identification data; and
    - the data describing particular regions contains more noise. In the example, this has been shown for the whole region, but the same applies when the noise is increased only in part of the system's operating region.

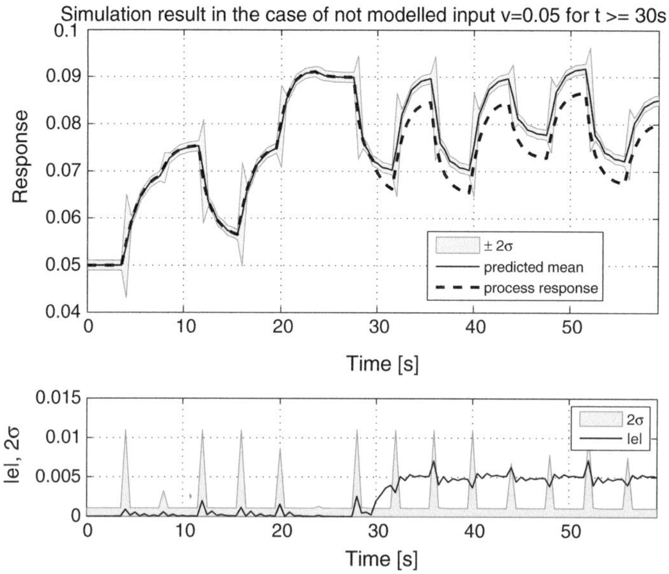

Fig. 2.28 Influence of the unmodelled input signal on the GP model prediction

These two causes cannot be easily distinguished without prior knowledge about the identified system.

3. When the unmodelled influence is introduced to the system, the model simulation response, including the variance, does not change.

References

1. Sjöberg, J., Zhang, Q., Ljung, L., Benveniste, A., Delyon, B., Glorennec, P.Y., Hjalmarsson, H., Juditsky, A.: Nonlinear black-box modelling in system identification: a unified overview. Automatica **31**(12), 1691–1724 (1995)
2. Ljung, L.: System Identification - Theory for the User, 2nd edn. Prentice Hall, Upper Saddle River, NJ (1999)
3. Bai, E.W.: Prediction error adjusted Gaussian process for nonlinear non-parametric system identification. In: 16th IFAC Symposium on System Identification, pp. 101–106. IFAC, Brussels (2012)
4. Johansen, T.A., Shorten, R., Murray-Smith, R.: On the interpretation and identification of dynamic Takagi-Sugeno fuzzy models. IEEE Trans. Fuzzy Syst. **8**, 297–313 (2000)
5. Murray-Smith, R., Johansen, T.A. (eds.): Multiple Model Approaches to Modelling and Control. Taylor and Francis, London (1997)
6. Murray-Smith, R., Johansen, T.A., Shorten, R.: On transient dynamics, off-equilibrium behaviour and identification in blended multiple model structures. In: Proceedings of European Control Conference, pp. BA-14. Karlsruhe (1999)
7. Suykens, J.A.K., Gestel, T.V., Brabanteer, J.D., Moor, B.D., Vandewalle, J.: Least Squares Support Vector Machines. World Scientific, Singapore (2002)

8. Snelson, E., Rasmussen, C.E., Ghahramani, Z.: Warped Gaussian processes. In: Thrun, S., Saul, L., Schölkopf, B. (eds.) Advances in Neural Information Processing Systems, vol. 16, pp. 337–344 (2004)
9. Lazáro-Gredilla, M.: Bayesian warped Gaussian processes. Advances in Neural Information Processing Systems, vol. 26. MIT Press, Cambridge, MA (2013)
10. Nelles, O.: Nonlinear System Identification. Springer, Berlin (2001)
11. Goodwin, G.C.: Identification: experiment design. In: Singh, M.G. (ed.) Systems and Control Encyclopedia, vol. 4 (I–L), pp. 2257–2264. Pergamon Press, Oxford (1987)
12. Norgaard, M., Ravn, O., Poulsen, N.K., Hansen, L.K.: Neural Networks for Modelling and Control of Dynamic Systems. A Practitioner's Handbook. Advanced Textbooks in Control and Signal Processing. Springer, London (2000)
13. Pronzato, L.: Optimal experimental design and some related control problems. Automatica **44**(2), 303–325 (2008)
14. Kocijan, J., Přikryl, J.: Soft sensor for faulty measurements detection and reconstruction in urban traffic. In: Proceedings 15th IEEE Mediterranean Electromechanical Conference (MELECON), pp. 172–177. Valletta (2010)
15. Haykin, S.: Neural Networks, A Comprehensive Foundation. Macmillan College Publishing Company, New York, NY (1994)
16. Ackermann, E.R., de Villiers, J.P., Cilliers, P.J.: Nonlinear dynamic systems modeling using Gaussian processes: predicting ionospheric total electron content over South Africa. J. Geophys. Res. A: Space Phys. **116**(10) (2011)
17. Kocijan, J., Girard, A., Banko, B., Murray-Smith, R.: Dynamic systems identification with Gaussian processes. In: Troch, I., Breitenecker, F. (eds.) Proceedings of 4th IMACS Symposium on Mathematical Modelling (MathMod), pp. 776–784. Vienna (2003)
18. Kocijan, J., Petelin, D.: Output-error model training for Gaussian process models. In: Dobnikar, A., Lotrič, U., Šter, B. (eds.) Adaptive and Natural Computing Algorithms. Lecture Notes in Computer Science, vol. 6594, pp. 312–321. Springer, Berlin (2011)
19. Frigola, R., Rasmussen, C.E.: Integrated pre-processing for Bayesian nonlinear system identification with Gaussian processes. In: 52nd IEEE Conference on Decision and Control (CDC) (2013)
20. Frigola, R., Lindsten, F., Schön, T.B., Rasmussen, C.E.: Bayesian inference and learning in Gaussian process state-space models with particle MCMC. In: L. Bottou, C. Burges, Z. Ghahramani, M. Welling, K. Weinberger (eds.) Advances in Neural Information Processing Systems, vol. 26, pp. 3156–3164 (2013)
21. Wang, J., Fleet, D., Hertzmann, A.: Gaussian process dynamical models. Adv. Neural Inf. Process. Syst. **18**, 1441–1448 (2005)
22. Levine, W.S. (ed.): The Control Handbook. CRC Press, IEEE Press, Boca Raton (1996)
23. Ko, J., Fox, D.: GP-Bayesfilters: Bayesian filtering using Gaussian process prediction and observation models. Auton. Robots **27**(1), 75–90 (2009)
24. Deisenroth, M.P., Turner, R.D., Huber, M.F., Hanebeck, U.D., Rasmussen, C.E.: Robust filtering and smoothing with Gaussian processes. IEEE Trans. Autom. Control **57**(7), 1865–1871 (2012). doi:10.1109/TAC.2011.2179426
25. Deisenroth, M.P., Huber, M.F., Hannebeck, U.D.: Analytic moment-based Gaussian process filtering. In: Proceedings of the 26th Annual International Conference on Machine Learning, pp. 225–232. Montreal (2009)
26. Tong, C.H., Furgale, P., Barfoot, T.D.: Gaussian process Gauss–Newton: non-parametric state estimation. In: 2012 Ninth Conference on Computer and Robot Vision, pp. 206–213. Toronto (2012)
27. Turner, R., Rasmussen, C.E.: Model based learning of sigma points in unscented Kalman filtering. Neurocomputing **80**, 47–53 (2012)
28. Wang, Y., Chaib-draa, B.: A marginalized particle Gaussian process regression. In: Neural Information Processing Systems, vol. 25 (2012)
29. Wang, Y., Chaib-draa, B.: An adaptive nonparametric particle filter for state estimation. In: 2012 IEEE International Conference on Robotics and Automation, pp. 4355–4360. IEEE (2012)

30. Peltola, V., Honkela, A.: Variational inference and learning for non-linear state-space models with state-dependent observation noise. In: Proceedings of the 2010 IEEE International Workshop on Machine Learning for Signal Processing, MLSP 2010, pp. 190–195 (2010)
31. Turner, R., Deisenroth, M.P., Rasmussen, C.E.: System identification in Gaussian process dynamical systems. Nonparametric Bayes Workshop at NIPS. Whistler (2009)
32. Turner, R., Deisenroth, M.P., Rasmussen, C.E.: State-space inference and learning with Gaussian processes. In: Proceedings of 13th International Conference on Artificial Intelligence and Statistics, vol. 9, pp. 868–875. Sardinia (2010)
33. Snelson, E., Ghahramani, Z.: Sparse Gaussian processes using pseudo-inputs. In: Weiss, Y., Schölkopf, B., Platt, J. (eds.) Advances in Neural Information Processing Systems, vol. 18, pp. 1257–1264. MIT Press, Cambridge, MA (2006)
34. Hartikainen, J., Riihimäki, J., Särkkä, S.: Sparse spatio-temporal Gaussian processes with general likelihoods. In: Honkela, T., Duch, W., Girolami, M., Kaski, S. (eds.) Artificial Neural Networks and Machine Learning - ICANN 2011. Lecture Notes in Computer Science, vol. 6791, pp. 193–200. Springer, Berlin (2011)
35. Hartikainen, J., Seppänen, M., Särkkä, S.: State-space inference for non-linear latent force models with application to satellite orbit prediction. In: Proceedings of the 29th International Conference on Machine Learning (ICML-12), pp. 903–910. Edinburgh (2012)
36. Särkkä, S., Hartikainen, J.: Infinite-dimensional kalman filtering approach to spatio-temporal Gaussian process regression. In: Proceedings of the 15th International Conference on Artificial Intelligence and Statistics (AISTATS). JMLR: W&CP, vol. 22, pp. 993–1001 (2012)
37. Särkkä, S., Solin, A., Hartikainen, J.: Spatiotemporal learning via infinite-dimensional Bayesian filtering and smoothing: a look at Gaussian process regression through Kalman filtering. IEEE Signal Process. Mag. **30**(4), 51–61 (2013). doi:10.1109/MSP.2013.2246292
38. Chiuso, A., Pillonetto, G., De Nicolao, G.: Subspace identification using predictor estimation via Gaussian regression. In: Proceedings of the IEEE Conference on Decision and Control (2008)
39. Blum, A.L., Langley, P.: Selection of relevant features and examples in machine learning. Artif. Intell. **97**(1–2), 245–271 (1997)
40. Guyon, I., Elisseeff, A.: An introduction to variable and feature selection. J. Mach. Learn. Res. **3**, 1157–1182 (2003)
41. Kohavi, R., John, G.H.: Wrappers for feature subset selection. Artif. Intell. **97**(1–2), 273–324 (1997)
42. May, R., Dandy, G., Maier, H.: Review of input variable selection methods for artificial neural networks (Chap). Artificial Neural Networks - Methodological Advances and Biomedical Applications, pp. 19–44. InTech, Rijeka (2011)
43. Lind, I., Ljung, L.: Regressor selection with the analysis of variance method. Automatica **41**, 693–700 (2005)
44. Bishop, C.M.: Pattern Recognition and Machine Learning. Springer Science + Business Media, New York, NY (2006)
45. He, X., Asada, H.: A new method for identifying orders of input-output models for nonlinear dynamical systems. In: Proceedings of the American Control Conference, San Francisco, CA, pp. 2520–2523 (1993)
46. Bomberger, J.D., Seborg, D.E.: Determination of model order for NARX models directly from input-output data. J. Process Control **8**(5–6), 459–468 (1998)
47. Broer, H., Takens, F.: Dynamical Systems and Chaos. Epsilon Uitgaven, Utrecht (2009)
48. Stark, J., Broomhead, D.S., Davies, M.E., Huke, J.: Delay embeddings for forced systems: II stochastic forcing. J. Nonlinear Sci. **13**(6), 519–577 (2003)
49. Rasmussen, C.E., Williams, C.K.I.: Gaussian Processes for Machine Learning. MIT Press, Cambridge, MA (2006)
50. Thompson, K.: Implementation of Gaussian process models for non-linear system identification. Ph.D. thesis, University of Glasgow, Glasgow (2009)
51. Deisenroth, M.P.: Efficient reinforcement learning using Gaussian processes. Ph.D. thesis, Karlsruhe Institute of Technology, Karlsruhe (2010)

52. Murray-Smith, R., Girard, A.: Gaussian process priors with ARMA noise models. In: Proceedings of Irish Signals and Systems Conference, pp. 147–152. Maynooth (2001)
53. Ažman, K., Kocijan, J.: Identifikacija dinamičnega sistema z znanim modelom šuma z modelom na osnovi Gaussovih procesov. In: Zajc, B., Trost, A. (eds.) Zbornik petnajste elektrotehniške in računalniške konference (ERK), pp. 289–292. Portorož (2006). (In Slovene)
54. MAC Multi-Agent Control: Probabilistic reasoning, optimal coordination, stability analysis and controller design for intelligent hybrid systems (2000–2004). Research Training Network, 5th EU framework
55. Neal, R.M.: Bayesian Learning for Neural Networks. Lecture Notes in Statistics, vol. 118. Springer, New York, NY (1996)
56. Stein, M.L.: Interpolation of Spatial Data. Springer, New York, NY (1999)
57. Duvenaud, D.: The kernel cookbook: advice on covariance functions (2013)
58. Álvarez, M.A., Luengo, D., Lawrence, N.D.: Latent force models. J. Mach. Learn. Res. - Proc. Track **5**, 9–16 (2009)
59. Álvarez, M.A., Peters, J., Schölkopf, B., Lawrence, N.D.: Switched latent force models for movement segmentation. In: Advances in Neural Information Processing Systems 23: 24th Annual Conference on Neural Information Processing Systems 2010. Proceedings of a meeting held 6–9 December 2010, Vancouver, pp. 55–63 (2010)
60. Honkela, A., Girardot, C., Gustafson, E.H., Liu, Y.H., Furlong, E.M.F., Lawrence, N.D., Rattray, M.: Model-based method for transcription factor target identification with limited data. Proc. Natl. Acad. Sci. USA **107**(17), 7793–7798 (2010)
61. Neo, K.K.S., Leithead, W.E., Zhang, Y.: Multi-frequency scale Gaussian regression for noisy time-series data. In: UKACC International Control Conference. Glasgow (2006)
62. Nguyen-Tuong, D., Peters, J.: Using model knowledge for learning inverse dynamics. In: Proceedings of IEEE International Conference on Robotics and Automation, pp. 2677–2682 (2010)
63. Zhang, Y., Leithead, W.E.: Exploiting Hessian matrix and trust region algorithm in hyperparameters estimation of Gaussian process. Appl. Math. Comput. **171**(2), 1264–1281 (2005)
64. Petelin, D., Filipič, B., Kocijan, J.: Optimization of Gaussian process models with evolutionary algorithms. In: Dobnikar, A., Lotrič, U., Šter, B. (eds.) Adaptive and Natural Computing Algorithms. Lecture Notes in Computer Science, vol. 6594, pp. 420–429. Springer, Berlin (2011)
65. Eiben, A.E., Smith, J.E.: Introduction to Evolutionary Computing. Natural Computing Series. Springer, Heidelberg (2003)
66. Hachino, T., Kadirkamanathan, V.: Multiple Gaussian process models for direct time series forecasting. IEEJ Trans. Electr. Electron. Eng. **6**(3), 245–252 (2011)
67. Hachino, T., Takata, H.: Identification of continuous-time nonlinear systems by using a Gaussian process model. IEEJ Trans. Electr. Electron. Eng. **3**(6), 620–628 (2008)
68. Storn, R., Price, K.: Differential evolution – a simple and efficient heuristic for global optimization over continuous spaces. J. Global Opt. **11**, 341–359 (1997)
69. Price, K., Storn, R., Lampinen, J.: Differential Evolution. Natural Computing Series. Springer, Heidelberg (2005)
70. Kennedy, J., Eberhart, R.: Particle swarm optimization. In: Proceedings of IEEE International Conference on Neural Networks, pp. 1942–1948. IEEE Press. (1995)
71. Kennedy, J., Eberhart, R., Shi, Y.: Swarm Intelligence. Morgan Kaufmann, San Francisco, CA (2001)
72. Hachino, T., Yamakawa, S.: Non-parametric identification of continuous-time Hammerstein systems using Gaussian process model and particle swarm optimization. Artif Life Robotics **17**(1), 35–40 (2012)
73. Birge, B.: Particle swarm optimization toolbox. http://www.mathworks.com/matlabcentral/fileexchange/7506-particle-swarm-optimization-toolbox
74. Pohlheim, H.: GEATbx - the genetic and evolutionary algorithm toolbox for Matlab. http://www.geatbx.com/

75. McHutchon, A., Rasmussen, C.E.: Gaussian process training with input noise. In: J. Shawe-Taylor, R. Zemel, P. Bartlett, F. Pereira, K. Weinberger (eds.) Advances in Neural Information Processing Systems, vol. 24, pp. 1341–1349 (2011)
76. Shi, J.Q., Choi, T.: Gaussian Process Regression Analysis for Functional Data. Chapman and Hall/CRC, Taylor & Francis group, Boca Raton, FL (2011)
77. Seeger, M.W., Kakade, S.M., Foster, D.P.: Information consistency of nonparametric Gaussian process methods. IEEE Trans. Inf. Theory **54**(5), 2376–2382 (2008)
78. Trobec, R., Vajteršic, M., Zinterhof, P. (eds.): Parallel Computing: Numerics, Applications, and Trends. Springer, London (2009)
79. Kurzak, J., Bader, D.A., Dongarra, J. (eds.): Scientific Computing with Multicore and Accelerators. Chapman & Hall/CRC Computational Science Series, 1st edn. CRC Press, Boca Raton, FL (2011)
80. Shen, J.P., Lipasti, M.H.: Modern Processor Design: Fundamentals of Superscalar Processors. McGraw-Hill Series in Electrical and Computer Engineering, 1st edn. McGraw-Hill, New York, NY (2004)
81. Kirk, D.B., Hwu, W.W.: Programming Massively Parallel Processors A Hands-on Approach, 1st edn. Morgan Kaufmann, San Francisco, CA (2010)
82. Tomov, S., Dongarra, J., Baboulin, M.: Towards dense linear algebra for hybrid GPU accelerated manycore systems. Parallel Comput. **36**(5), 232–240 (2010)
83. NVIDIA Corporation, Santa Clara, CA: Cuda Programming Guide Version 2.3.1 (2009)
84. Advanced Micro Devices, Inc.: AMD Accelerated Parallel Processing OpenCL Programming Guide. Sunnyvale, CA (2011)
85. Srinivasan, B.V., Duraiswami, R.: Scaling kernel machine learning algorithm via the use of GPUs. In: GPU Technology Conference, NVIDIA Research Summit. NVIDIA (2009)
86. Srinivasan, B.V., Hu, Q., Duraiswami, R.: GPUML: graphical processors for speeding up kernel machines. In: Workshop on High Performance Analytics - Algorithms, Implementations, and Applications, SIAM Conference on Data Mining. SIAM (2010)
87. Musizza, B., Petelin, D., Kocijan, J.: Accelerated learning of Gaussian process models. In: Proceedings of the 7th EUROSIM Congress on Modelling and Simulation (2010)
88. Blackford, L., Petitet, A., Pozo, R., Remington, K., Whaley, R., Demmel, J., Dongarra, J., Duff, I., Hammarling, S., Henry, G., et al.: An updated set of basic linear algebra subprograms (BLAS). ACM Trans. Math. Softw. **28**(2), 135–151 (2002)
89. Volkov, V., Demmel, J.: LU, QR and Cholesky factorizations using vector capabilities of GPUs. Technical Report UCB/EECS-2008-49, EECS Department, University of California, Berkeley, CA (2008). http://www.eecs.berkeley.edu/Pubs/TechRpts/2008/EECS-2008-49.html
90. Humphrey, J.R., Price, D.K., Spagnoli, K.E., Paolini, A.L., J.Kelmelis, E.: CULA: hybrid GPU accelerated linear algebra routines. In: Kelmelis, E.J. (ed.) Proceeding of SPIE: Modeling and Simulation for Defense Systems and Applications V, vol. 7705. SPIE (2010)
91. Anderson, E., Bai, Z., Bischof, C., Blackford, S., Demmel, J., Dongarra, J., Du Croz, J., Greenbaum, A., Hammarling, S., McKenney, A., Sorensen, D.: LAPACK Users' Guide, 3rd edn. Society for Industrial and Applied Mathematics, Philadelphia, PA (1999)
92. Smola, A.J., Bartlett, P.L.: Sparse greedy Gaussian process regression. Advances in Neural Information Processing Systems, vol. 13, pp. 619–625. MIT Press, Cambridge, MA (2001)
93. Wahba, G.: Spline Models for Observational Data. Society for Industrial and Applied Mathematics, Philadelphia, PA (1990)
94. Williams, C.K.I., Seeger, M.: Using the Nyström method to speed up kernel machines. Advances in Neural Information Processing Systems, pp. 682–688. MIT Press, Cambridge, MA (2001)
95. Quiñonero-Candela, J., Rasmussen, C.E.: A unifying view of sparse approximate Gaussian process regression. J. Mach. Learn. Res. **6**, 1939–1959 (2005)
96. Quiñonero-Candela, J., Rasmussen, C.E., Williams, C.K.I.: Approximation methods for Gaussian process regression (Chap). Large Scale Learning Machines, pp. 203–223. MIT Press, Cambridge, MA (2007)

97. Chalupka, K., Williams, C.K.I., Murray, I.: A framework for evaluating approximation methods for Gaussian process regression. J. Mach. Learn. Res. **14**, 333–350 (2013)
98. Murray, I.: Gaussian processes and fast matrix-vector multiplies. In: Presented at the Numerical Mathematics in Machine Learning workshop at the 26th International Conference on Machine Learning (ICML 2009), Montreal (2009). http://citeseerx.ist.psu.edu/viewdoc/summary?doi=10.1.1.153.7688
99. Leithead, W.E., Zhang, Y., Leith, D.J.: Time-series Gaussian process regression based on Toeplitz computation of O(N2) operations and O(N) level storage. In: Joint 44th IEEE Conference on Decision and Control and European Control Conference (CDC-ECC). Sevilla (2005)
100. Leithead, W.E., Zhang, Y.: O(N-2)-operation approximation of covariance matrix inverse in Gaussian process regression based on quasi-Newton BFGS method. Commun. Stat.-Simul. Comput. **36**(2), 367–380 (2007)
101. Zhang, Y., Leithead, W.E.: Approximate implementation of the logarithm of the matrix determinant in Gaussian process regression. J. Stat. Comput. Simul. **77**(4), 329–348 (2007)
102. Lawrence, N.D., Seeger, M., Herbrich, R.: Fast sparse Gaussian process methods: the informative vector machine. In: Becker, S., Thrun, S., Obermayer, K. (eds.) Advances in Neural Information Processing Systems, vol. 15, pp. 609–616. MIT Press, Cambridge, MA (2003)
103. Csató, L., Opper, M.: Sparse online Gaussian processes. Neural Comput. **14**(3), 641–668 (2002)
104. Sathiya Keerthi, S., Chu, W.: A matching pursuit approach to sparse Gaussian process regression. In: Weiss, Y., Schölkopf, B., Platt, J. (eds.) Advances in Neural Information Processing Systems, vol. 18, pp. 643–650. MIT Press, Cambridge, MA (2006)
105. Seeger, M., Williams, C.K.I., Lawrence, N.D.: Fast forward selection to speed up sparse Gaussian process regression. In: Ninth International Workshop on Artificial Intelligence and Statistics, Society for Artificial Intelligence and Statistics (2003)
106. Lazáro-Gredilla M., Quiñonero-Candela J., Figueiras-Vidal A.: Sparse spectral sampling. Technical report, Microsoft Research, Cambridge (2007)
107. Lazáro-Gredilla, M., Quiñonero-Candela, J., Rasmussen, C.E., Figueiras-Vidal, A.R.: Sparse spectrum Gaussian process regression. J. Mach. Learn. Res. **11**, 1865–1881 (2010)
108. Titsias, M.: Variational learning of inducing variables in sparse Gaussian processes. In: The 12th International Conference on Artificial Intelligence and Statistics (AISTATS), vol. 5, pp. 567–574 (2009)
109. Ni, W., Tan, S.K., Ng, W.J., Brown, S.D.: Moving-window GPR for nonlinear dynamic system modeling with dual updating and dual preprocessing. Ind. Eng. Chem. Res. **51**(18), 6416–6428 (2012)
110. Oba, S., Sato, M., Ishii, S.: On-line learning methods for Gaussian processes. In: Dorffner, G., Bischof, H., Hornik, K. (eds.) Artificial Neural Networks (ICANN 2001). Lecture Notes in Computer Science, vol. 2130, pp. 292–299. Springer, Berlin (2001). doi:10.1007/3-540-44668-0
111. Grbić, R., Slišković, D., Kadlec, P.: Adaptive soft sensor for online prediction based on moving window Gaussian process regression. In: 2012 11th International Conference on Machine Learning and Applications, pp. 428–433 (2012)
112. Ranganathan, A., Yang, M.H., Ho, J.: Online sparse Gaussian process regression and its applications. IEEE Trans. Image Process. **20**, 391–404 (2011)
113. Nguyen-Tuong, D., Seeger, M., Peters, J.: Real-time local GP model learning. From Motor Learning to Interaction Learning in Robots, pp. 193–207. Springer, Heidelberg (2010)
114. Tresp, V.: A Bayesian committee machine. Neural Comput. **12**, 2719–2741 (2000)
115. Shi, J.Q., Murray-Smith, R., Titterington, D.M.: Hierarchical Gaussian process mixtures for regression. Statist. Comput. **15**(1), 31–41 (2005)
116. Gregorčič, G., Lightbody, G.: Local model identification with Gaussian processes. IEEE Trans. Neural Netw. **18**(5), 1404–1423 (2007)
117. Petelin, D., Kocijan, J.: Control system with evolving Gaussian process model. In: Proceedings of IEEE Symposium Series on Computational Intelligence, SSCI 2011. IEEE, Paris (2011)

118. Angelov, P., Filev, D.P., Kasabov, N.: Evolving Intelligent Systems: Methodology and Applications. IEEE Press Series on Computational Intelligence. Wiley-IEEE Press, New York, NY (2010)
119. Åström, K.J., Wittenmark, B.: Computer Controlled Systems: Theory and Design. Prentice Hall, Upper Saddle River, NJ (1984)
120. Isermann, R., Lachman, K.H., Matko, D.: Adaptive Control Systems. Systems and Control Engineering. Prentice Hall International, Upper Saddle River, NJ (1992)
121. Fritzke, B.: Growing cell structures—a self-organizing network for unsupervised and supervised learning. Neural Netw. **7**(9), 1441–1460 (1994)
122. Angelov, P., Buswell, R.: Evolving rule-based models: a tool for intelligent adaptation. In: Proceedings of the Joint 9th NAFIPS International Conference, pp. 1062–1066. IEEE Press (2001)
123. Kasabov, N.K.: Evolving Connectionist Systems: Methods and Applications in Bioinformatics, Brain Study and Intelligent Machines. Springer, New York, NY (2002)
124. Abusnina, A., Kudenko, D.: Adaptive soft sensor based on moving Gaussian process window, pp. 1051–1056. IEEE (2013)
125. Petelin, D., Grancharova, A., Kocijan, J.: Evolving Gaussian process models for the prediction of ozone concentration in the air. Simul. Modell. Pract. Theory **33**(1), 68–80 (2013)
126. Deisenroth, M.P., Rasmussen, C.E.: PILCO: A model-based and data-efficient approach to policy search. In: Proceedings of the 28th International Conference on Machine Learning (ICML 2011). Bellevue, WA (2011)
127. Ni, W., Tan, S.K., Ng, W.J.: Recursive GPR for nonlinear dynamic process modeling. Chem. Eng. J. **173**(2), 636–643 (2011)
128. Ni, W., Wang, K., Chen, T., Ng, W.J., Tan, S.K.: GPR model with signal preprocessing and bias update for dynamic processes modeling. Control Eng. Pract. **20**(12), 1281–1292 (2012)
129. Duvenaud, D., Lloyd, J.R., Grosse, R., Tenenbaum, J.B., Ghahramani, Z.: Structure discovery in nonparametric regression through compositional kernel search. In: Proceedings of the 30th International Conference on Machine Learning (2013)
130. Seeger, M.: Low rank updates for the Cholesky decomposition. Technical report, University of California, Berkeley, CA (2008)
131. Murray-Smith, D.J.: Methods for the external validation of continuous system simulation models: a review. Math. Comput. Modell. Dyn. Syst. **4**(1), 5–31 (1998)
132. Cawley, G.C., Talbot, N.L.C.: On over-fitting in model selection and subsequent selection bias in performance evaluation. J. Mach. Learn. Res. **11**, 2079–2107 (2010)
133. Hvala, N., Strmčnik, S., Šel, D., Milanić, S., Banko, B.: Influence of model validation on proper selection of process models — an industrial case study. Comput. Chem. Eng. **29**, 1507–1522 (2005)
134. Kocijan, J., Girard, A., Banko, B., Murray-Smith, R.: Dynamic systems identification with Gaussian processes. Math. Comput. Modell. Dyn. Syst. **11**(4), 411–424 (2005)
135. Girard, A.: Approximate methods for propagation of uncertainty with Gaussian process models. Ph.D. thesis, University of Glasgow, Glasgow (2004). http://citeseerx.ist.psu.edu/viewdoc/summary?doi=10.1.1.63.8313
136. Girard, A., Rasmussen, C.E., Candela, J.Q., Murray-Smith, R.: Gaussian process priors with uncertain inputs - application to multiple-step ahead time series forecasting. In: Becker, S., Thrun, S., Obermayer, K. (eds.) Advances in Neural Information Processing Systems, vol. 15, pp. 542–552. MIT Press, Cambridge, MA (2003)
137. Groot, P., Lucas, P., van den Bosch, P.: Multiple-step time series forecasting with sparse Gaussian processes. In: Proceedings of the 23rd Benelux Conference on Artificial Intelligence (BNAIC 2011), pp. 105–112. Ghent (2011)
138. Gutjahr, T., Ulmer, H., Ament, C.: Sparse Gaussian processes with uncertain inputs for multi-step ahead prediction. In: 16th IFAC Symposium on System Identification, pp. 107–112. Brussels, (2012)
139. Girard, A., Rasmussen, C., Murray-Smith, R.: Gaussian process priors with uncertain inputs: multiple-step ahead prediction. Technical report DCS TR-2002-119, University of Glasgow, Glasgow (2002)

140. Kocijan, J., Likar, B.: Gas-liquid separator modelling and simulation with Gaussian-process models. Simul. Modell. Pract. Theory **16**(8), 910–922 (2008)
141. Ažman, K., Kocijan, J.: Application of Gaussian processes for black-box modelling of biosystems. ISA Trans. **46**, 443–457 (2007)
142. Cho, J., Principe, J.C., Erdogmus, D., Motter, M.A.: Quasi-sliding model control strategy based on multiple-linear models. Neurocomputing **70**, 960–974 (2007)
143. Gauthier, J.P., Hammouri, H., Othman, S.: A simple observer for nonlinear systems applications to bioreactors. IEEE Trans. Autom. Control **6**, 875–880 (1992)
144. Lind, I.: Regressor selection in system identification using ANOVA. Licentiate thesis, University of Linköping, Linköping (2001)

Chapter 3
Incorporation of Prior Knowledge

This chapter is devoted to the topic of incorporating prior knowledge into GP models.

The theory of dynamic systems modelling is mainly devoted to white-box models, i.e. first-principles models, or black-box models that are models identified from data. Moreover, engineering practice is usually between these two extreme cases. In practice, grey-box models are frequently used, with these grey-box models having very different levels of prior knowledge incorporated. The rest of the missing information is obtained from the data.

From this point of view, it is important to know how a certain method can allow prior knowledge to be incorporated. Various forms of prior knowledge can be incorporated into GP models.

3.1 Different Prior Knowledge and Its Incorporation

A GP model, as we have seen in the previous chapter, needs prior knowledge, prior in short, in the different levels of inference to make the modelling procedure most effective. The first prior we have to place is the knowledge of the distributions over the function we expect to model in the form of mean function and covariance function. This prior over models can rule out or in certain models that might or might not be optimal for the data at hand.

There are further kinds of prior knowledge used in the modelling process that are not inherent only to GP modelling, but also general with respect to the identification procedure. The selection of experiments, input datasets, preprocessing and modelling methods are only a few examples of prior knowledge that come together with the modelling process itself.

J. Kocijan, *Modelling and Control of Dynamic Systems Using Gaussian Process Models*, Advances in Industrial Control,
DOI 10.1007/978-3-319-21021-6_3

Yet another possibility is a combination of GP models with other kinds of models to provide different forms of hybrid models. GP models may complement white-box models as a supplement for parts of a model, e.g. in [1] or [2], as a model of stochastic input, known as latent force model (LFM), e.g. in [3] or [4], or as a model of residuals from white-box models, e.g. in [5] or [6].

The most of this chapter is devoted to kinds of prior knowledge, which have not yet been elaborated up to this point, but can be important for a nonlinear systems modelling process in practice.

A trained GP model carries information about the observed system in the following elements:

- input–output data $\mathcal{D} = \{(\mathbf{Z}, \mathbf{y})\}$, describing the input–output behaviour of the system,
- the covariance function, which expresses the correlation between the data,
- the mean value of the input–output modelling function.

While the mean value selection has been elaborated in Sect. 2.4.4, this chapter is devoted to two other possibilities for prior knowledge incorporation into the GP model. The first possibility is either to change or to add extra data to the input–output data set $\mathcal{D}$. The second possibility is to appropriately change the covariance function, so it expresses our different, stronger, prior beliefs about the system. This enables the incorporation of useful knowledge regarding the dynamics of a system into a probabilistic learning framework.

Further possibilities for prior knowledge incorporation exist, and examples are given in [7].

3.1.1 Changing Input–Output Data

The first mentioned possibility for prior information incorporation is to change the input–output data $\mathcal{D}$ in which the behaviour of the unknown system is contained in explicit form, i.e. the system's output as a function of the corresponding values of regressors. There are several possibilities:

- Extra data points $\{(\mathbf{z}_i, y_i)\}$ can be added, reflecting some prior knowledge. Examples of these are static characteristics, e.g. equilibrium curves or hyperplanes, or some boundary conditions, like hard constraints where a process has reached its limits.
- A new regressor can be added to already-selected regressors, increasing the input dimension of the model. Into this regressor, additional information about every training data point in the data $\mathcal{D}$ is encoded, e.g. the state of the hysteresis of the system for a particular training point $(\mathbf{z}_i, y_i)$, as was given in [8, 9]. The selection of regressors may be used to exercise the prior knowledge in nonlinear systems' identification in general, not just for GP models.

Example 3.1 (*Incorporation of known static characteristic*) This example is intended to show how easily the prior knowledge about a static characteristic can be incorporated into a GP model. The first-order nonlinear system [10]:

$$y(k+1) = \frac{y(k)}{1 + y^2(k)} + u(k)^3 \tag{3.1}$$

is utilised to illustrate the modelling. The sampling time $T_s = 1$s. The regressors $u(k)$ and $y(k)$ are used for the modelling. An input signal of 30 points is generated first with a random generator using no hold time. The input signal from which regressors are generated is shown in Fig. 3.1 in the top-left figure. The validating simulation response of the GP model of the system described with Eq. (3.1) and identified based on these points is depicted in Fig. 3.1 in the top-right figure. It is clear that the model has problems matching the behaviour of the original system, especially in the stationary states.

Thirteen points of a known static input–output characteristic are then simply added to the input and target vectors. These can be seen in Fig. 3.1 in the bottom-left figure.

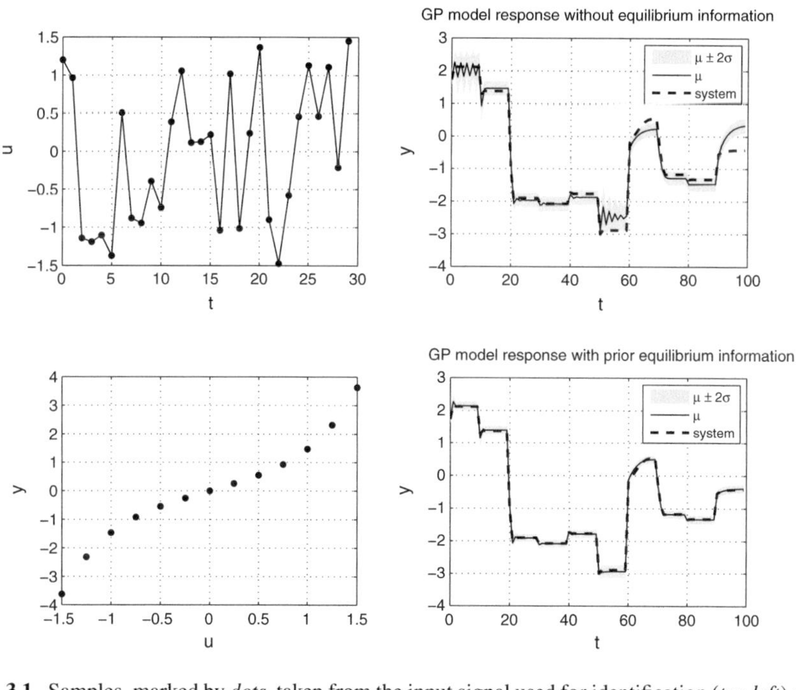

Fig. 3.1 Samples, marked by *dots*, taken from the input signal used for identification (*top left*) and simulation response of the GP model to validation of the input signal (*top right*); samples, marked by *dots*, from the static input–output characteristic that are concatenated to input and target data (*bottom left*) and the simulation response of the GP model with the included prior knowledge to the validation input signal (*bottom right*)

The validation simulation of the GP model obtained with prior information about the static characteristic is depicted in Fig. 3.1 in the bottom-right figure. The GP model now behaves much better near the given stationary states.

3.1.2 Changing the Covariance Function

The role of the covariance function is to correlate the data constituting the GP model. Our *a priori* knowledge is expressed through the choice of the covariance-function family. The squared exponential function described with Eq. (2.13) or (2.14), for example, is the most widely, sometimes blindly, used covariance function employed for functional representation, as it represents common prior beliefs like the stationarity and the smoothness of the prior GP model, and it is fairly easy to use. However, if it is already known that the unknown system has some other properties that we would like to express with the covariance function, like periodicity, non-stationarity, it could be a good idea to choose the covariance function from a different function family, see, e.g. [11] or [12] for the latter case.

Another possibility is to change the nature of the data in the input–output data $\mathcal{D}$ so that it represents the derivative instead of the functional information. As the derivative of the GP remains a GP [13], this is allowed if we appropriately adapt the covariance function for the derivative data. This can be generalised to any linear operation and is presented in a general form in [14]. We restrict our attention to the derivative information since this is the most useful in the context of dynamic systems.

An example of the incorporation of derivatives is the knowledge of the state variables that are time derivatives of other state variables. Such an example is described in [15].

Another example of including data with the derivative information is the inclusion of linear local models. These linear local models can be combined with data representing the signals of a modelled system. Such a GP model could be a useful tool for combining local models, as it can replace the local models with the system's response samples in the regions where the local models are difficult to identify, e.g. in the off-equilibrium regions of the dynamic system [16]. Another advantage of incorporating the linear local models is a reduction of the size of the GP model, which could reduce the training time of the model, as described in Sect. 3.3.1.

There is also a possibility that the noise present at the output of the system is not white and Gaussian. If the parameters of the noise model are known, the 'noise part' of the covariance function can be adopted accordingly, as in [17].

3.1.3 Combination with the Presumed Structure

In a case we are familiar with, or where we would like to impose the structure of the model, in the sense of explicit equations describing the model, two possibilities are presented in this chapter.

The first one is the application of a Gaussian process framework for the modelling of the Wiener and Hammerstein models, and the second one is the application of GP models to develop a linear parameter-varying model, i.e. a linear model with varying parameters that are described by GP models. Section 3.2 describes the Wiener and Hammerstein models in the GP framework, and Sect. 3.3.2 describes the linear parameter-varying models with GP models of varying parameters.

Other combinations of embedded, parallel or serial connections of first-principles models and GP models are possible. An overview of the combinations of white-box and black-box models, in general, is given in [18].

3.2 Wiener and Hammerstein GP Models

This section shows two methods for modelling particular types of nonlinear dynamic systems as GP models, namely, the Wiener and Hammerstein models.

Block-oriented structures where a nonlinear system model is presented as a combination of static-nonlinear and dynamic-linear submodels are well known in dynamic systems identification, e.g. [18, 19]. Depending on the position of the static nonlinearity and the dynamic-linear model, these structures are named the Wiener model, the Hammerstein model or combinations like the Wiener–Hammerstein, the Wiener-Hammerstein-Wiener model and similar. Prior knowledge in the form of a known structure like the Wiener and Hammerstein models is attractive to the engineering community.

There is a vast amount of literature on the topic of block-oriented, nonlinear systems identification. A selection of topics and an overview of the references can be found in [20]. The idea of using nonparametric models, in particular kernel methods, like support-vector machines, in block-oriented structures is not new and is documented in [20]. Moreover, regardless of the well-established methods relating to this topic, new research and application results emerge continuously, because of the facilitated analysis and control design of the nonlinear systems that are frequently found in practice.

GP models are suited to the incorporation of structural prior knowledge about the modelled system in the form of a block-oriented structure. Published examples are listed as follows. GP models for modelling Wiener–Hammerstein models are described in [21]. The identification of a Wiener model with a GP model is described in [22] or [23], with a study of the identification consistency in [24]. Hammerstein models with GP models can be found in [25, 26] and for continuous-time systems in [27].

The difficulty associated with Hammerstein and Wiener system identification is the interaction between the linear and the nonlinear parts. If the nonlinear part is identified at a separate stage, then the difficulty is alleviated. The following two subsections present two different methods. The first one for modelling the Wiener model is a one-stage method and the second one, for modelling the Hammerstein model, is based on two separate stages, which is a principle that might be attractive for engineering practice as a complement to more general frameworks for the GP modelling of block-oriented models like those in the references listed above.

3.2.1 GP Modelling Used in the Wiener Model

The structure of the Wiener model is composed of a linear dynamic block that is followed by a nonlinear static block, as depicted in Fig. 3.2. This structure is matched with a relatively small number of processes. One of them is the pH titration process from chemical engineering. Another possibility is a dynamic system where a nonlinearity exists in the sensor's static characteristic.

No straightforward linear parametrisation exists for the structure of the Wiener model [18], and when identified in one stage, methods different from the linear identification are necessary in general.

In this section, the method from [23] is summarised, where the model to be identified is a combination of a linear state-space model and a nonparametric model for the nonlinear block. The parameters of a linear model that are considered as random variables and the nonlinear part are identified at the same time using a Bayesian approach, i.e. the posterior density estimation algorithm. The method provides a posterior probability-density function estimate $p(\boldsymbol{\theta}|\mathcal{D})$ of the parameters $\boldsymbol{\theta}$ of both blocks given the measured data $\mathcal{D}$ containing the output measurements $\{y_i|i = 1, \dots N\}$ and the input measurements $\{u_i|i = 1, \dots N\}$.

The noise enters the system, not only at the output of the nonlinear block, but also internally to the linear dynamic part of the system. The method deals with a wide range of nonlinear mappings, and there is no assumption that the nonlinearity is an invertible and monotonic function. The Wiener model is presented as a system with one input and one output, but the generalisation for multiple-input, multiple-output models is straightforward.

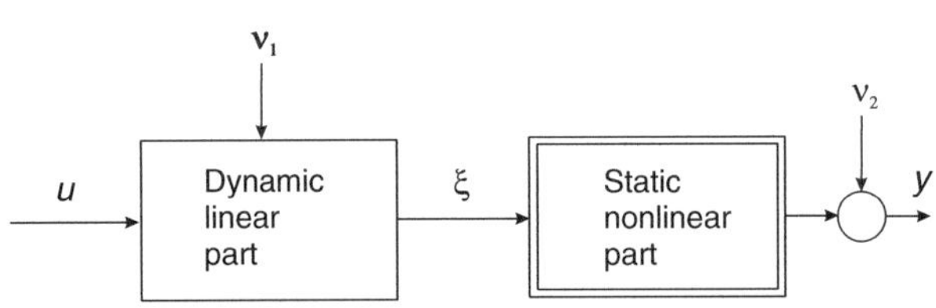

Fig. 3.2 Principial scheme of the Wiener model with a linear dynamic block and a nonlinear static block. The system noise ν_1 and the measurement noise ν_2 are not measured

The Wiener model is described in the state-space form as

$$\mathbf{x}(k+1) = \mathbf{A}\mathbf{x}(k) + \mathbf{b}u(k) + \boldsymbol{\nu}_1(k), \tag{3.2}$$

$$\xi = \mathbf{c}\mathbf{x}(k), \tag{3.3}$$

$$y(k) = f_s(\xi) + \nu_2(k), \tag{3.4}$$

where $\boldsymbol{\nu}_1 \sim \mathcal{N}(\mathbf{0}, \boldsymbol{\Sigma}_{\nu 1})$ is the process noise, $\nu_2 \sim \mathcal{N}(0, \sigma_{\nu 2}^2)$ is the measurement noise, $\mathbf{x}, \boldsymbol{\nu}_1 \in \mathbb{R}^n$, $u, \xi, y \in \mathbb{R}$ are the input signal, the output from the linear block and the output signal from the nonlinear block, respectively.

The elements of the vector $\mathbf{c}$ are fixed to $\mathbf{c} = \begin{bmatrix} 1 \ 0 \ \dots \ 0 \end{bmatrix}$ without any loss of generality. The system parameters $\mathbf{A}, \mathbf{b}, \boldsymbol{\Sigma}_{\nu 1}, \sigma_{\nu 2}^2$ and the hyperparameters of the nonlinear mapping f_s need to be identified with a Bayesian approach by considering the parameters as random variables.

Based on the way that prior values are set to the system parameters $\mathbf{A}$ and $\mathbf{b}$, the utility of ARD may be exploited. See [23] for more details.

The nonlinear mapping f_s is modelled as a GP model that is described by its mean and covariance function. The linear mean function $m_f(\xi) = \xi$, or any other function, can be used with respect to the shape of the nonlinearity. One of the smooth covariance functions, e.g. the squared exponential or the Matérn, is suggested because the used data is affected by stochastic disturbances and, as a result, smooth regression functions are favoured to avoid over-fitting. In the case of non-smooth nonlinearities and a smooth covariance function, the nonlinearity is approximated with a smooth function.

The posterior density of $\boldsymbol{\theta}$, which contains the system parameters of the linear part and the hyperparameters of the nonlinear part, has to be found. The joint posterior density of the parameters and the system states $\mathbf{x}$ is computed as

$$p(\boldsymbol{\theta}, \mathbf{x}|\mathcal{D}) = p(\mathbf{x}|\boldsymbol{\theta}, \mathcal{D})p(\boldsymbol{\theta}|\mathcal{D}). \tag{3.5}$$

The density $p(\boldsymbol{\theta}|\mathcal{D})$ is obtained by a straightforward marginalisation of Eq. (3.5).

As the posterior density $p(\boldsymbol{\theta}, \mathbf{x}|\mathcal{D})$ is analytically intractable, we have to use one of the possible approximations. The proposed solution is to use an MCMC sampler [28], a numerical approximation method, to solve the inference problem.

The standard Gibbs sampler, which targets some joint density by alternately sampling from its conditionals [29], is problematic in our case when it comes to the state inference. Consequently, a particle Gibbs sampler [30], a particle MCMC method, is suggested. Furthermore, it should be noted that from the practitioner's point of view, if you understand the principle of MCMC, it is not necessary to understand all the technical details of the particle MCMC to be able to use it as a component in this identification procedure.

The general algorithm for the identification of the Wiener system with particle MCMC is as follows:

Algorithm: WIENER($\mathcal{D}$)

---

initialise the values of the system parameters and the GP model
set the initial trajectory of **x**
for $i \leftarrow 1$ **to** $maxSample$
  do
  - run the algorithm for the sampling system parameters for the selected prior
  - sample the posteriori distribution of the hyperparameters
  - sample the posteriori distribution $f_s[i]$
  - set the vector of samples of the system parameters and the GP model
  - run the conditional particle filter targeting the distribution of **x**
  - sample a selected trajectory of **x** for this iteration

The sampling steps in the algorithm are performed according to the principle of the Gibbs sampling, except for the last step, which is performed according to the particle Gibbs sampling [30].

The convergence analyses for this numerically intensive method are given in [23] and with fewer assumptions in [24].

In the next example, the presented method is used for the identification of the GP Wiener model. Even though the nonlinear system used for the example does not exhibit problems like a non-monotonic nonlinear mapping function, it serves the purpose to illustrate the utility of the identification method.

Example 3.2 (*GP Wiener model*) The nonlinear system [31] that is used in this section for the illustration of the Wiener model's identification in the framework of the GP model is composed of the linear dynamic part

$$\mathbf{x}(k+1) = \begin{bmatrix} 1.414 & -0.6065 \\ 1 & 0 \end{bmatrix} \mathbf{x}(k) + \begin{bmatrix} 0.5 \\ 0 \end{bmatrix} u(k), \tag{3.6}$$

$$\xi(k) = \begin{bmatrix} 0.2088 & 0.1966 \end{bmatrix} \mathbf{x}(k) \tag{3.7}$$

and the subsequent static nonlinearity

$$y = \frac{\xi}{\sqrt{0.1 + 0.9\xi^2}}. \tag{3.8}$$

The sampling period of the signals is one time unit. The input signal u is a stochastic signal with a hold time, i.e. the period of time for which the signal stays constant, of 10 time units. White Gaussian noises $\mathcal{N}(\mathbf{0}, 0.03^2\mathbf{I})$ and $\mathcal{N}(0, 0.02^2)$ are added to the states of the linear part and to the output of the process as the process and measurement noise, respectively.

The identification data contains 729 data points and the particle MCMC method uses 20,000 iterations and 15 particles. The model order is set to 2 and was obtained using the backward-elimination method (Sect. 2.3.2), but other order-selection methods including exploiting the ARD utility, as described in [23], could be used.

The nonlinear part of the system is modelled with a GP model containing the linear mean function $m_f(\xi) = \xi$. The GP model should contain the covariance function that reflects prior knowledge about the static nonlinearity. Because we are dealing with stochastic disturbances at the output, the suggestion is that the covariance function for the functional part of the GP model is a smooth covariance function in order to avoid any modelling of the process-inherent noise. The squared exponential covariance function described with Eq. (2.14) for the function part summed with Eq. (2.11) for the noise part is used in our case, but other stationary covariance functions might also be used for the function part

$$C(\xi_i, \xi_j) = \sigma_f^2 \exp\left(-\frac{1}{2l^2}(\xi_i - \xi_j)^2\right) + \sigma_n^2 \delta_{ij}. \tag{3.9}$$

The identification input and output signals are depicted in Fig. 3.3.

The predictions obtained from the model or its blocks can be found by averaging over the posterior. In our case, the Wiener model parameters of the linear part and the GP model of the nonlinear part are presented as the first and second moments, i.e. the mean values and covariances that match the first and second moments of the Gaussian mixtures of posterior realisations for each of the linear model parameters and the GP model. The predictions for the models with the parameters and the GP model obtained with the moment matching can be seen in Figs. 3.4 and 3.5.

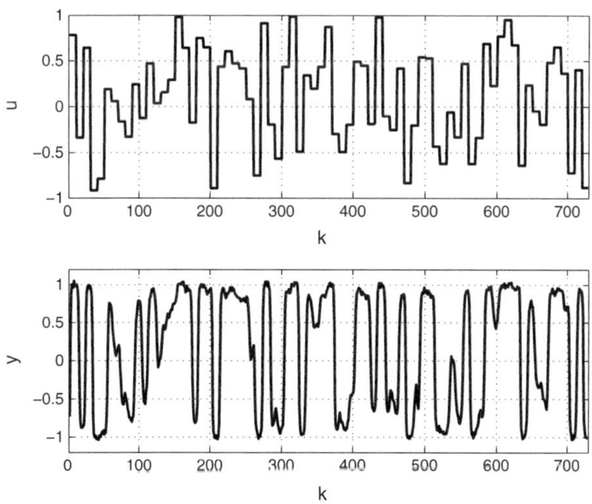

Fig. 3.3 Input and output signals used for the identification of the Wiener model

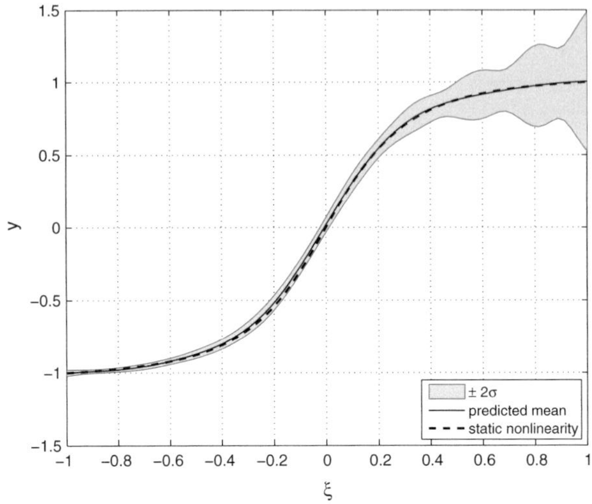

Fig. 3.4 The static-nonlinear part of the Wiener model scheme obtained from Gaussian mixtures for every input point with the matching of moments

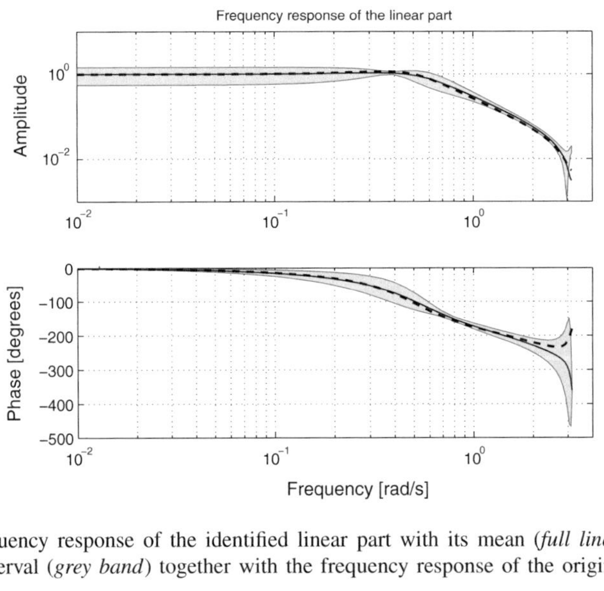

Fig. 3.5 Frequency response of the identified linear part with its mean (*full line*) and a 95 % confidence interval (*grey band*) together with the frequency response of the original linear part (*dashed line*)

The identified nonlinear part of the Wiener model with a good matching to the original nonlinearity is shown in Fig. 3.4.

The Bode plot of the identified linear part, together with the original linear part, is depicted in Fig. 3.5. As is clear from Figs. 3.4 and 3.5, the obtained model shows a satisfactory performance.

In the case of a validation with the simulation response of the model with an input signal that is different from the one used for the identification, we have various options. We can use the matched mean values and covariances for the linear model parameters and the matched posterior mean functions and covariance functions of the GP model to be used in the dynamic system simulation, taking account of the uncertainties as described in Sect. 2.6 or, alternatively, use the simulation with the Monte Carlo method.

3.2.2 GP Modelling Used in the Hammerstein Model

The Hammerstein structure consists of a nonlinear static block followed by a linear dynamic block, as illustrated in Fig. 3.6. It is a frequently applied, nonlinear dynamic systems modelling approach. This kind of model can be used where the actuator dominates the system's behaviour with its nonlinear static characteristic.

The structure of the Hammerstein model can be linearly parameterised, which can be reflected in the choice of regressors when modelling with the linear model. The idea behind this approach is to represent a static nonlinearity with a polynomial approximation and, in this case, the overall input–output relationship is linear in the parameters [18]. In the case of a GP model identification with a linear covariance function, this approach requires the manual or automated selection of polynomial regressors, which are at the same time also regressors of the complete GP Hammerstein model. The GP model with a linear covariance function can, consequently, be used to model the input–output relationship. The identification procedure is mainly composed of the regressors' selection, which might be a lengthy operation, and the input–output identification, which are both tightly interconnected.

In the case that the nonlinearity requires a complicated polynomial representation, a two-stage procedure might be an alternative choice. The two-stage procedure is similar to engineering practice and consists initially of the identification of a static nonlinearity, which is followed by the identification of the dynamic part. This is the case we describe in the continuation.

There is no assumption necessary that the nonlinearity is an invertible or monotonic function. The white-noise disturbance with a normal distribution is presumed to be at the output of the nonlinear block. The gain of the linear dynamic part is assumed to be 1. If the situation is otherwise, the gain is joined with the static nonlinearity.

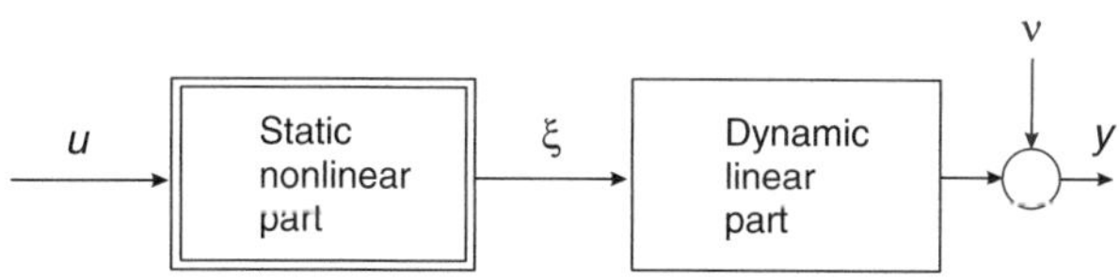

Fig. 3.6 Principial scheme of the Hammerstein model with a linear dynamic block and a nonlinear static block. The measurement noise ν is not measured

Modelling of the Static Part

The idea behind the method is that the GP-NFIR model of the entire system is identified and used as the model of the static nonlinearity with appropriate values for the regressors.

First, the GP-NFIR model of the entire system is made with identification data pairs $\mathcal{D} = \{(\mathbf{z}_i, y_i)|i = 1, \ldots, N\}$. The regressors are delayed samples of the input signal u.

$$\hat{y}(k) = f(u(k-1), u(k-2), \ldots, u(k-m)) + \nu, \tag{3.10}$$

where f represents the GP model of the entire system to be modelled without using any knowledge of the block structure. Assuming that the number of regressors m and the number of data N are large enough, the input–output mapping f can be consistently described with such a model. Note that the number of regressors can be relatively large to obtain a satisfactory model, but the model can be obtained relatively quickly. This is one of the reasons why FIR models can often be found in signal-processing applications.

Next, this GP-NFIR model can be considered as a model of the static nonlinearity when using regressors that correspond to a constant signal at the input

$$u = u(k-1) = u(k-2) = \cdots = u(k-m). \tag{3.11}$$

Again, as was the case with the illustrative Example 3.2, the nonlinear system was identified with a GP model with the composite covariance function described with Eq. (3.9). The constant mean function $m_f = 0$ is selected.

How is the GP model affected in the case of a nonlinearity, for which it gets different values of the output data for the same values of the input data, e.g. hysteresis? The GP model can be interpreted as a linear smoother [32] and it averages the obtained information, and therefore in the case of ambiguous output data, the GP model will predict the distribution with the mean value corresponding to the average of the output data.

Modelling of the Dynamic Part

When the GP model of static nonlinearity is obtained, the intermediate signal $\hat{\xi}$ can be inferred from the input signal with this model. The output of the GP model is the predictive distribution that will form the input for the linear dynamic part of the Hammerstein model.

These predictive distributions can be considered as an uncertain input signal for the linear dynamic part. Therefore, we need to consider the learning of the dynamic GP model using a linear covariance function with stochastic inputs.

The assumption is made that the input variables are independent and normally distributed. The derivation for the general covariance function is given first, adopted from [33].

We recall that the GP prior on f_d, with a zero mean and covariance function $C(\mathbf{z}_i, \mathbf{z}_j)$, implies that $E(y_i) = 0$ and $\mathrm{cov}(y_i, y_j) = C(\mathbf{z}_i, \mathbf{z}_j)$ (Eq. 1.10).

In the situation where stochastic inputs are present, we have

$$y_i = f_d(\mathbf{z}_i), \tag{3.12}$$

with the regression vector

$$\mathbf{z}_i = \boldsymbol{\mu}_{\mathbf{z}_i} + \boldsymbol{\nu}_{\mathbf{z}_i}, \tag{3.13}$$

where

$$\boldsymbol{\nu}_{\mathbf{z}_i} \sim \mathcal{N}(\mathbf{0}, \boldsymbol{\Sigma}_{\mathbf{z}_i}).$$

Given $\mathbf{z}_i \sim \mathcal{N}(\boldsymbol{\mu}_{\mathbf{z}_i}, \boldsymbol{\Sigma}_{\mathbf{z}_i})$, although the process in the general case is no longer Gaussian, the mean and covariance function of the random process can still be determined. According to the law of iterated expectations, we can write [33]

$$E(y_i|\boldsymbol{\mu}_{\mathbf{z}_i}) = E(E(y_i|\mathbf{z}_i)) = 0. \tag{3.14}$$

The law of conditional variances says that

$$\mathrm{var}(y_i|\boldsymbol{\mu}_{\mathbf{z}_i}) = E(\mathrm{var}(y_i|\mathbf{z}_i)) + \mathrm{var}(E(y_i|\mathbf{z}_i)) = E(\mathrm{var}(y_i|\mathbf{z}_i)). \tag{3.15}$$

Extending this result to the covariances leads to [33]

$$\mathrm{cov}(y_i, y_j|\boldsymbol{\mu}_{\mathbf{z}_i}, \boldsymbol{\mu}_{\mathbf{z}_j}) = \int\int C(\mathbf{z}_i, \mathbf{z}_j) p(\mathbf{z}_i, \mathbf{z}_j) d\mathbf{z}_i d\mathbf{z}_j, \tag{3.16}$$

where a noise variation for each input, i.e.

$$p(\mathbf{z}_i) = \mathcal{N}_{\mathbf{z}_i}(\boldsymbol{\mu}_{\mathbf{z}_i}, \boldsymbol{\Sigma}_{\mathbf{z}_i}), \tag{3.17}$$

$$p(\mathbf{z}_j) = \mathcal{N}_{\mathbf{z}_j}(\boldsymbol{\mu}_{\mathbf{z}_j}, \boldsymbol{\Sigma}_{\mathbf{z}_j}), \tag{3.18}$$

is allowed.

Let $C(\boldsymbol{\mu}_{\mathbf{z}_i}, \boldsymbol{\mu}_{\mathbf{z}_j})$ denote the covariance function with stochastic inputs giving the covariance between y_i and y_j. Assuming the input variables are independent, given their characteristics, it can be defined as

$$C(\boldsymbol{\mu}_{\mathbf{z}_i}, \boldsymbol{\mu}_{\mathbf{z}_j}) = \int\int C(\mathbf{z}_i, \mathbf{z}_j) p(\mathbf{z}_i) p(\mathbf{z}_j) d\mathbf{z}_i d\mathbf{z}_j. \tag{3.19}$$

This integral is not solvable for all possible covariance functions. But it can be solved for the linear covariance function described with Eq. (2.22) as follows. The noise part is, for reasons of convenience, omitted from the expressions.

Equation (3.19) with the linear covariance function can be written as

$$C(\mu_{\mathbf{z}_i}, \mu_{\mathbf{z}_j}) = \int\int \mathbf{z}_i^{\mathrm{T}} \boldsymbol{\Lambda}^{-1} \mathbf{z}_j p(\mathbf{z}_i) p(\mathbf{z}_j) d\mathbf{z}_i d\mathbf{z}_j. \tag{3.20}$$

It can be integrated over $\mathbf{z}_i$ first

$$\int \mathbf{z}_i p^{\mathrm{T}}(\mathbf{z}_i) d\mathbf{z}_i \Psi = E_{\mathbf{z}_i}(\mathbf{z}_i)\Psi = \mu_{\mathbf{z}_i}^{\mathrm{T}} \Psi, \tag{3.21}$$

where $\Psi = \int \boldsymbol{\Lambda}^{-1} \mathbf{z}_j p(\mathbf{z}_j) d\mathbf{z}_j$,
and the obtained result is integrated over $\mathbf{z}_j$

$$\int \mu_{\mathbf{z}_i}^{\mathrm{T}} \boldsymbol{\Lambda}^{-1} \mathbf{z}_j p(\mathbf{z}_j) d\mathbf{z}_j = \mu_{\mathbf{z}_i}^{\mathrm{T}} \boldsymbol{\Lambda}^{-1} \int \mathbf{z}_j p(\mathbf{z}_j) d\mathbf{z}_j \tag{3.22}$$

$$= \mu_{\mathbf{z}_i}^{\mathrm{T}} \boldsymbol{\Lambda}^{-1} \mu_{\mathbf{z}_j}. \tag{3.23}$$

The obtained result is the same as for data learning without the input uncertainty, Eq. (2.22). This means that the same covariance function and learning procedure can be pursued in the case of stochastic inputs using their mean values.

The obtained Hammerstein model has to be validated with a simulation. This is also carried out in two stages:

- The input data samples are treated, as described with Eq. (3.11), to form the input data for the static model of the nonlinear part. The output predictions of the GP model are random variables that will be used as the input data for the dynamic part of the Hammerstein model.
- The response of the dynamic part is simulated in such a way that the lagged samples of the output signals are fed back and used as regressors, together with the lagged predictive distributions from the nonlinear static model. This can be considered as making iterative predictions for uncertain input values, represented by the stochastic inputs, as described in Sect. 2.6, with final expressions in Appendix B.

Next, we have an illustrative example where the presented procedures are used for an identification of the GP Hammerstein model.

Example 3.3 (*GP Hammerstein model*) The nonlinear system that is used in this section for an illustration of the Hammerstein model's identification with a GP model is composed of the static nonlinearity

$$\xi = \frac{u}{\sqrt{0.1 + 0.9u^2}} \tag{3.24}$$

and the consequent linear dynamic part

$$y(k+1) = 1.4138y(k) - 0.6065y(k-1) \\ + 0.1044\xi(k) + 0.0883\xi(k-1). \quad (3.25)$$

The sampling period is one time unit. The input signal $u(k)$ is similar to the one in Example 3.2, i.e. a random signal with a hold time, i.e. the period of time for which the signal stays constant, of 10 time units. White Gaussian noise with the distribution $\mathcal{N}(0, 0.02^2)$ is added to the output of the process as the measurement noise.

The two-stage identification procedure is divided into the following steps: the modelling of the static part, the compensation of the static nonlinearity and the generation of the input signal for the linear system identification; and the identification of the dynamic-linear part.

(a) The modelling of static part

The nonlinearity is assumed to be monotone. The static nonlinearity is identified utilising GP-NFIR model identification with the GP model that has the composite covariance function described with Eq. (3.9). The regression vector is composed of 20 lagged values of the input-signal samples with the forward selection of regressors [34].

Figure 3.7 shows an identified direct static part with the used measurements sampled from the output and input signals, also shown in Fig. 3.7.

(b) The generation of the input signal for the linear system identification

The input data for the identification of the linear dynamic part, i.e. samples from ξ, are obtained as predictions of the static GP model with samples from the input signal u at its input. The predicted signal $\hat{\xi}$ is depicted in Fig. 3.8. The outputs of the static GP model are predictive distributions that are determined with the mean values and the corresponding variances.

(c) The identification of the dynamic-linear part

The signals shown in Figs. 3.8 and 3.9 are used for training the GP model with a linear covariance function.

The Bode plot of the identified linear part, together with the original linear part, is depicted in Fig. 3.10.

The obtained Hammerstein model may alternatively be validated with a simulation using an input signal that is different to the one used for the identification, as described in Sect. 2.6, but this is beyond the scope of this example.

The section presented two possible methods for the identification of the Wiener and Hammerstein models using Gaussian processes. These models can accommodate noisy data as well as the uncertainties of the identified model, which is all reflected in the shape of the predicted output distributions. The GP Wiener and GP Hammerstein models can be used for the design of robust, nonlinear control and other designs where these kinds of nonlinear models with information about the uncertainty can be utilised.

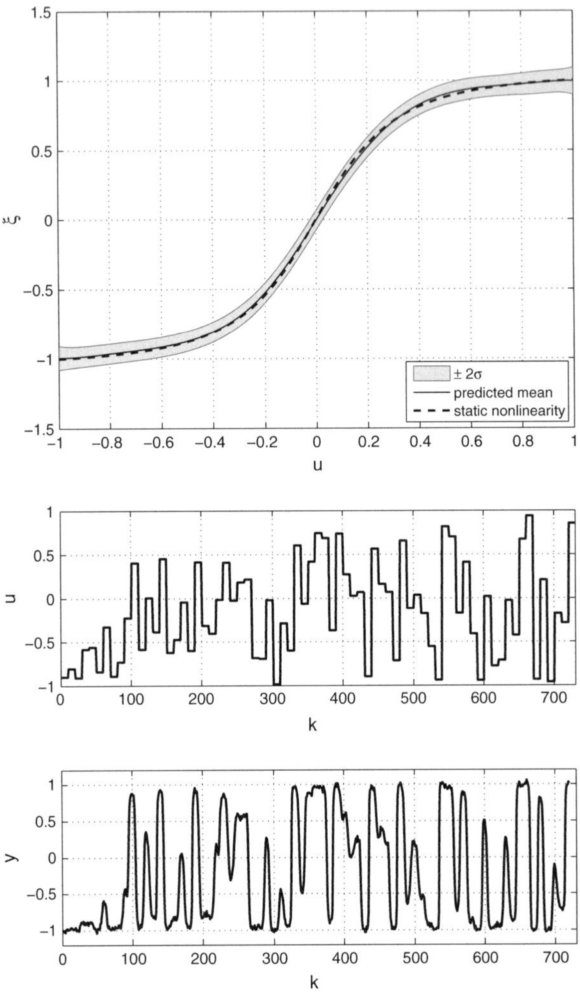

Fig. 3.7 Static part of the Hammerstein model scheme, *upper* part of figure; with input and output signals used for modelling, which are same as those in Fig. 3.3, *lower* part of figure

3.3 Incorporation of Local Models

The identification of nonlinear dynamic systems from measured data has received plenty of attention in the past few decades and numerous methods have been developed. Initially, methods with adaptive basis functions [18], like artificial neural networks and fuzzy logic models, and later methods with fixed basis functions [18], so-called kernel methods, appeared. A number of them can be viewed as universal approximators. The main practical disadvantages of these black-box modelling methods are [35]:

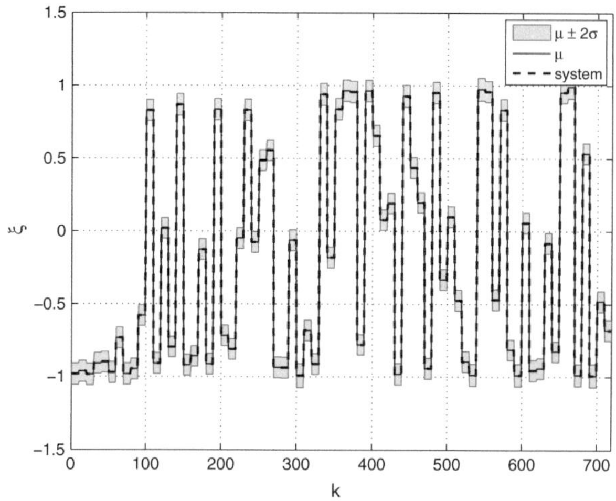

Fig. 3.8 Output signal from the static part as the response on the input signal u. Mean values from the output signal are used for the identification of the linear part of the Hammerstein model scheme

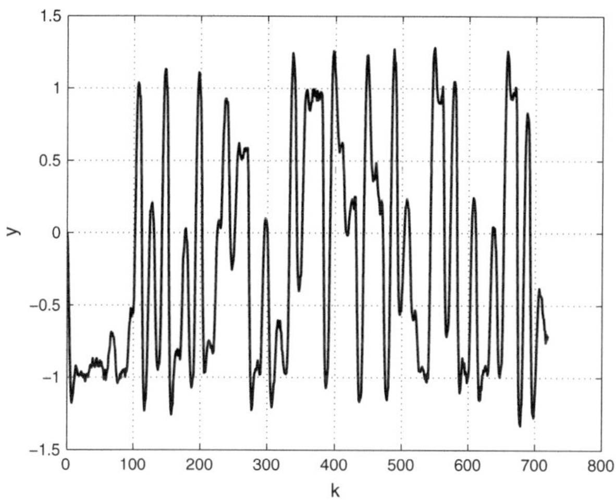

Fig. 3.9 Output signal for the identification of the linear part of the Hammerstein model scheme which is the same as the nonlinear systems output signal y

- the lack of transparency, i.e. a model structure does not reflect the physical properties of the system,
- the curse of dimensionality.

An alternative method to circumvent the disadvantages of the global black-box model of the system is to employ a network of local models, wherein each model describes some particular operating region. Such an approach is referred to as a local model network— LMN, [36].

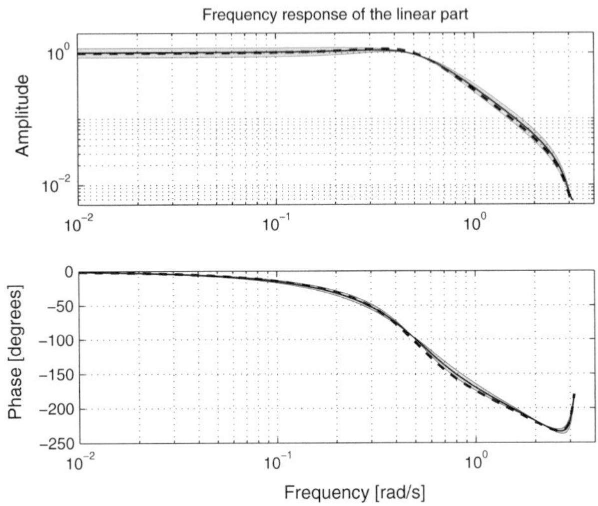

Fig. 3.10 Frequency response of the identified linear part with its mean (*full line*) and a 95 % confidence interval (*grey band*) together with the frequency response of the original linear part (*dashed line*)

LMNs are attractive for the so-called *divide and conquer* control design [36]. In this approach, the global behaviour is represented by a network of simple local models, where each local model describes some particular operating region, and the global behaviour is realised by blending the dynamics of the local models. An accurate representation of nonlinear dynamics with locally valid linear models is important from the control-design point of view, because local controllers can be designed for each of the corresponding local models and blended to a local controller network or blended gain-scheduled controller, e.g. [37]. In the case that the nonlinear dynamics cannot be interpreted with locally obtained information at every operating point, then the LMN can be seen as a black-box model.

The number of unknown parameters in the LMN is typically smaller than in neural networks with a comparable quality of fit. However, some of the inconveniences still related to LMN are [16, 35, 38]:

- the problem of describing off-equilibrium dynamics,
- the global/local optimisation issue,
- the scheduling vector selection.

The problem of modelling off-equilibrium dynamics [16, 35, 38] with local models originates in a system's 'tendency' towards equilibrium. As a consequence, there is usually a lack of measured data in the regions away from equilibrium, which makes the construction of valid local models for those regions rather difficult. This problem can be highlighted, e.g. in the process industries, where a lot of data can only be taken in particular operating regions of the system, so that only those regions can be satisfactorily modelled with local models. For the rest of the tentative operating regions, the lack of data prevents the construction of valid local models.

Two approaches to the optimisation of the LMN parameters are possible [38]:

- the global learning approach, where all the LMN parameters are optimised using all the identification data in a single regression operation,
- the local learning approach, where the parameters for each local model representing a corresponding subspace are estimated separately using the corresponding data.

LMN optimisation with a global learning approach usually provides a globally better fit as the local model parameters in the off-equilibrium regions are used to increase the level of validity for the associated local models, but these parameters no longer represent the system's local dynamics [16].

In contrast, when a local learning approach is used, the local models' parameters do represent local dynamic behaviour, which results in more transparent local models. Such models are more applicable for use in analysis and control design. Their drawback is that they are valid in smaller operating regions, which results in non-modelled regions of the system, leading again to the problem of describing off-equilibrium dynamics.

LMN approaches, regardless of the blending realisation, also encounter the issue of scheduling vector selection. The scheduling vector—usually a subset of the model's regressors—is the vector defining the current region of operation and assists the blending mechanism to accurately match the nonlinear dynamics. With a reduction of the dimension of the scheduling vector, the regions in which the individual local models try to match the nonlinear dynamics increase, but unfortunately at the cost of a decreased accuracy with respect to the distance from the regions where these local models were obtained. Furthermore, the blending based on the reduced scheduling vector can result in a non-smooth and sometimes discontinuous LMN [38].

More on the LMN approaches to system identification can be found e.g. in [36] and [18] and more on the problems associated with this approach in [16, 35] and [38].

The GP model is a possible alternative that solves some of the mentioned problems [16]. Such a model smooths the information given as the identification data. The model's output is predicted by weighting targets with respect to the distance between the input data used for identification and a new input data. The identification of GP models, however, does suffer from an increasing computational burden with an increasing number of data points being used for the modelling. One of the ways to lessen the computational burden is to combine the LMN and GP models. An overview of other methods used to reduce the computational burden is discussed in Sect. 2.5.2.

Much of the computational burden can be removed by the introduction of local models in the GP model. A local model, typically parameterised with only a few parameters, can successfully describe a subset of the training points, reflecting the local dynamics of the system. Thus the introduction of local models into the GP model can result in a reduced computational burden associated with the optimisation of the hyperparameters.

When comparing the GP model with incorporated local models to the LMN, several benefits of the GP model can be noticed. Some typical LMN problems, e.g. off-equilibrium dynamics, global/local optimisation issues and scheduling vector selection, are avoided, and also confidence in the model's prediction is given using the GP model.

The combination of local models and GP models can be pursued in two directions. The first is the *mixture of GP models*, also called the *mixture of GP experts*. As described in Sect. 2.5.2, this is a model where the system's model is composed of locally valid GP models. The idea follows that of using weighted basis functions, as in Eq. (2.3), for approximating the nonlinearity of the system to be modelled. It uses the divide-and-conquer strategy that is common to LMN methods and can be written as

$$p(y) = \sum_{i=1}^{l} w_i p_i(y), \tag{3.26}$$

where $0 \leq w_i \leq 1$, $\sum_{i=1}^{l} w_i = 1$, and $p_i(y)$ is a probability density function for the variable y.

The theoretical background for the GP mixture models is described in, e.g. [39]. In the context of dynamic systems, GP mixture models can be implemented in various ways. A very common situation is local GP models with a linear covariance function, as in, e.g. [40–42], or any other kinds of covariance functions, as in, e.g. [43–45]. The weights w_i can also be determined differently, as in, e.g. [41, 46], or the strategy of switching among local GP models is used, e.g. [47, 48]. Local GP models can be used for modelling the local parts of nonlinear mapping, but they can also represent models of batches in the process industry, e.g. [49].

The second way of using local models is the incorporation of local models into a GP model in the form of prior knowledge. In the following sections, two possible ways to incorporate GP models into one GP model are given: local models incorporated into a Gaussian process model, i.e. a LMGP model, and a fixed-structure Gaussian process model, i.e. a FSGP model.

3.3.1 Local Models Incorporated into a GP Model

Since differentiation is a linear operation, the derivative of a GP remains a GP [13]. Consequently, within the Gaussian process modelling framework, the derivatives can be used together with functional values, thus providing a way to include linear local models into the GP model. This topic is elaborated in [13, 50–52], with an application for identification in [53] and for dynamic systems control in [54]. The description here is mostly adapted from [52]. The GP model with incorporated local models will be referred to as an LMGP (local models incorporated into a Gaussian processes) model.

Consider the autoregressive model of a nth-order dynamic system $y(k) = f(\mathbf{z}(k)) + \nu$, where the regression vector $\mathbf{z}(k)$ is composed of previous values of the output values y up to a given lag n and the control input values u up to a given lag m, with the regression vector dimension $D = n + m$:

$$\mathbf{z}(k) = [y(k-1), y(k-2), \ldots, y(k-n), u(k-1), u(k-2), \ldots, u(k-m)]^{\mathrm{T}}. \quad (3.27)$$

The goal is to model the dynamic system $y(k) = f(\mathbf{z}(k)) + \nu$ using a GP model with the finite number $N_{eq} \in \mathbb{N}$ of incorporated linear local models. Let us assume that the point $\mathbf{z}_i; i \in \{1, \ldots, N_{eq}\}$ is one of equilibrium points of a stable, generally nonlinear, system $y(k) = f(\mathbf{z}(k))+\nu$. We would like to present the system's dynamic behaviour in the vicinity of the point $\mathbf{z}_i$ with an approximation in the form of a linear local model $\mathcal{M}_i$:

$$y(\mathbf{z}) = f(\mathbf{z}_i) + \boldsymbol{\theta}_i^T (\mathbf{z} - \mathbf{z}_i) + \nu, \quad (3.28)$$

where

$$\boldsymbol{\theta}_i^T = \left[\mathbf{a}_i^T, \mathbf{b}_i^T\right], \quad (3.29)$$

$$\mathbf{a}_i = \left[\frac{\partial f}{\partial y_{k-1}}, \ldots, \frac{\partial f}{\partial y_{k-n}}\right]_i^T, \quad (3.30)$$

$$\mathbf{b}_i = \left[\frac{\partial f}{\partial u_{k-1}}, \ldots, \frac{\partial f}{\partial u_{k-m}}\right]_i^T, \quad (3.31)$$

$\boldsymbol{\theta}_i$ are the parameters of the linear local model $\mathcal{M}_i$ centred at $\mathbf{z}_i$ and $f(\mathbf{z}_i)$ is the function value in the selected point $\mathbf{z}_i$.

Two different types of information are used to construct the linear local model $\mathcal{M}_i$:

- the functional values (functional observation in [13])—values of the system's output $f(\mathbf{z}_i)$ in the centre of the model $\mathbf{z}_i$,
- the derivatives (derivative observation in [13])—vector of partial derivatives of the system's output $f(\mathbf{z})$ with respect to the components $z_{di}; d = 1, \ldots, D$ of the vector of the regressor $\mathbf{z}_i$:

$$\boldsymbol{\theta}_i = \left[\frac{\partial f}{\partial y_{k-1}}, \ldots, \frac{\partial f}{\partial y_{k-n}}, \frac{\partial f}{\partial u_{k-1}}, \ldots, \frac{\partial f}{\partial u_{k-m}}\right]_i^{\mathrm{T}}.$$

Local models can be derived using any standard linear regression method that gives a consistent and unbiased solution, see e.g. [55]. Local models' order is the same as the order of LMGP model.

In order to include the derivatives into the GP model, only the functional part of covariance function must be changed appropriately. The following results are derived for the frequently used squared exponential covariance functions, but other covariance functions may be used.

When using the squared exponential covariance function described with Eq. (2.14) between two data points, one can find that the covariance function between a data point and the derivative is:

$$\frac{\partial}{\partial z_{di}} C_f(\mathbf{z}_i, \mathbf{z}_j) = \mathrm{cov}\left[\frac{\partial f(\mathbf{z}_i)}{\partial z_{di}}, f(\mathbf{z}_j)\right]$$
$$= -v w_d (z_{di} - z_{dj}) \exp\left[-\frac{1}{2}\sum_{g=1}^{D} w_g (z_{gi} - z_{gj})^2\right], \tag{3.32}$$

where C_f represents the functional part of covariance function, $i, j = 1, \ldots, N_{eq}$ and $d, e = 1, \ldots, D$. In the same manner, the covariance function between two derivatives reads:

$$\frac{\partial^2}{\partial z_{di} \partial z_{ej}} C_f(\mathbf{z}_i, \mathbf{z}_j) = \mathrm{cov}\left[\frac{\partial f(\mathbf{z}_i)}{\partial z_{di}}, \frac{\partial f(\mathbf{z}_j)}{\partial z_{ej}}\right]$$
$$= v w_e (\delta_{e,d} - w_d (z_{ei} - z_{ej})(z_{di} - z_{dj}))$$
$$\times \exp\left[-\frac{1}{2}\sum_{g=1}^{D} w_g (z_{gi} - z_{gj})^2\right], \tag{3.33}$$

where $\delta_{e,d}$ is the Kronecker operator between the indices d and e:

$$\delta_{e,d} = \begin{cases} 1, & e = d \\ 0, & \text{otherwise.} \end{cases} \tag{3.34}$$

The problem of off-equilibrium dynamics dictates the following approach to LMGP model composition. Regions of the system where enough data is given for the identification of local models—usually in the vicinity of the equilibrium curve—are modelled with local models. Regions of the system that are lacking enough data to construct local models are modelled with individual samples of the system's response. This knowledge is together incorporated into the GP model, as illustrated in Fig. 3.11 for the first-order example. The GP model 'smooths' this information and is able to make a robust prediction of the system's response even where the data describing the system is sparse.

With the introduction of the local models into the GP model, the derivatives are added to the vector of the targets $\mathbf{y}$ of the GP model. The values of the regressors corresponding to the included derivatives are added to the matrix of regressors $\mathbf{Z}$.

As in Eq. (3.27), n is the order of the system and $D = m + n$ is the number of regressors. The following notation is used in continuation: the subscript oeq denotes data representing out-of-equilibrium behaviour (response), and the subscript eq denotes data representing the equilibrium behaviour of the system in the form of local models.

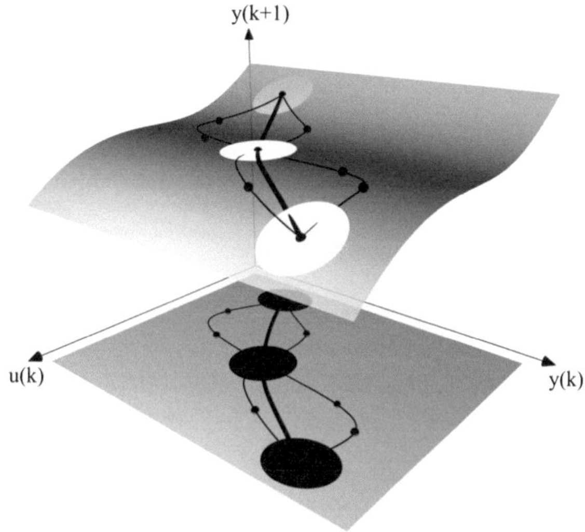

Fig. 3.11 Target data of the LMGP model consists of local models and data points—an illustration of the approach for a first-order system. The *ovals* represent the local models' proximity regions and the *dots* represent the samples of the off-equilibrium system's response

Given N_{eq} local models and N_{oeq} samples of the system's response, describing the system's behaviour, one of the possible ways to compose the input/target data $\{(\mathbf{Z}, \mathbf{y})\}$ for the identification, or training, of hyperparameters is:

$$\mathbf{Z} = \left[\mathbf{Z}_{oeq}, \mathbf{Z}_{eq}, \mathbf{Z}_{eq}, \ldots, \mathbf{Z}_{eq}\right], \quad \mathbf{y} = \begin{bmatrix} \mathbf{Y}_{oeq1} \\ \mathbf{Y}_{eq1} \\ \boldsymbol{\theta}_1 \\ \vdots \\ \boldsymbol{\theta}_D \end{bmatrix}, \tag{3.35}$$

$$\mathbf{Z}_{oeq} = [\mathbf{Y}_{oeq}\ \mathbf{U}_{oeq}]^{\mathrm{T}}, \tag{3.36}$$

$$\mathbf{Z}_{eq} = [\mathbf{Y}_{eq}\ \mathbf{U}_{eq}]^{\mathrm{T}}, \tag{3.37}$$

where

$\mathbf{Y}_{oeq1}$ is a $N_{oeq} \times 1$ target vector of the system's out-of-equilibria response points;
$\mathbf{Z}_{oeq}$ is a $D \times N_{oeq}$ matrix of the appropriate regressors corresponding to the target vector $\mathbf{Y}_{oeq1}$;
$\mathbf{Y}_{eq1}$ is a $N_{eq} \times 1$ target vector of the system's response points in the centres of the local models;
$\mathbf{Z}_{eq}$ is a $D \times N_{eq}$ matrix of appropriate regressors corresponding to the target vector $\mathbf{Y}_{eq1}$;
$\boldsymbol{\theta}_1$ is a $N_{eq} \times 1$ vector of the derivatives $\frac{\partial f}{\partial y_{k-1}}$ at the matrix of regressors $\mathbf{Z}_{eq}$ (vector of derivatives $\frac{\partial f}{\partial y_{k-1}}$ for all N_{eq} incorporated local models);

$\boldsymbol{\theta}_n$ is a $N_{eq} \times 1$ vector of derivatives $\frac{\partial f}{\partial y_{k-n}}$ at the matrix of regressors $\mathbf{Z}_{eq}$;
$\boldsymbol{\theta}_{n+1}$ is a $N_{eq} \times 1$ vector of derivatives $\frac{\partial f}{\partial u_{k-1}}$ at the matrix of regressors $\mathbf{Z}_{eq}$;
$\boldsymbol{\theta}_D$ is a $N_{eq} \times 1$ vector of derivatives $\frac{\partial f}{\partial u_{k-m}}$ at the matrix of regressors $\mathbf{Z}_{eq}$;

Let $N = N_{eq} + N_{oeq}$ be the number of functional data points (input-target data points), where N_{eq} is the number of identified local models to be incorporated into the GP model. For the nth-order system, there are a total of D vectors of the derivatives $\boldsymbol{\theta}_i$, $i = 1, \ldots, D$ with the length N_{eq}—one for each regressor $\mathbf{z}_i$. Thus, the size of the matrix of regressors $\mathbf{Z}$ is $D \times (N + D \cdot N_{eq})$ and the length of the target vector $\mathbf{y}$ is $(N + D \cdot N_{eq})$.

When the identification data $\mathcal{D} = \{(\mathbf{Z}, \mathbf{y})\}$ is composed as presented, the covariance matrix $\mathbf{K}$, the vector of covariances between validation, also test, input data and identification data $\mathbf{k}(\mathbf{z}^*)$ and the autocovariance of the validation input data $\kappa(\mathbf{z}^*)$ need to be:

$$\mathbf{K} = \begin{bmatrix} \boldsymbol{\Sigma}_{f11} & \boldsymbol{\Sigma}_{f12} \\ \boldsymbol{\Sigma}_{f21} & \boldsymbol{\Sigma}_{f22} \end{bmatrix} + \sigma_n^2 \mathbf{I}, \tag{3.38}$$

where

$$\boldsymbol{\Sigma}_{f11} = \big[\big[C_f(\mathbf{z}_i, \mathbf{z}_j)\big]\big], \tag{3.39}$$

$$\boldsymbol{\Sigma}_{f12} = \Big[\Big[\mathrm{cov}[f(\mathbf{z}_i), \tfrac{\partial f(\mathbf{z}_j)}{\partial z_{ej}}]\Big]_{e=1} \ldots \Big[\mathrm{cov}[f(\mathbf{z}_i), \tfrac{\partial f(\mathbf{z}_j)}{\partial z_{ej}}]\Big]_{e=D}\Big], \tag{3.40}$$

$$\boldsymbol{\Sigma}_{f21} = \begin{bmatrix} \Big[\mathrm{cov}[\tfrac{\partial f(\mathbf{z}_i)}{\partial z_{di}}, f(\mathbf{z}_j)]\Big]_{d=1} \\ \vdots \\ \Big[\mathrm{cov}[\tfrac{\partial f(\mathbf{z}_i)}{\partial z_{di}}, f(\mathbf{z}_j)]\Big]_{d=D} \end{bmatrix}, \tag{3.41}$$

$$\boldsymbol{\Sigma}_{f22} = \begin{bmatrix} \Big[\mathrm{cov}[\tfrac{\partial f(\mathbf{z}_i)}{\partial z_{di}}, \tfrac{\partial f(\mathbf{z}_j)}{\partial z_{ej}}]\Big]_{d=1,e=1} & \ldots & \Big[\mathrm{cov}[\tfrac{\partial f(\mathbf{z}_i)}{\partial z_{di}}, \tfrac{\partial f(\mathbf{z}_j)}{\partial z_{ej}}]\Big]_{d=1,e=D} \\ \vdots & \vdots & \vdots \\ \Big[\mathrm{cov}[\tfrac{\partial f(\mathbf{z}_i)}{\partial z_{di}}, \tfrac{\partial f(\mathbf{z}_j)}{\partial z_{ej}}]\Big]_{d=D,e=1} & \ldots & \Big[\mathrm{cov}[\tfrac{\partial f(\mathbf{z}_i)}{\partial z_{di}}, \tfrac{\partial f(\mathbf{z}_j)}{\partial z_{ej}}]\Big]_{d=D,e=D} \end{bmatrix}, \tag{3.42}$$

$$\mathbf{k}(\mathbf{z}^*) = \begin{bmatrix} \big[C(\mathbf{z}_i, \mathbf{z}^*)\big] \\ \Big[\mathrm{cov}[\tfrac{\partial f(\mathbf{z}_i)}{\partial z_{di}}, f(\mathbf{z}^*)]\Big]_{d=1} \\ \vdots \\ \Big[\mathrm{cov}[\tfrac{\partial f(\mathbf{z}_i)}{\partial z_{di}}, f(\mathbf{z}^*)]\Big]_{d=D} \end{bmatrix}, \tag{3.43}$$

$$\kappa(\mathbf{z}) = \big[C(\mathbf{z}^*, \mathbf{z}^*)\big] = \sigma_f^2 + \sigma_n^2, \tag{3.44}$$

respectively.

The information about the system's behaviour in the vicinity of the point $\mathbf{z}_i$, that in a conventional GP model is presented as a (large) set of functional values as in [56, 57], is now compressed in the parameters of the linear local model $\mathcal{M}_i$. In this way, the behaviour of the system around each equilibrium point is represented by fewer data points, which can effectively reduce the computational burden.

We have to be careful to ensure that all the local models are formed using the same state representation. The values of the regressors described with Eq. (3.27) are used as state coordinates in our case, but other choices are possible as well.

The target data used for identification can contain noise information. Where this information is available, it is added to the corresponding diagonal elements of the covariance matrix [13]; where not, the hyperparameter describing the white-noise variance is trained. Prediction with the LMGP model is done in exactly the same way as with the conventional GP model, apart from the fact that the covariance functions are selected differently.

The dynamic response of the LMGP model in off-equilibrium regions is represented by data points (response samples) and therefore represents the global and not the local dynamic behaviour in these regions. On the other hand, the incorporated local models on the equilibrium curve encapsulate the system's local dynamics and as the parameters of these local models do not change with optimisation, the dynamics of the system remains well modelled. The LMGP model does not have a scheduling variable and also does not suffer very much from partitioning, as the local models need only be put over the equilibrium curve in the necessary density.

Besides not suffering from some of the problems of the LMN approach, the *confidence measure*, i.e. variance, in the LMGP model's predictions, depending on the input data, is also provided. This confidence measure can be seen as the criterion for model quality in the corresponding region of the system.

Example 3.4 (*LMGP modelling example*) The GP model with an incorporated LM approach is presented on the identification of the following discrete, nonlinear, second-order dynamic system [58]:

$$\begin{aligned} y(k) = \; & 0.893y(k-1) + 0.0371y^2(k-1) - 0.05y(k-2) \\ & - 0.05u(k-1)y(k-1) + 0.157u(k-1) + \nu(k), \end{aligned} \tag{3.45}$$

where the output signal is corrupted with white Gaussian noise $\nu(k) \sim \mathcal{N}(0, \sigma_\nu^2)$ with the variance $\sigma_\nu^2 = 4 \times 10^{-4}$.

Our task will be to model the region bounded by the input values spanning between $u_{\mathrm{min}} = -2$ in $u_{\mathrm{max}} = 4$ for the purpose of a multistep-ahead prediction. The static characteristic of the system from Eq. (3.45) in the region of interest is depicted in Fig. 3.12. The nonlinearity of the system is also shown in Fig. 3.13, where the system's response to the alternating step signal with a growing magnitude is presented.

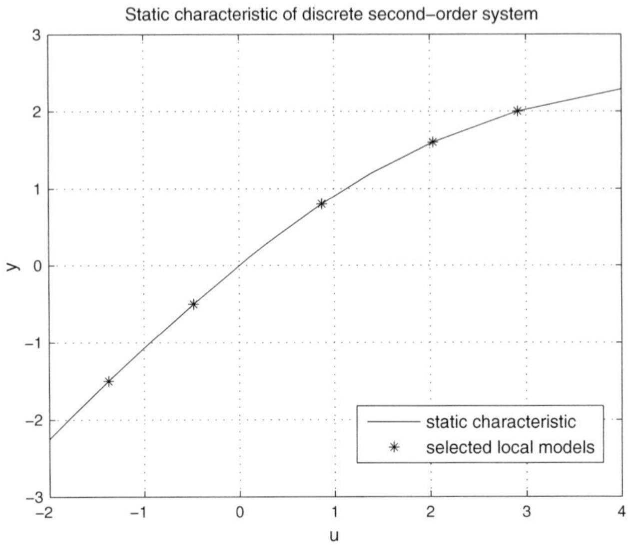

Fig. 3.12 Static characteristic of the second-order dynamic system

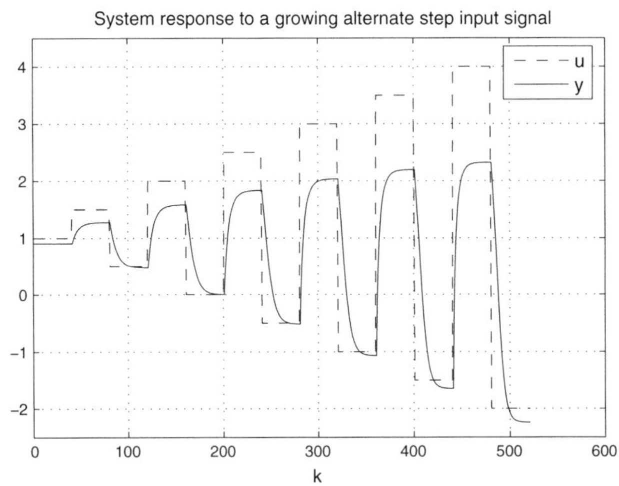

Fig. 3.13 Response of the second-order dynamic system to a growing alternate step input signal

As stated in Sect. 3.1, two different types of data represent the unknown system in the LMGP model:

- the local models, describing the system's dynamics in their centres and vicinity, with centres lying on the equilibrium curve,
- the samples of the system's response, which describe the system's regions not described by the LM (usually transient regions between equilibrium states).

These two different representations require two different measurement types.

To obtain the off-equilibrium data, which describe the system's behaviour in transient regions, the system must be excited with such an input signal that values of the regressors cover as much operating space $\mathbb{R}^D$ as possible. In our example, a pseudo-random binary signal (PRBS) is used as an excitation signal, except that the magnitude of the input value could occupy any random value between $u_{\min}$ and $u_{\max}$ when changed. The identification data for the LMGP model is later sampled from the input and the system's response.

To obtain the local model's parameters, the system is first driven into the equilibrium point with a static input signal. After the settlement of the system's response, a PRBS with a small magnitude ΔU is added to the input to stimulate the system's dynamic response around the equilibrium point.

To obtain the equilibrium dynamics of the system described with Eq. (3.45), five, approximately evenly distributed, local models on the equilibrium curve are identified. Their centres on the equilibrium curve can be seen in Fig. 3.12. The PRBS signal with the switching time $T_{sw} = 4$ steps and the magnitude of the perturbation $\Delta U = 0.3$ are selected so that the local models can be identified, despite the noise. The models are identified using the *instrumental-variable (IV)* method [55].

An example of the identified local model's response in the equilibrium point $(U_{eq}, Y_{eq}) = (0.415, 0.4)$ is presented in Fig. 3.14, and it can be seen that the model perfectly captures the dynamics of the system. It should be, however, taken into account that here the identified system is ideal and of known order.

Each of the five local models contribute one functional value (the value of system's response at the equilibrium point) and four derivatives (one for each regressor) to the identification data. Thus, together with fourteen points, sampled out of the system's off-equilibrium response, the LMGP model is formed using 39 identification points.

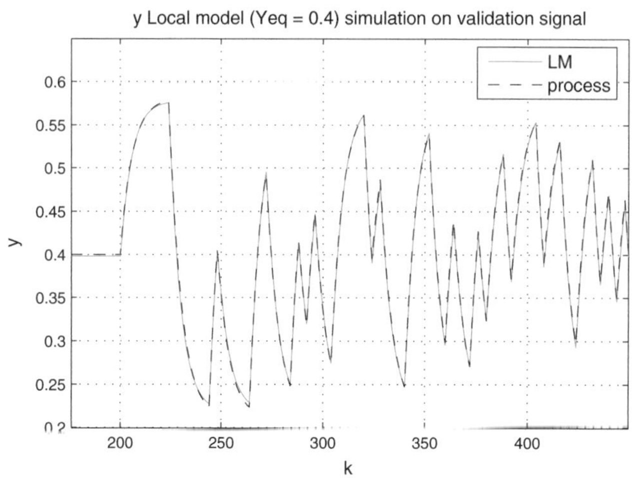

Fig. 3.14 Example of the identified local model's response in the equilibrium point $Y_{eq} = 0.4$

The estimates of local models' parameter variances, gained through identification, are added to the corresponding diagonal elements of the covariance matrix, defined with Eq. (3.38).

The acquired LMGP model is validated with data not used in the identification. Two different input signals are used for simulation, both random input signals, but the first with the switching time step $T_{sw} = 4$ and the second with the switching time step $T_{sw} = 20$ and both with the range between $u_{\min} = -2$ and $u_{\max} = 4$. The validation signal 1 is driving the LMGP model mainly in the region away from equilibria. The validation signal 2, where the pulses of which the input signal is composed are of longer duration, on the other hand, allows the model to reach the steady state.

The idea behind this choice is to show acceptable behaviour for the model, whether it is operating near to or far from the equilibrium points. The results of the naive simulation where the input signal is the validation signal 1 is presented in Fig. 3.15, and the corresponding simulation error with the accompanying 95 % confidence interval of the model's prediction is shown in Fig. 3.16. Note that for illustration purposes, these two figures represent only a segment of the whole simulation result. From Figs. 3.15 and 3.16, it can deduce that the description of the system behaviour in the off-equilibrium regions is satisfactory, even though only fourteen samples of the system's response are used. The model could be further improved by adding more samples of the system's response to the identification data.

The segment of the result of the naive simulation on validation signal 2 is depicted in Figs. 3.17 and 3.18, where the model's output signal is compared to the system's output signal and absolute simulation error together with 95 % confidence interval

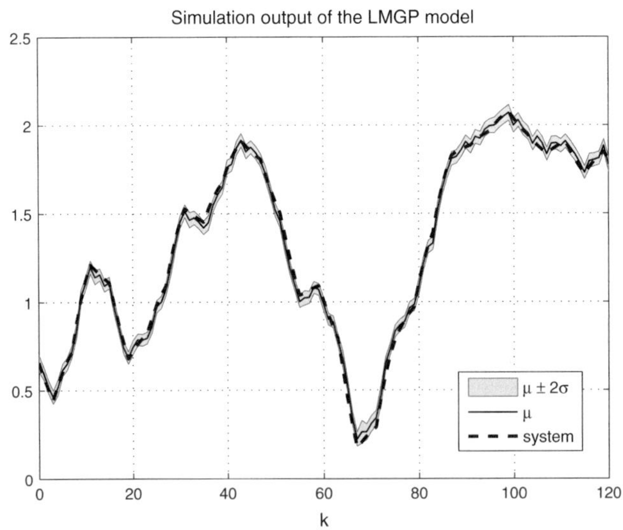

Fig. 3.15 Simulation of identified LMGP model on validation signal 1

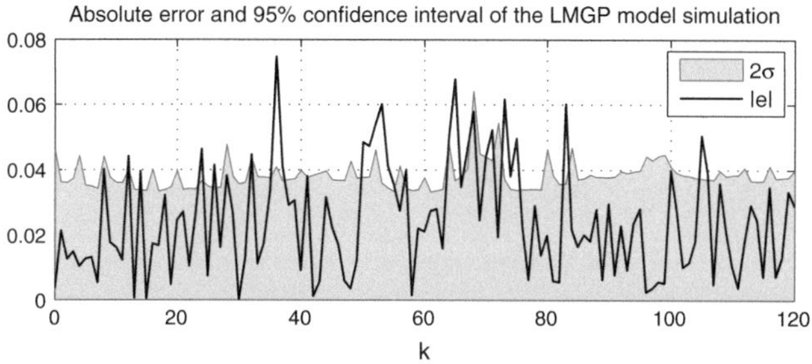

Fig. 3.16 Absolute error of the LMGP model simulation on the validation signal 1 together with the 95 % confidence interval

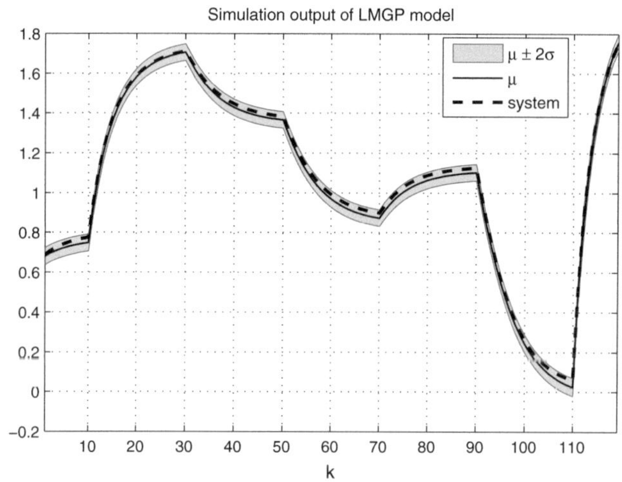

Fig. 3.17 Simulation of identified LMGP model on validation signal 2

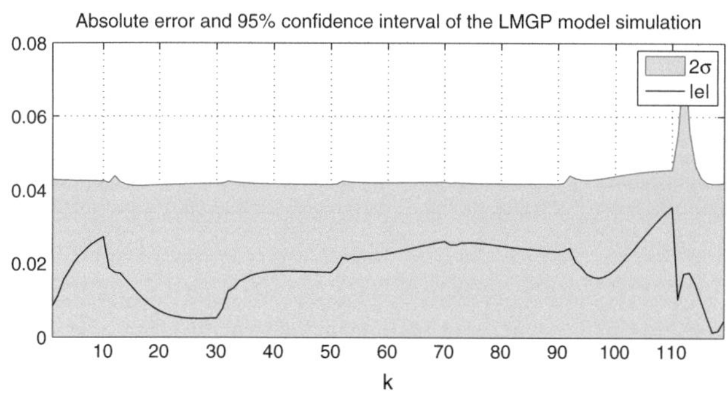

Fig. 3.18 Absolute error of the LMGP model simulation on the validation signal 2 together with the 95 % confidence interval

Table 3.1 Two performance measures of the simulation for the validation of the illustrative example, i.e. the standardised mean square error SMSE and mean standardised log loss MSLL

Validation signal	SMSE	MSLL
1	0.0038	–2.07
2	0.0020	–3.10

are shown correspondingly. These results show that the model matches the response of the process also in steady states.

Also, two performance measures, i.e. the standardised mean square error SMSE described with Eq. (2.54) and the mean standardised log loss MSLL described with Eq. (2.57), are applied for the validation of the simulation error for both validation studies. The results of both performance measures applied to the validation signals can be seen in Table 3.1.

In the LMGP model framework, the three exposed problems of the LMN approach have a diminished influence. The system in the off-equilibrium regions is represented by the data points and the interpolation of predictions between them is smooth as one of the attributes of the GP model. The problem of changing the model's local dynamics properties to provide a better global fit was solved within the Gaussian processes framework, i.e. the information describing the system's local dynamics does not change with optimisation. The problem of scheduling vector selection drops out as there is no scheduling vector. The problem of region partitioning is reduced as the local models are put only on the equilibrium curve and not over the whole operating region, as in the case of the LMN. Also, the values of the covariance function hyperparameters can be used as an indication of the influence along the corresponding regressor components.

3.3.2 *Fixed-Structure GP Model*

In this section, a parametric approach with a fixed linear model structure and varying parameters, i.e. a linear parameter-varying model or LPV model, based on the GP models is described. It is called the fixed-structure Gaussian process (FSGP) model [59]. The FSGP model is a model with a predetermined linear structure where the varying and probabilistic parameters are represented by GP models. As such, the FSGP model opens up possibilities that are different from other GP-based models of dynamic systems from the control-design point of view. It is the prior information about the model's structure that distinguishes the FSGP model from other GP-based models which are nonparametric. The idea of approximating the functional dependence of varying parameters is not new [60]. For example, an approach using radial basis function neural networks can be found in [61]. In the FSGP model, the varying parameters are represented by GP models, which brings some benefits in comparison with other GP-based approaches and other LPV approaches.

There are several reasons to use GP models for modelling the varying parameters of the LPV model:

- They tend to achieve acceptable modelling results, even with relatively small training datasets.
- The GP model gives a measure of the prediction confidence, which is dependent on the density of the training data and the covariance function.
- When local models are blended, with GP models predicting the local models' parameters, the GP models also provide information about the dependence of the parameters on the individual regressors.

The FSGP model therefore addresses problems such as nonparametricity of the general GP model, the confidence of the varying parameters' predictions and the small number of data for the identification of these varying parameters.

The FSGP method [59] can be used as a simple but still effective and, in the sense of prediction, potentially fast engineering tool. As the model posses a known structure with parameters, it can be used for a wider range of control-design methods, not only model-based predictive control, e.g. gain-scheduling control, as will be shown in Chap. 4.

This section is divided into two subsections in which the FSGP method for continuous-time and discrete-time systems is shown.

Modelling the Nonlinear Continuous-Time System

The FSGP approach to modelling a nonlinear system is a combination of velocity-based linearisation (VBL) [62] and modelling with GPs. The derivations are given for the systems with one input and one output, but the generalisation for multiple-input multiple-output systems is straightforward. The application of this method for the modelling of a process-engineering plant is in [63].

Consider the continuous nonlinear system written in the state space

$$\begin{aligned} \dot{\mathbf{x}}(t) &= f_t(\mathbf{x}(t), u(t)), \\ y(t) &= g_t(\mathbf{x}(t), u(t)). \end{aligned} \tag{3.46}$$

We would like to model the system described with Eq. (3.46) using a LPV model of a known and predetermined structure, where the varying parameters are modelled with GP models, thus providing not only the values of the model parameters but also the corresponding measure of the uncertainty. Such a model, when frozen at any operating point, would result in a linear local model, which is distinguishable from, e.g. a LMGP model, which is a nonparametric model represented by data pairs and the covariances among these data. The modelling method specifically addresses issues such as the blending of local models, the scheduling variable selection and modelling the nonlinear dynamics at a distance from the equilibrium regions.

One way to deal with the issue of accurately modelling off-equilibrium behaviour based on local linear models is representing the nonlinear system with VBL. VBL, in contrast to the conventional Taylor-series-expansion approach, enables the representation of the system at every operating point and not just in the equilibrium regions. Nevertheless, the blending and scheduling mechanisms still need to be determined.

The system described with Eq. (3.46) may be reformulated, without any loss of generality, in the form denoted as extended local linear equivalence (ELLE) [62].

$$\begin{aligned} \dot{\mathbf{x}}(t) &= \tilde{\mathbf{A}}_t \mathbf{x}(t) + \tilde{\mathbf{b}}_t u(t) + F_t(\varrho), \\ y(t) &= \tilde{\mathbf{c}}_t \mathbf{x}(t) + \tilde{d}_t u(t) + G_t(\varrho), \end{aligned} \tag{3.47}$$

where $\mathbf{x}(t) \in \mathbb{R}^n$, $u(t) \in \mathbb{R}$, $\tilde{\mathbf{A}}_t, \tilde{\mathbf{b}}_t, \tilde{\mathbf{c}}_t, \tilde{d}_t$ are appropriately dimensioned constant matrices, $F_t(\cdot)$ and $G_t(\cdot)$ are smooth nonlinear functions and $\varrho = \varrho(\mathbf{x}(t), u(t)) \in \mathbb{R}^q$, $q \leq n+1$, embodies the nonlinear dependence of the dynamics on the state and input with $\nabla_x \varrho$, $\nabla_u \varrho$ constant [64]. Index t denotes the continuous-time model.

Differentiating Eq. (3.47) an alternative representation of the nonlinear system is obtained [62]

$$\begin{aligned} \dot{\mathbf{x}}(t) &= \omega(t), \\ \dot{\omega}(t) &= \mathbf{A}_t(\varrho)\omega(t) + \mathbf{b}_t(\varrho)\dot{u}(t), \\ \dot{y}(t) &= \mathbf{c}_t(\varrho)\omega(t) + d_t(\varrho)\dot{u}(t) \end{aligned} \tag{3.48}$$

where

$$\mathbf{A}_t(\varrho) = \tilde{\mathbf{A}}_t + \frac{\partial F_t}{\partial \mathbf{x}}(\mathbf{x}(t), u(t)), \tag{3.49}$$

$$\mathbf{b}_t(\varrho) = \tilde{\mathbf{b}}_t + \frac{\partial F_t}{\partial u}(\mathbf{x}(t), u(t)), \tag{3.50}$$

$$\mathbf{c}_t(\varrho) = \tilde{\mathbf{c}}_t + \frac{\partial G_t}{\partial \mathbf{x}}(\mathbf{x}(t), u(t)), \tag{3.51}$$

$$d_t(\varrho) = \tilde{d}_t + \frac{\partial G_t}{\partial u}(\mathbf{x}(t), u(t)). \tag{3.52}$$

At every operating point ϱ_0, the elements of $\mathbf{A}_t(\varrho_0)$, $\mathbf{b}_t(\varrho_0)$, $\mathbf{c}_t(\varrho_0)$ and $d_t(\varrho_0)$ are the parameters of the local models, identified in the close vicinity of the operating point ϱ_0. We will not focus on the details of how the local models are obtained, which can be found, e.g. in [36] and references therein. Nevertheless, the identified linear local models need to be of the same order, they must describe the corresponding region satisfactorily well, and they must be located at equilibrium as well as off-equilibrium points. The off-equilibrium models are necessary as they uniquely define the system [65] and also provide the GP model with the training data describing the entire operating region.

It is important to note that a local linear input–output model only specifies the parameters up to a co-ordinate transformation [65].

$\mathbf{A}_t(\varrho)$, $\mathbf{b}_t(\varrho)$, $\mathbf{c}_t(\varrho)$ and $d_t(\varrho)$ are smooth functions of the variable ϱ, and this can be modelled with interpolations between the parameters of the identified local linear models. GP models are proposed to model each element of $\mathbf{A}_t(\varrho)$, $\mathbf{b}_t(\varrho)$, $\mathbf{c}_t(\varrho)$ and

$d_t(\varrho)$ based on local model parameters as learning data, because modelling with GP models gives acceptable results, even with relatively small training data sets and gives a measure of the confidence for the model predictions based on data density. This is done under the assumption that the parameters are independent of each other.

The smoothing property of the GP models is used to interpolate or to blend the values of the local model parameters for the operating points lying between the points where the linear local models were identified. The input data into the GP models are the coordinates of the current operating point ϱ_0. Each of the GP models' predictions, at the current input is the predictive distribution of the corresponding non-constant LPV model parameter, expressed with the mean value denoted as $\boldsymbol{\mu}^i_{\text{FSGP}}$ and the associated variance denoted as $\boldsymbol{\sigma}^{2i}_{\text{FSGP}}$, e.g. for the element denoted b_i its predictive distribution is $p(b_i) = \mathcal{N}(\mu_{b_i}, \sigma^2_{b_i})$.

The mean, and therefore the most likely values $\boldsymbol{\mu}^i_{\text{FSGP}}$ of GP models that model the elements of $\mathbf{A}_t(\varrho) = [\hat{a}_{ij}]$, $\mathbf{b}_t(\varrho) = [\hat{b}_i]$, $\mathbf{c}_t(\varrho) = [\hat{c}_j]$ and $d_t(\varrho) = \hat{d}$ are used for the global model simulation. The calculated variances $\boldsymbol{\sigma}^{2i}_{\text{FSGP}}$ express the confidence in the predicted mean values of the parameters, depending on the amount of information available for the modelling. This information can be effectively used, e.g. for control purposes. It can be used to retain the system in better modelled regions, i.e. the regions with a smaller parameter variance.

The FSGP modelling procedure, therefore, consists of two stages.

1. The first stage is the identification of the linear local models at the equilibrium and off-equilibrium points. The results of the first stage are coefficients (parameters) of the linear local models, and at the same time, derivatives of the nonlinear functions $f_t(\cdot)$ and $g_t(\cdot)$ from Eq. (3.46).
2. In the second stage, sets of values corresponding to each of the linear local model parameters are used for the training of the GP models. The training is pursued as described in Chap. 2. Through the relevance detection property [32] of the GP modelling method the relevant regressors, i.e. the state values and the input values, to which the parameters are functionally linked, are revealed via the values of the hyperparameters. This is how the issue of scheduling vector ϱ selection is solved.

The nonlinear system model is implemented using a VBL [64].

The obtained nonlinear model can be viewed as a parametric model with probabilistic and variable parameters $\mathbf{w}(\varrho(t)) = [a_{11}(\varrho(t)), \ldots, d(\varrho(t))]^{\mathrm{T}}$—a FSGP model. Each element of the vector of parameters $\mathbf{w}(\varrho(t))$ is $[w_i(\varrho(t))] \sim \mathcal{GP}(E(w_i(\varrho)), \operatorname{cov}(w_i(\varrho_j), w_i(\varrho_l)))$. It depends on the vector of scheduling variables $\varrho(t)$, which can consist of all the states and inputs or a subset of them.

Modelling the Nonlinear Discrete-Time System

In, this section the same procedure will be repeated for discrete-time systems. The method is elaborated in [59, 66]. The nonlinear system model is again realised using a VBL. The realisation in [64] was originally developed as a framework for continuous systems, but it can be extended to discrete systems [67]. Since linear local models are usually identified as discrete-time models, the following method is more appealing for practice.

Consider a sampled, nonlinear, dynamic system, which can be represented in the state space as:

$$\begin{aligned} \mathbf{x}(k+1) &= f\left(\mathbf{x}(k), u(k)\right), \\ y(k) &= g\left(\mathbf{x}(k), u(k)\right), \end{aligned} \tag{3.53}$$

where k denotes the discrete-time instants, $\mathbf{x}(k)$ is the state vector, u is the input signal and y is the output signal. We suppose that the sampling time T_s used for the representation of the original system is chosen to be small enough so that the system described with Eq. (3.53) captures all the nonlinear dynamics of the original system.

We would like to model the system described with Eq. (3.53) that is a discretised representation of the original system. We start with the discrete-time version because the proposed modelling is based on combining information obtained from identified local models, which are based on sampled signals and consequently written in discrete-time form.

To combine local information in the global model, including the off-equilibrium regions, a VBL is the appropriate approach [64]. However, the VBL approach can be applied for continuous systems only. Consequently, a description of a discrete-time system of Eq. (3.53) that can be treated with the VBL approach is required. The system described with Eq. (3.53) with a zero-order hold, representing a model of a digital/analogue converter, can be seen as a continuous delayed system [67, 68]:

$$\begin{aligned} \mathbf{x}(t+T_s) &= f\left(\mathbf{x}(t), u(t)\right), \\ y(t) &= g\left(\mathbf{x}(t), u(t)\right). \end{aligned} \tag{3.54}$$

The system of Eq. (3.54) has an equivalent response to the system described with Eq. (3.53) in sampled instances and is in a form that enables an analysis based on a VBL.

The system of Eq. (3.54) can be reformulated, without any loss of generality, in the form denoted as ELLE [65]

$$\begin{aligned} \mathbf{x}(t+T_s) &= \tilde{\mathbf{A}}\mathbf{x}(t) + \tilde{\mathbf{b}}u(t) + F(\varrho), \\ y(t) &= \tilde{\mathbf{c}}\mathbf{x}(t) + \tilde{d}u(t) + G(\varrho), \end{aligned} \tag{3.55}$$

where $\mathbf{x}(t) \in \mathbb{R}^n$, $u(t) \in \mathbb{R}$ and $\tilde{\mathbf{A}}, \tilde{\mathbf{b}}, \tilde{\mathbf{c}}, \tilde{d}$ are an appropriately dimensioned constant matrix, two vectors and a scalar, respectively, $F(\cdot)$ and $G(\cdot)$ are nonlinear functions and $\varrho = \varrho(\mathbf{x}(t), u(t)) \in \mathbb{R}^q$, $q \leq n+1$, embodies the nonlinear dependence of the dynamics on the state and the input with $\nabla_x \varrho$, $\nabla_u \varrho$ constant [65].

Relative to the operating point $\varrho(t) = \varrho_0$, where $(\mathbf{x}, u) = (\mathbf{x}_0, u_0)$, the first equation of Eq. (3.55) can be rewritten using

$$\delta\mathbf{x}(t) = \mathbf{x}(t) - \mathbf{x}_0, \tag{3.56}$$

$$\delta u(t) = u(t) - u_0, \tag{3.57}$$

as:

$$\mathbf{x}(t + T_s) = \tilde{\mathbf{A}}\,(\mathbf{x}_0 + \delta\mathbf{x}(t)) + \tilde{\mathbf{b}}\,(u_0 + \delta u(t)) + F(\varrho_0). \tag{3.58}$$

Close to the operating point ϱ_0 and assuming the local linearity it follows that:

$$\begin{aligned}
&\mathbf{x}(t + T_s) - \mathbf{x}_0 + \mathbf{x}_0 \\
&\quad = \tilde{\mathbf{A}}\mathbf{x}_0 + \tilde{\mathbf{A}}\delta\mathbf{x}(t) + \tilde{\mathbf{b}}u_0 + \tilde{\mathbf{b}}\delta u(t) + \mathbf{F}_0 + \mathbf{F}_{\mathbf{x}0}\delta\mathbf{x}(t) + \mathbf{F}_{u0}\delta u(t),
\end{aligned} \tag{3.59}$$

$$\begin{aligned}
\delta\mathbf{x}(t + T_s) + \mathbf{x}_0 = {} & \left(\tilde{\mathbf{A}}\mathbf{x}_0 + \tilde{\mathbf{b}}u_0 + \mathbf{F}_0\right) + \left(\tilde{\mathbf{A}} + \mathbf{F}_{\mathbf{x}0}\right)\delta\mathbf{x}(t) \\
& + \left(\tilde{\mathbf{b}} + \mathbf{F}_{u0}\right)\delta u(t),
\end{aligned} \tag{3.60}$$

$$\begin{aligned}
\delta\mathbf{x}(t + T_s) = {} & (\mathbf{f}_0 - \mathbf{x}_0) + \left(\tilde{\mathbf{A}} + \mathbf{F}_{\mathbf{x}0}\right)\delta\mathbf{x}(t) \\
& + \left(\tilde{\mathbf{b}} + \mathbf{F}_{u0}\right)\delta u(t),
\end{aligned} \tag{3.61}$$

where $\mathbf{f}_0 = f(\mathbf{x}_0, u_0)$, $\mathbf{F}_0 = F(\mathbf{x}_0, u_0)$, $\mathbf{F}_{\mathbf{x}0} = \frac{\partial F}{\partial \mathbf{x}}(\mathbf{x}_0, u_0)$ and $\mathbf{F}_{u0} = \frac{\partial F}{\partial u}(\mathbf{x}_0, u_0)$.

When differentiating Eq. (3.61) with regards to time, we obtain:

$$\dot{\mathbf{x}}(t + T_s) = \left(\tilde{\mathbf{A}} + \mathbf{F}_{\mathbf{x}0}\right)\dot{\mathbf{x}}(t) + \left(\tilde{\mathbf{b}} + \mathbf{F}_{u0}\right)\dot{u}(t), \tag{3.62}$$

$$\dot{\mathbf{x}}(t + T_s) = \mathbf{A}(\varrho_0)\dot{\mathbf{x}}(t) + \mathbf{b}(\varrho_0)\dot{u}(t), \tag{3.63}$$

where $\mathbf{A}(\varrho_0) = (\tilde{\mathbf{A}} + \mathbf{F}_{\mathbf{x}0})$ and $\mathbf{b}(\varrho_0) = (\tilde{\mathbf{b}} + \mathbf{F}_{u0})$. Similarly, the second equation of Eq. (3.55) can be rewritten as:

$$\dot{y}(t) = \mathbf{c}(\varrho_0)\dot{\mathbf{x}}(t) + d(\varrho_0)\dot{u}(t), \tag{3.64}$$

with $\mathbf{c}(\varrho_0) = (\tilde{\mathbf{c}} + \mathbf{G}_{\mathbf{x}0})$ and $d(\varrho_0) = (\tilde{d} + G_{u0})$, where $\mathbf{G}_{\mathbf{x}0} = \frac{\partial G}{\partial \mathbf{x}}(\mathbf{x}_0, u_0)$ and $G_{u0} = \frac{\partial G}{\partial u}(\mathbf{x}_0, u_0)$.

Equations (3.63) and (3.64) are valid for any operating point determined with the scheduling vector ϱ. Therefore, the system described with Eq. (3.55) can be generally written as:

$$\begin{aligned}
\dot{\mathbf{x}}(t + T_s) &= \mathbf{A}(\varrho)\dot{\mathbf{x}}(t) + \mathbf{b}(\varrho)\dot{u}(t), \\
\dot{y}(t) &= \mathbf{c}(\varrho)\dot{\mathbf{x}}(t) + d(\varrho)\dot{u}(t),
\end{aligned} \tag{3.65}$$

where

$$\mathbf{A}(\varrho) = \tilde{\mathbf{A}} + \frac{\partial F}{\partial \mathbf{x}}(\mathbf{x}(t), u(t)), \quad (3.66)$$

$$\mathbf{b}(\varrho) = \tilde{\mathbf{b}} + \frac{\partial F}{\partial u}(\mathbf{x}(t), u(t)), \quad (3.67)$$

$$\mathbf{c}(\varrho) = \tilde{\mathbf{c}} + \frac{\partial G}{\partial \mathbf{x}}(\mathbf{x}(t), u(t)), \quad (3.68)$$

$$d(\varrho) = \tilde{d} + \frac{\partial G}{\partial u}(\mathbf{x}(t), u(t)). \quad (3.69)$$

Afterwards, the method follows the modelling procedure for continuous systems.

The model described with Eq. (3.65) in the appropriate state space canonical form can be, if convenient, reformulated in the input–output form for a standard linear equation as

$$\dot{y}(t + T_s) = \mathbf{z}_{\mathrm{FSGP}}\mathbf{w} \quad (3.70)$$

where $\mathbf{z}_{\mathrm{FSGP}} = [\dot{y}(t), \ldots, \dot{y}(t - nT_s), \dot{u}(t), \ldots, \dot{u}(t - nT_s)]^{\mathrm{T}}$ is the regression vector of the FSGP model and $\mathbf{w}(\varrho(t)) = [w_i(\varrho(t))]; i = 1, \ldots, 2n$ is the vector of parameters modelled with GP models.

The FSGP model prediction $\dot{\hat{y}}(t + T_s)$ can be calculated in the same way as for linear models [18]; therefore, as the mean prediction

$$E(\dot{\hat{y}}(t + T_s)) = \mathbf{z}_{\mathrm{FSGP}}^{\mathrm{T}} E(\mathbf{w}) \quad (3.71)$$

and the corresponding variance

$$\mathrm{var}(\dot{\hat{y}}(t + T_s)) = \mathbf{z}_{\mathrm{FSGP}}^{\mathrm{T}} \mathrm{cov}(\mathbf{w}) \mathbf{z}_{\mathrm{FSGP}}. \quad (3.72)$$

Nevertheless, we have to keep in mind that the parameters contained in $\mathbf{w}(\varrho(t))$ are considered independent, which means that the model tends to be overconfident in its predictions.

Caution needs to be exercised with the implementation of a FSGP model when it is simulated. The global, discrete-time, local model-based, FSGP model containing a linear parameter-varying system with a GP model for each of the varying parameters is simulated as shown in Figs. 3.19 and 3.20 and used for the model validation.

The simulation of the FSGP model is based on a VBL approach [64] and the principal block scheme can be seen in Fig. 3.19. The FSGP model is simulated as a continuous-time-delayed system with sampling of the output signal.

The central block in Fig. 3.19 contains the linear parameter-varying system with a GP model for each of the varying parameters presented in more detail in Fig. 3.20.

The input signal derivative in Fig. 3.19 is due to the use of the VBL approach, which is necessary for the modelling of the nonlinear dynamics at a distance from equilibria based on local information. It is important to point out that the FSGP model generally serves as an analysis tool and the derivative is not implemented in,

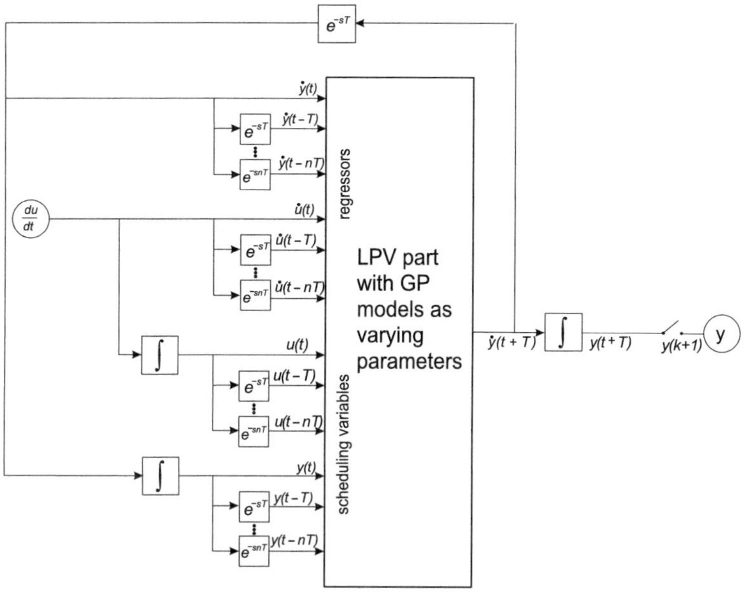

Fig. 3.19 Block scheme for the simulation of the FSGP model

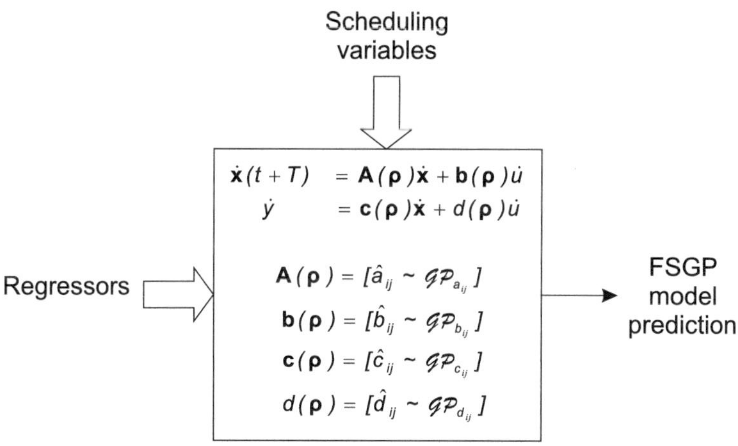

Fig. 3.20 LPV part with GP models as varying parameters—the masked part from Fig. 3.19

e.g. closed-loop control. However, the use of signal derivative can be circumvented, as shown in [64], if there is a practical problem with noisy input signal. Nevertheless, the model input signals are often noise free.

Example 3.5 (*FSGP modelling example*) The modelling procedure is illustrated with the same second-order discrete nonlinear system that is used for the LMGP modelling Example 3.4:

$$
\begin{aligned}
y(k+1) &= f\left(y(k), y(k-1), u(k)\right) \\
&= k_1 y(k) + k_2 y^2(k) + k_3 y(k-1) + k_3 u(k) y(k) + k_4 u(k), \quad (3.73)
\end{aligned}
$$

with the constants $k_1 = 0.893, k_2 = 0.0371, k_3 = -0.05, k_4 = 0.157$ and the sampling time $T_s = 1\,\text{s}$. The process model is supposed to be used for the control design, based on the information available from the FSGP model.

The system described with Eq. (3.73) represented in continuous-time-delayed form is as follows:

$$
\begin{aligned}
y(t+T_s) &= f\left(y(t), y(t-T_s), u(t)\right) \\
&= k_1 y(t) + k_2 y^2(t) + k_3 y(t-T_s) + k_3 u(t) y(t) + k_4 u(t) \quad (3.74)
\end{aligned}
$$

and in the ELLE form of Eq. (3.55):

$$
\begin{aligned}
\mathbf{x}(t+T_s) &= \begin{bmatrix} k_1 & k_3 \\ 1 & 0 \end{bmatrix} \mathbf{x}(t) + \begin{bmatrix} k_4 \\ 0 \end{bmatrix} u(t) + k_2 x^2(t) + k_3 u(t) x(t), \\
y(t) &= [\,1 \;\; 0\,]\, \mathbf{x}(t),
\end{aligned} \quad (3.75)
$$

where $\mathbf{x}(t) = [x(t) x(t-T_s)]^{\mathsf{T}} = [y(t)\; y(t-T_s)]^{\mathsf{T}}$. Following the VBL approach, the system can be further written in the forms described with Eqs. (3.63) and (3.64) as a second-order system with varying parameters:

$$
\begin{aligned}
\dot{\mathbf{x}}(t+T_s) &= \begin{bmatrix} a_1(\varrho) & a_2(\varrho) \\ 1 & 0 \end{bmatrix} \dot{\mathbf{x}}(t) + \begin{bmatrix} b_1(\varrho) \\ b_2(\varrho) \end{bmatrix} \dot{u}(t), \\
\dot{y}(t) &= [\,1 \;\; 0\,]\, \dot{\mathbf{x}}(t),
\end{aligned} \quad (3.76)
$$

where the parameters $a_1(\varrho) = \frac{\partial x(t+T_s)}{\partial x(t)}$ and $b_1(\varrho) = \frac{\partial x(t+T_s)}{\partial u(t)}$ depend on the vector of the scheduling variables $\varrho = [y(t)\; u(t)]^T$:

$$
\begin{aligned}
a_1(\varrho) &= k_1 + 2k_2\, y(t) + k_3\, u(t), \\
b_1(\varrho) &= k_3\, y(t) + k_4,
\end{aligned} \quad (3.77)
$$

while the parameters $a_2 = \frac{\partial x(t+T_s)}{\partial x(t-T_s)} = k_3$ and $b_2 = \frac{\partial x(t)}{\partial u(t)} = 0$ are constant across the whole operating region.

Our aim is to model the system described with Eq. (3.74) using the FSGP model and use the information from this model for the control design. The FSGP model will consist of the second-order model structure with the modelled varying parameters $\hat{a}_1(\varrho)$ and $\hat{b}_1(\varrho)$, while the other parameters are constant

$$\dot{\mathbf{x}}(t+T_s) = \begin{bmatrix} \hat{a}_1(\varrho) & a_2 \\ 1 & 0 \end{bmatrix} \dot{\mathbf{x}}(t) + \begin{bmatrix} \hat{b}_1(\varrho) \\ b_2 \end{bmatrix} \dot{u}(t),$$
$$\dot{y}(t) = [\,1\ 0\,]\,\dot{\mathbf{x}}(t). \tag{3.78}$$

The two varying parameters are modelled with two GP models, while the constant parameters a_2 and b_2 are set to their corresponding values. The training points, i.e. the values of the local model parameters, are in general obtained with the identification of linear local models. Any suitable linear model identification algorithm that gives consistent and unbiased results can be used for the identification, e.g. the IV method [55].

The identification of the linear local models necessary for composing the FSGP model is performed in the region determined by $1 < u < 3$ and $0.9 < y < 2$. Two kinds of local models are collected: linear local models on the equilibrium curve and linear local models at a distance from the equilibrium curve.

The equilibrium models are obtained with the identification of models using the IV method in the vicinity of arbitrarily selected equilibrium points (Fig. 3.21) using a pseudo-random binary signal for the excitation of the process. The choice of the method, the excitation signal and the equilibrium points is arbitrary and usually depends on the process itself or on the available data. Twenty-one local linear models are identified on the equilibrium curve in our case.

The models at a distance from the equilibrium curve are obtained from the identification data obtained in regions away from equilibrium. These data are gathered from the random input signal that also excited the system at a distance from equilibrium.

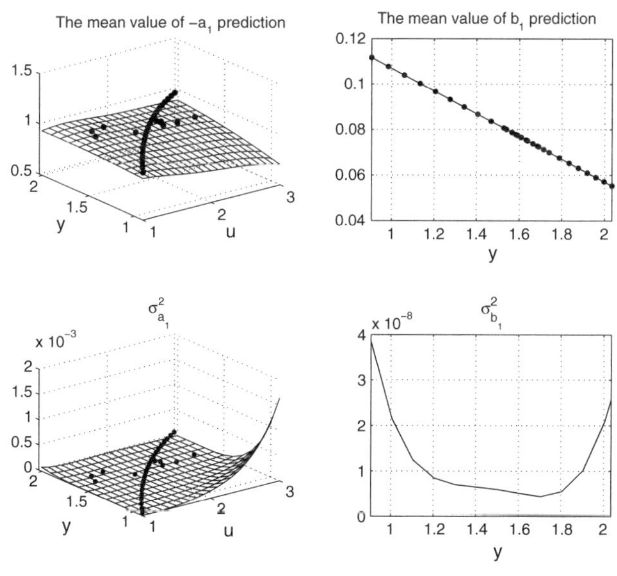

Fig. 3.21 Predictions of $\hat{a}_1$ (*left*) and $\hat{b}_1$ (*right*) and the associated variances

Nine local linear models are identified at a distance from the equilibrium curve in our case.

Two GP models that represent the varying parameters each encompass 30 values of the operating point values and the corresponding local model parameters as training data. Modelling the GP models also makes it possible to take into account the variances of the identified local model parameters obtained during the identification of the local models. In our case, we utilise only the mean values. The GP model representing the varying parameter $a_1(\varrho)$ and the GP model representing the varying parameter $b_1(\varrho)$ are trained with the scheduling vector ϱ being the regressor vector. A squared exponential covariance function described by Eq. (2.14) is used for the GP models, because the functions of the varying parameters are presumed to be smooth.

The obtained varying parameters models' predictions represented by the mean values and variances can be seen in Fig. 3.21.

A global FSGP model with GP models representing the varying parameters is put together and validated with a computer simulation. The process and the FSGP model are excited inside and outside the region where the local models were identified and where the model is to be consistent.

The responses of the system and the FSGP model, with a zero-order hold on the output, on a staircase input signal are depicted in Fig. 3.22. The variances that are associated with each of the predicted parameters are given in Fig. 3.23.

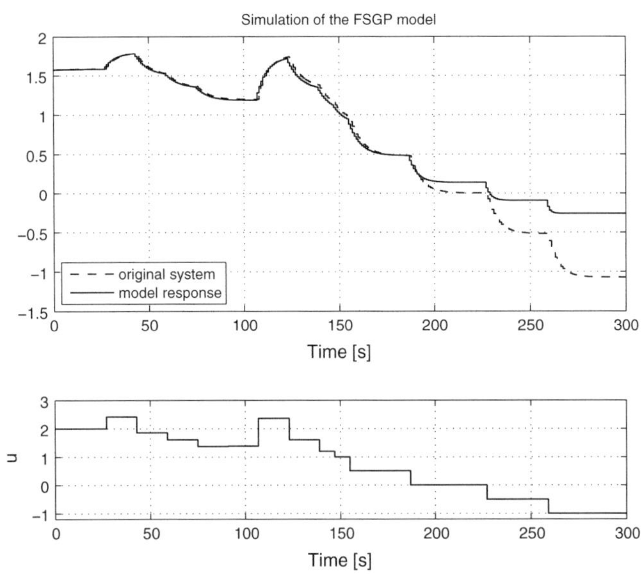

Fig. 3.22 Comparison of simulated responses of the system and the FSGP model with a zero-order hold on the output on the validation signal. The region before 150 s is the trained region and that after 150 s is the untrained region

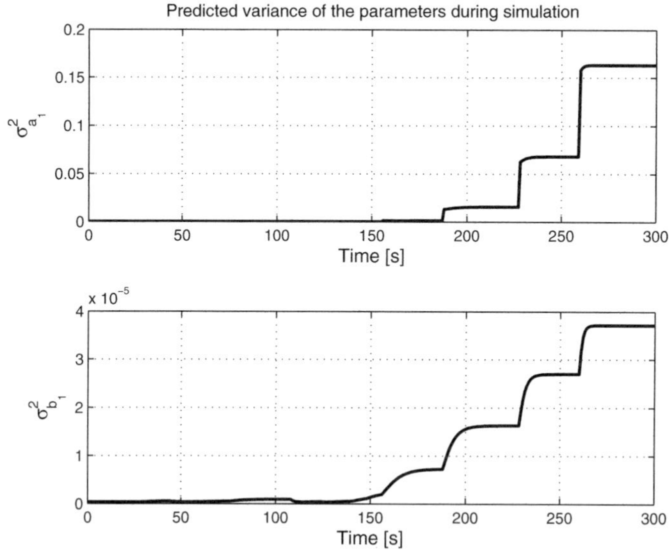

Fig. 3.23 Predicted variance of the parameters $\hat{a}_1$ (*top*) and $\hat{b}_1$ (*bottom*) during a simulation. The region before 150 s is the trained region and that after 150 s is the untrained region

It is clear from Fig. 3.22 that when the FSGP model predicts within the region where its components are identified, the accuracy of the FSGP model's predictions is noticeably better than outside of this region. The uncertainty of the predictions can be detected via the increase in the variances of the parameters that can be seen in Fig. 3.23.

A possible application of the developed model is control design, which is discussed in Chap. 4.

References

1. Hansen, J., Murray-Smith, R., Johansen, T.A.: Nonparametric identification of linearizations and uncertainty using Gaussian process models–application to robust wheel slip control. In: Joint 44th IEEE Conference on Decision and Control and European Control Conference (CDC-ECC), pp. 7994–7999. Sevilla (2005)
2. Schmitt, K., Madsen, J., Anitescu, M., Negrut, D.: A Gaussian process based approach for handling uncertainty in vehicle dynamics simulation. Int. Mech. Eng. Congr. Expos. (IMECE) **11**, 617–628 (2008)
3. Álvarez, M.A., Luengo, D., Lawrence, N.D.: Latent force models. J. Mach. Learn. Res. Proc. Track **5**, 9–16 (2009)
4. Hartikainen, J., Särkkä, S.: Sequential inference for latent force models. In: Proceedings of the 27th Conference on Uncertainty in Artificial Intelligence, UAI 2011, pp. 311–318 (2011)

5. Ko, J., Klein, D.J., Fox, D., Haehnel, D.: Gaussian processes and reinforcement learning for identification and control of an autonomous blimp. In: Proceedings of the International Conference on Robotics and Automation, pp. 742–747. Rome (2007)
6. Chen, N., Qian, Z., Meng, X., Nabney, I.: Short-term wind power forecasting using Gaussian processes. In: International Joint Conference on Artificial Intelligence IJCAI'13, pp. 1771–1777 (2013)
7. Murray-Smith, R., Pearlmutter, B.A.: Transformations of Gaussian process priors. In: Winkler, J., Niranjan, M., Lawrence, N. (eds.) Deterministic and Statistical Methods in Machine Learning. Lecture Notes in Computer Science, vol. 3635, pp. 110–123. Springer, Heidelberg (2005)
8. Ažman, K., Kocijan, J.: Identifikacija dinamičnega sistema s histerezo z modelom na osnovi Gaussovih procesov. In: B. Zajc, A. Trost (eds.) Zbornik štirinajste elektrotehniške in računalniške konference (ERK 2005), vol. A, pp. 253–256. Portorož, Slovenia (2005). (In Slovene)
9. Ažman, K.: Identifikacija dinamičnih sistemov z Gaussovimi procesi. Ph.D. thesis, University of Ljubljana, Ljubljana (2007). (In Slovene)
10. Narendra, K.S., Parthasarathy, K.: Identification and control of dynamical systems using neural networks. IEEE Trans. Neural Netw. **1**(1), 4–27 (1990)
11. Gibbs, M.N.: Bayesian Gaussian processes for regression and classification. Ph.D. thesis, University of Cambridge, Cambridge (1997)
12. Leith, D.J., Leithead, W.E., Neo, K.S.: Gaussian regression based on models with two stochastic processes. In: Proceedings of IFAC 16th World Congress 2005, pp. 1–6. Prague (2005)
13. Solak, E., Murray-Smith, R., Leithead, W.E., Leith, D.J., Rasmussen, C.E.: Derivative observations in Gaussian process models of dynamic systems. In: Becker, S., Thrun, S., Obermayer, K. (eds.) Advances in Neural Information Processing Systems, vol. 15, pp. 529–536. MIT Press, Cambridge, MA (2003)
14. Särkkä, S.: Linear operators and stochastic partial differential equations in Gaussian process regression. In: Proceedings of the International Conference on Artificial Neural Networks (ICANN). Espoo (2011)
15. Hall, J., Rasmussen, C., Maciejowski, J.: Modelling and control of nonlinear systems using Gaussian processes with partial model information. In: Conference on Decision and Control (CDC) (2012)
16. Murray-Smith, R., Johansen, T.A., Shorten, R.: On transient dynamics, off-equilibrium behaviour and identification in blended multiple model structures. In: Proceedings of European Control Conference, pp. BA-14. Karlsruhe (1999)
17. Murray-Smith, R., Girard, A.: Gaussian process priors with ARMA noise models. In: Proceedings of Irish Signals and Systems Conference, pp. 147–152. Maynooth (2001)
18. Nelles, O.: Nonlinear System Identification. Springer, Heidelberg (2001)
19. Haber, R., Unbehauen, H.: Structure identification of nonlinear dynamic systems: A survey on input/output approaches. Automatica **26**(4), 651–677 (1990)
20. Giri, F., Bai, E.W. (eds.) Block-oriented Nonlinear System Identification. Lecture Notes in Control and Information Sciences, vol. 404, Springer, Heidelberg (2010)
21. Leith, D.J., Leithead, W.E., Murray-Smith, R.: Nonlinear structure identification with application to Wiener-Hammerstein systems. In: Proceedings of 13th IFAC Symposium on System Identification, Rotterdam (2003)
22. Lindsten, F., Schön, T.B., Jordan, M.I.: A semiparametric bayesian approach to Wiener system identification. In: 16th IFAC Symposium on System Identification, pp. 1137–1142. Brussels (2012)
23. Lindsten, F., Schön, T.B., Jordan, M.I.: Bayesian semiparametric Wiener system identification. Automatica **49**(7), 2053–2063 (2013). doi:10.1016/j.automatica.2013.03.021
24. Pillonetto, G.: Consistent identification of Wiener systems: a machine learning viewpoint. Automatica **49**, 2704–2712 (2013)
25. Kocijan, J.: Incorporating knowledge about model structure in the identification of Gaussian-process models. In: Recent Advances in Telecommunications, Signals and Systems, pp. 124–129. Lemesos (2013)

26. Hachino, T., Hashiguchi, Y., Takata, H., Fukushima, S., Igarashi, Y.: Local Gaussian process models for identification of discrete-time Hammerstein systems. ICIC Express Lett. **8**, 173–179 (2014)
27. Hachino, T., Yamakawa, S.: Non-parametric identification of continuous-time Hammerstein systems using Gaussian process model and particle swarm optimization. Artif. Life Robot. **17**(1), 35–40 (2012)
28. Martinez, W., Martinez, A.: Computational Statistics Handbook with MATLAB, 2nd edn. Chapman & Hall/CRC Computer Science & Data Analysis. Taylor and Francis, London (2007)
29. Robert, C., Casella, G.: Monte Carlo Statistical Methods Springer Texts in Statistics. Springer, New York, NY (2004)
30. Andrieu, C., Doucet, A., Holenstein, R.: Particle Markov chain Monte Carlo methods. J. R. Stat. Soc. Ser. B (Stat. Methodol.) **72**(3), 269–342 (2010)
31. Tötterman, S., Toivonen, H.T.: Support vector method for identification of Wiener models. J. Process Control **19**(7), 1174–1181 (2009)
32. Rasmussen, C.E., Williams, C.K.I.: Gaussian Process. Mach. Learn. MIT Press, Cambridge, MA (2006)
33. Girard, A.: Approximate methods for propagation of uncertainty with Gaussian process models. Ph.D. thesis, University of Glasgow, Glasgow (2004). http://citeseerx.ist.psu.edu/viewdoc/summary?doi=10.1.1.63.8313
34. May, R., Dandy, G., Maier, H.: Artificial Neural Networks - Methodological Advances and Biomedical Applications. In: Chapter Review of Input Variable Selection Methods for Artificial Neural Networks, pp. 19–44. InTech, Rijeka (2011)
35. Johansen, T.A., Shorten, R., Murray-Smith, R.: On the interpretation and identification of dynamic Takagi-Sugeno fuzzy models. IEEE Trans. Fuzzy Syst. **8**, 297–313 (2000)
36. Murray-Smith, R., Johansen, T.A. (eds.): Multiple Model Approaches to Modelling and Control. Taylor and Francis, London (1997)
37. Kocijan, J., Žunič, G., Strmčnik, S., Vrančić, D.: Fuzzy gain-scheduling control of a gas-liquid separation plant implemented on a PLC. Int. J. Control **75**, 1082–1091 (2002)
38. Gregorčič, G., Lightbody, G.: From multiple model networks to the Gaussian processes prior model. In: Proceedings of IFAC ICONS Conference, pp. 149–154. Faro (2003)
39. Shi, J.Q., Choi, T.: Gaussian Process Regression Analysis for Functional Data. Chapman and Hall/CRC, Taylor and Francis group, Boca Raton, FL (2011)
40. Gregorčič, G., Lightbody, G.: An affine Gaussian process approach for nonlinear system identification. Syst. Sci. J. **29**(2), 47–63 (2003)
41. Gregorčič, G., Lightbody, G.: Local model identification with Gaussian processes. IEEE Trans. Neural Netw. **18**(5), 1404–1423 (2007)
42. Yu, J.: Online quality prediction of nonlinear and non-Gaussian chemical processes with shifting dynamics using finite mixture model based Gaussian process regression approach. Chem. Eng. Sci. **82**, 22–30 (2012)
43. Shi, J.Q., Murray-Smith, R., Titterington, D.M.: Hierarchical Gaussian process mixtures for regression. Stat. Comput. **15**(1), 31–41 (2005)
44. Grbić, R., Slišković, D., Kadlec, P.: Adaptive soft sensor for online prediction and process monitoring based on a mixture of Gaussian process models. Comput. Chem. Eng. **58**, 84–97 (2013)
45. Zhao, X., Fu, Y., Liu, Y.: Human motion tracking by temporal-spatial local Gaussian process experts. IEEE Trans. Image Process. **20**(4), 1141–1151 (2011)
46. Nguyen-Tuong, D., Seeger, M., Peters, J.: Real-time local GP model learning. In: Chapter, From Motor Learning to Interaction Learning in Robots, vol. 264, pp. 193–207.Springer, Heidelberg (2010)
47. Bukkapatnam, S.T.S., Cheng, C.: Forecasting the evolution of nonlinear and nonstationary systems using recurrence-based local Gaussian process models. Phys. Rev. E **82**(5) (2010)
48. Hu, M., Sun, Z.: Multimodel nonlinear predictive control with Gaussian process model. Lect. Notes Electr. Eng. **238**, 1895–1902 (2014)

49. Yu, J., Chen, K., Rashid, M.: A Bayesian model averaging based multi-kernel Gaussian process regression framework for nonlinear state estimation and quality prediction of multiphase batch processes with transient dynamics and uncertainty. Chem. Eng. Sci. **93**, 96–109 (2013)
50. Leith, D.J., Leithead, W.E., Solak, E., Murray-Smith, R.: Divide and conquer identification using Gaussian processes. In: Proceedings of the 41st Conference on Decision and Control, pp. 624–629. Las Vegas, AZ (2002)
51. Ažman, K., Kocijan, J.: Comprising prior knowledge in dynamic Gaussian process models. In: Rachev, B., Smrikarov, A. (eds.) International Conference on Computer Systems and Technologies —CompSysTech 2005, pp. IIIB.2–1 – IIIB.2–6. Varna (2005)
52. Ažman, K., Kocijan, J.: Dynamical systems identification using Gaussian process models with incorporated local models. Eng. Appl. Artif. Intell. **24**(2), 398–408 (2011)
53. Kocijan, J., Girard, A.: Incorporating linear local models in Gaussian process model. In: Proceedings of IFAC 16th World Congress, p. 6. Prague (2005)
54. Ažman, K., Kocijan, J.: Non-linear model predictive control for models with local information and uncertainties. Trans. Inst. Meas. Control **30**(5) (2008)
55. Ljung, L.: System Identification—Theory for the User, 2nd edn. Prentice Hall, Upper Saddle River, NJ (1999)
56. Kocijan, J., Girard, A., Banko, B., Murray-Smith, R.: Dynamic systems identification with Gaussian processes. Math. Comput. Model. Dyn. Syst. **11**(4), 411–424 (2005)
57. Kocijan, J., Banko, B., Likar, B., Girard, A., Murray-Smith, R., Rasmussen, C.E.: A case based comparison of identification with neural networks and Gaussian process models. In: Proceedings of IFAC ICONS conference, pp. 137–142. Faro (2003)
58. Matko, D., Škrjanc, I., Mušič, G.: Robustness of fuzzy control and its application to a thermal plant. Math. Comput. Simul. **51**, 245–255 (2000)
59. Ažman, K., Kocijan, J.: Fixed-structure Gaussian process model. Int. J. Syst. Sci. **40**(12), 1253–1262 (2009)
60. Leith, D.J., Leithead, W.E.: Survey of gain-scheduling analysis and design. Int. J. Control **73**, 1001–1025 (2000)
61. Peng, H., Ozaki, T., Haggan-Ozaki, V., Toyoda, Y.: A parameter optimization method for radial basis function type models. IEEE Trans. Neural Netw. (2) (2003)
62. Leith, D.J., Leithead, W.E.: Gain scheduled and nonlinear systems: dynamic analysis by velocity-based linearisation families. Int. J. Control **70**(2), 289–317 (1998)
63. Kocijan, J., Ažman, K.: Applications of varying parameters modelling with Gaussian processes. In: 2nd IFAC International Conference on Intelligent Control Systems and Signal Processing (ICONS), Istanbul (2009)
64. Leith, D.J., Leithead, W.E.: Analytic framework for blended multiple model systems using linear local models. Int. J. Control **72**, 605–619 (1999)
65. Leith, D.J., Leithead, W.E.: Global reconstruction of nonlinear systems from families of linear systems. In: Proceedings of 15th IFAC World Congress 2002, p. 6. Barcelona (2002)
66. Ažman, K., Kocijan, J.: Identifikacija dinamičnega sistema z znanim modelom šuma z modelom na osnovi Gaussovih procesov. In: Zajc, B., Trost, A. (eds.) Zbornik petnajste elektrotehniške in računalniške konference (ERK), pp. 289–292. Portorož (2006). (In Slovene)
67. Nyström, R.H., Åkesson, B.M., Toivonen, H.T.: Gain-scheduling controllers based on velocity-form linear parameter-varying models applied to an example process. Ind. Eng. Chem. Res. **41**, 220–229 (2002)
68. Rosenvasser, Y.N., Polyakov, E.Y., Lampe, B.P.: Application of Laplace transformation for digital redesign of continuous control systems. IEEE Trans. Autom. Control **44**(4), 883–886 (1999)

Chapter 4
Control with GP Models

Previous chapters dealt with the GP models used for the identification of dynamic systems. Dynamic systems identification is frequently used for control design, and GP models identification is no exception. This chapter gives a comprehensive overview of the control methods based on GP models.

Control is the activity that makes a system behave in the desired way. A general overview of the field of automatic control can be found in encyclopaedic books like [1, 2]. There are different possible divisions of control types. We are resorting to those common to control engineering community. The control of dynamic systems is, in general, divided to open- and closed-loop control. Closed-loop control utilises the difference between the desired and the measured output and applies it as feedback to bring the process output close to the desired value. Closed-loop control is further divided into two basic types: disturbance-rejection control and reference-tracking control. While the former is proposed for mitigating various disturbances that prevent a system from behaving in the desired way, the latter is focused on following the specified reference values.

The methods that are described in the following sections are meant for the control of nonlinear and uncertain dynamic systems, i.e. the systems for which GP models are very suitable. The requirements for a controlled system that is most of the time in a closed-loop can be expressed in many different ways, as will become clear from the examples in this chapter and in the chapter describing applications, later in the book.

What is not addressed in this chapter is the minimum-entropy control [3]. This is a control method where the desired probability density function, which is not necessarily Gaussian, of the system output is controlled in a closed-loop manner. GP models have not yet been used for such a type of control in general.

The nonlinear systems analysis that goes hand-in-hand with control design is addressed in many well-known references, e.g. [4, 5], and others, but the analysis

J. Kocijan, *Modelling and Control of Dynamic Systems Using Gaussian Process Models*, Advances in Industrial Control,
DOI 10.1007/978-3-319-21021-6_4

of nonparametric models or models containing nonparametric parts that would be applicable in the context of GP models is relatively scarce, e.g. [6, 7].

Closed-loop *stability* and *performance* are the two properties that are analysed most frequently when closed-loop control is designed. In particular, closed-loop stability is a very important requirement and of interest for control design.

On Closed-Loop Stability

It is important to mention some concepts about a system's stability that are used later in the text. The reader is referred to, e.g. [4, 5] for a more formal and elaborate explanation. It should be noted that there is no universal method for the analysis of all nonlinear control systems.

The analysis of a system's stability is based on the Lyapunov stability theory, which deals with deterministic systems and signals. Stability in the sense of Lyapunov means that the system's trajectory can be kept arbitrarily close to the origin by starting sufficiently close to it. If the trajectory converges to the origin we talk about asymptotic stability, and when it converges to the origin faster than an exponential function, we talk about exponential stability.

It is also important to distinguish between local and global stability. The idea behind *local stability* is that the stability properties of a nonlinear system in the close vicinity of an equilibrium point are more or less the same as those of its linearised approximation.

The *global stability* of a system's equilibrium point, on the other hand, means that asymptotic or exponential stability holds for any initial state of the system.

The Lyapunov theory can be divided into the indirect and direct analysis methods. Lyapunov's indirect or linearisation method is concerned with the local stability of a nonlinear system. The concept behind it is that a nonlinear system should behave in a similar way to its linearised approximation for a small range deviations around the point of linearisation. It serves as the justification for using a linear control technique in practice and shows that stable design by linear control guarantees the stability of the nonlinear system locally [5].

The direct method or Lyapunov analysis is a generalisation of the energy concepts associated with a mechanical system. The idea is to construct an energy-like scalar field, known as the Lyapunov function, for the system to see whether the function decreases with time. The direct method is used for an analysis of global stability.

Another useful concept is that of input-output stability. The bounded-input, bounded-output stable systems are those where the bounded input in the system causes the bounded output of the system. Furthermore, the passivity formalism is used when combinations of subsystems are analysed and this assists with the construction of Lyapunov functions for feedback-control purposes.

Nevertheless, the stability results for nonlinear closed-loop systems can mainly be provided for parametric models. GP modelling is a computational-intelligence-based method. Because the GP model itself is probabilistic and nonparametric, the standard analytical methods for the closed-loop analysis of deterministic systems do not apply in general. Nevertheless, for certain control structures some general directions for stability analysis can be given, as will be seen in this chapter.

The analysis of probability of closed-loop stability seems to be a more appropriate tool for a closed-loop system containing a GP model. Nevertheless, because of the nonparametric model and of the nonlinear characteristics, at present only numerical evaluations seem to be applicable for a stability evaluation, e.g. Monte Carlo simulations.

It is important to note that until now there were no published research results available on the topic of closed-loop stability when a GP model of the process to be controlled is involved, except the empirical ones using computer simulations. This is despite the fact that a lot of control applications have been investigated worldwide. So far, computer simulations have been the most frequently used, general-purpose analysis tool for closed-loop systems containing GP models in publications.

The same lack of analytical tools for particular types of models was also the case in the early days of the now well-established computational intelligence methods like fuzzy control and neural networks. The discovery of applicable analytical methods for closed-loop systems containing fuzzy or neural methods was lagging behind after many successful control applications.

Therefore, a stability analysis, as the most important issue in closed-loop control design, has been many times illustrated with computer simulations, which are now a common tool in engineering design.

Chapter Outline

Control design always depends on the model of the system to be controlled. Different types of models also mean different types of control methods, as will be seen in the following sections. The chapter is devoted to control methods utilising GP models that were already published in the literature. Some of the methods are described in more detail; for other methods the reader is referred to the literature for the detailed description. The single-input, single-output cases are elaborated in this chapter for the sake of an easier understanding of the principles, but they can be generalised to the multiple-input, multiple-output cases as in [8] for some of the described control methods.

It is clear from the literature that both deterministic and stochastic treatments of control systems using GP models are used according to the authors' convenience and problem setting. Consequently, no attempt to fit the described control methods within one framework is made, rather we try to follow the method description from its literature source.

Descriptions of control applications where GP models are used only to complement or assist conventional control design methods can also be found. Examples of such design methods are: control systems where GP models are used for modelling static nonlinearities, e.g. a friction curve in [9], or deviations from a compensated nominal model, e.g. in GP-assisted model-reference adaptive control [6, 7, 10], or proportional-integral control with a GP-based soft sensor in [11]. These and other, not listed, hybrid methods are not classified in the covered control GP-based principles, but are no less interesting for engineering practice. They show that the use of GP models does not necessarily mean adopting new design methods, just improving the existing ones, where convenient.

The following control principles based on GP models are covered in this chapter:

- inverse dynamics control,
- optimal control,
- model predictive control,
- adaptive control,
- gain-scheduling control,
- model identification adaptive control,
- iterative learning for control.

4.1 Control with an Inverse Dynamics Model

The concept of the inverse model [12] is that such a model of the process is developed to be connected in series with the process and therefore to control the system in an open loop. This kind of approach is not usually meant as an effective control solution, but mainly as a demonstration of a particular machine-learning method.

The basic principle, in brief, is as follows. If the system to be controlled can be described by an input-output model

$$y(k+1) = h(y(k), \ldots, y(k-n+1), \ldots, u(k), \ldots, u(k-m)) \tag{4.1}$$

then the corresponding inverse model is

$$\hat{u}(k) = h^{-1}(y(k+1), y(k), \ldots, y(k-n+1), \ldots, u(k-1), \ldots, u(k-m)) \tag{4.2}$$

Assuming that this inverse system has been obtained, it can be used to generate a control input signal that approaches the desired process output signal, when the reference input signal is given to the inverse model. This means that samples of y in Eq. (4.2) are replaced by the reference values r.

The general principle is illustrated in Fig. 4.1.

The training of a realisable inverse model requires input-output stable process responses. The input-output stability of open-loop control is ensured only if both the inverse model and the process are input-output stable. This is because signals are always constrained in magnitude, which disables the open-loop control of unstable systems. Even in the case of a computer simulation, the input and output values cannot be infinitely large.

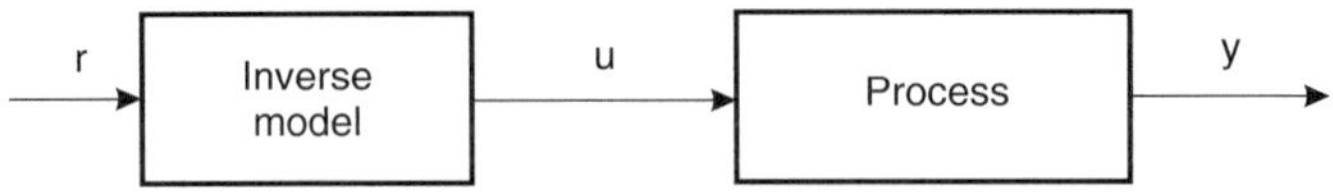

Fig. 4.1 General block scheme of direct inverse control

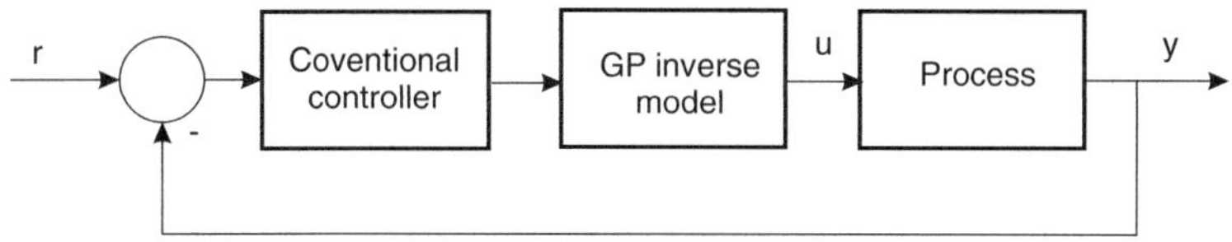

Fig. 4.2 General block scheme of inverse dynamics control

When the mentioned assumption is satisfied, the inverse model can be modelled from the appropriately selected output and input values of the process following Eq. (4.2). The resulting cascade system ideally becomes a unit delay. In the case of a dead time in the process to be controlled, it should be removed from the data for the inverse model training. The resulting cascade system in this case ideally becomes a delay corresponding to the process dead time plus the unit delay.

There is a further list of assumptions and constraints that needs to be satisfied for such a system to be implemented in practice. Actually, there are so many of them that it seems that only the computer simulation of such an open-loop control system is possible.

The assumptions necessary for open-loop control to be operational are: no disturbances in the system, no uncertainties and changes in the process and an open-loop controller that is the perfect inverse of the process in the region of its operation. It is not enough that the inverse model exists. To be realisable it also needs to be causal. Since it is not realistic for all these assumptions to be fulfilled, the inverse system is usually implemented in combination with closed-loop control or as part of the adaptive system.

A realisable control method that uses an inverse model for cancelling the nonlinearities of the process to be controlled is *inverse dynamics control*, e.g. [13, 14] illustrated in Fig. 4.2.

This is a closed-loop method that contains a conventional controller to deal with the mismatches between the nonlinearity compensator and the process as well as with the process disturbances. Such a method is used for the system control in robotics. The requirements for the stability and causality of the compensator, i.e. the GP inverse process model, must be fulfilled during the identification of the inverse model, which can be identified either offline or online. The controller in the loop must be designed appropriately to ensure the closed-loop stability. Assuming the inverse model is perfect, the cascade system of the nonlinearity compensator and the process ideally becomes a delay. The closed-loop stability under this assumption may be ensured with linear stabilising control.

An application with the GP model of the inverse process dynamics that is identified offline is given in [14, 15] for a robot-control investigation. These applications of referenced inverse GP models do not use the entire information from the prediction distribution, rather they focus on the mean value of the prediction. But the potential of the GP models for this kind of control is not utilised entirely, e.g. information about the variances could be used for maintaining or indicating the region of nominal closed-loop performance.

Feedforward control that eliminates the process nonlinearities is another control method that is used mainly in robotics. The principle is shown in Fig. 4.3.

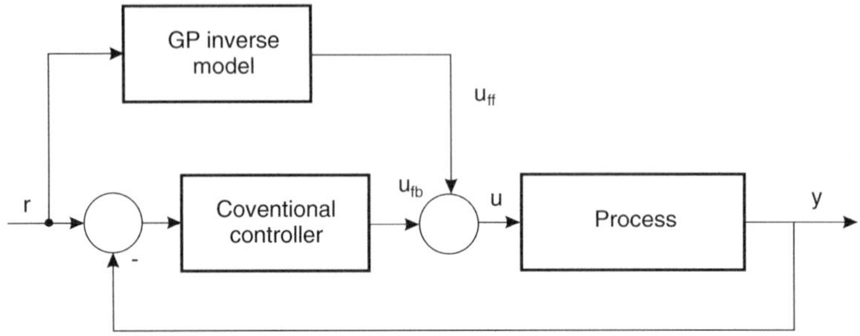

Fig. 4.3 General block scheme of the existing control system with the GP inverse model as a feedforward to improve the closed-loop performance

The control signal consists of the feedforward and feedback component $u = u_{ff} + u_{fb}$. The feedback loop with a conventional, frequently linear, controller is required to maintain the stability and the disturbance rejection for this control system, designed for set-point tracking. The feedforward part is an inverse model that compensates for the process nonlinearities, and provided that it is stable, it will not change the stability properties of the entire system. The inverse model has to be as accurate as possible in the region of operation to ensure the required performance. The closed-loop performance is deteriorated in the case of unmodelled nonlinearities. The feedforward is generally considered as a function of the desired set-point, in the case of robotic control that would mean the desired robot trajectories.

The concept has some practical advantages [12]. First, since it is assumed that a stabilising controller is available in advance, the data necessary for the inverse model can be collected from a previously assembled, closed-loop system without the feedforward component. Second, the feedforward signal can be introduced gradually during the control system's implementation as a caution. Third, in the case of avoiding inverse dynamics, static feedforward may be used with the feedback controller, compensating for an erroneous feedforward signal.

The inverse model can be identified offline or online. The case when the inverse GP model is identified offline and used in such a control set-up is described in [14, 16]. Again, like in the case of inverse dynamics control, only the mean value of the prediction is utilised. The adaptive cases are mentioned in Sect. 4.4.

Example 4.1 Comparison of the tracking performance

In this example all three described control methods that are based on the inverse process model are illustrated with the first-order nonlinear system.

Consider the nonlinear dynamic system [17] described by

$$y(k+1) = \frac{y(k)}{1 + y^2(k)} + (u(k) + d(k))^3 \qquad (4.3)$$

where u is the system's input signal, d is an unmeasured input disturbance, not correlated to the input signal u, y is the system's output signal. The closed-loop control requirement is to follow the set-point closely.

First, the inverse GP model is identified. The input signal u to the system was a random signal in the range $[-1.5, 1.5]$. About 2000 samples from the input and output signals were used to identify the inverse model. The regression vector of the inverse GP model is as follows: $[u(k-1),\ y(k+1),\ y(k)]^{\mathrm{T}}$; $k = 2, \ldots, N$ and the target is $[u(k)]$; $k = 2, \ldots, N$.

The obtained results are as expected. While direct inverse control (Fig. 4.4) works well in the region where the inverse model was identified, it cannot cope with the input disturbance, because the controller does not receive any information about its occurrence and effect.

The conventional controllers in the case of the inverse dynamics and the feed-forward control are proportional-integral controllers with the appropriately selected constants. The controlled systems are tested for the set-point tracking of rectangular pulses of about a half unit positive and negative magnitude for 100 time instants. A step-like disturbance of considerable magnitude, which is 0.2, i.e. about 40 % of the system input and output signal, is added to input d from the time instant $k = 50$ on, in order to test the closed-loop systems. The responses and input signals into the system for the described types of control are given in Figs. 4.4, 4.5 and 4.6.

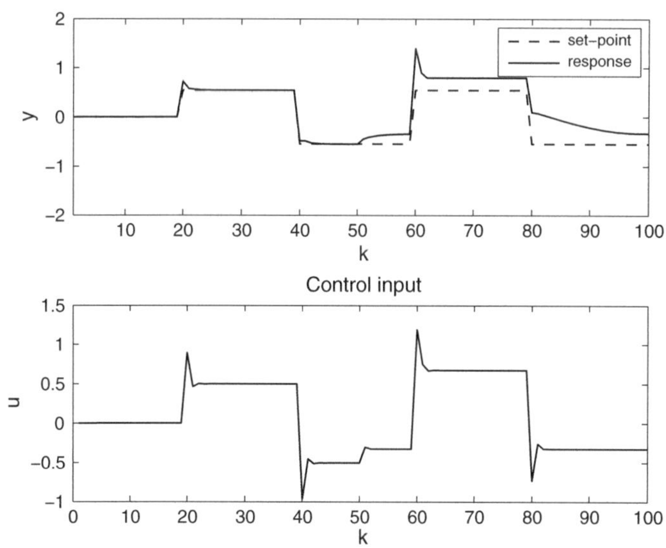

Fig. 4.4 Open-loop response with the direct inverse model control on a pulse change of the set-point signal and an additive step disturbance at the time instant $k = 50$ (*upper figure*) and the corresponding system input signal (*bottom figure*)

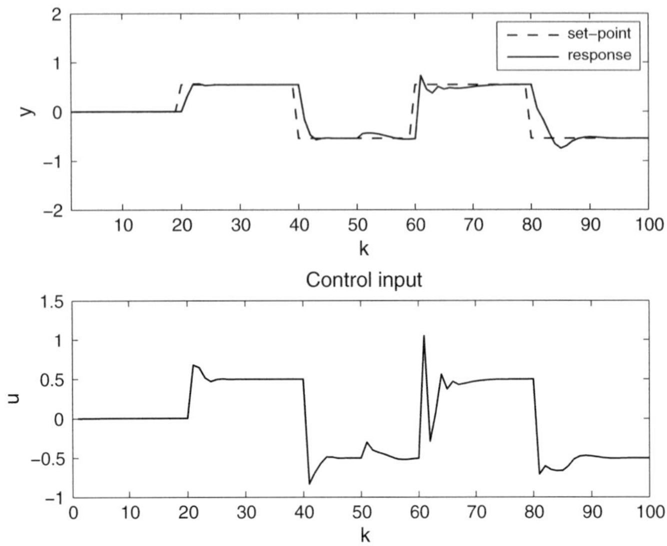

Fig. 4.5 Closed-loop response with inverse dynamics control on a pulse change of the set-point signal and an additive step disturbance at the time instant $k = 50$ (*upper figure*) and the corresponding system input signal (*bottom figure*)

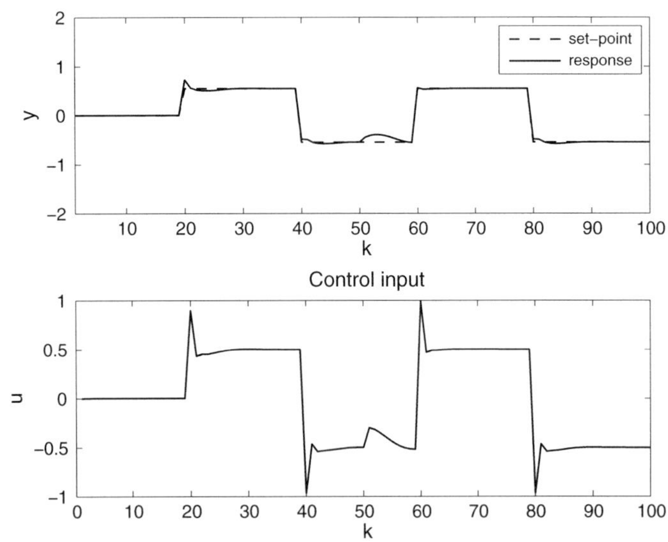

Fig. 4.6 Closed-loop response with feedforward control on a pulse change of the set-point signal and an additive step disturbance at the time instant $k = 50$ (*upper figure*) and the corresponding system input signal (*bottom figure*)

A better example is the response from the inverse dynamics control (Fig. 4.5), but it should be emphasised that some effort needs to be put into finding suitable constants for a conventional proportional-integral controller that assists the loop and ensures closed-loop stability.

The closed-loop control system with the inverse GP model in feedforward (Fig. 4.6) shows the best tracking and disturbance-rejection performance, which explains the relative popularity of this method in practice for selected applications.

It should be noted that these simulation cases contain absolutely no added noise, which would cause even more notable differences in the responses. Note that only the mean values, i.e. the most likely ones, are taken as output values of the inverse GP model and no other information about the predicted output distribution is taken into account.

4.2 Optimal Control

The general idea of optimal control [1, 18] is to design the control for a dynamic system by minimising the so-called performance index, constrained by the system dynamics. This performance index depends on the system variables and might include various measures, like the operating error, the control effort and others. The smaller performance index is, the smaller, under some assumptions, are the system variables, which also means closed-loop stability. This kind of philosophy is very general and encompasses a very broad set of control strategies. The formal description of a continuous problem, its discrete-time version, and the stochastic problem description are given next. As the formal descriptions originate from a continuous problem, its description will be more detailed. The problems are, for the sake of generality, presented for systems with multiple inputs and states.

The process model is given in the state space for a nonlinear, time-varying and dynamic system:

$$\dot{\mathbf{x}} = \mathbf{f}(\mathbf{x}, \mathbf{u}, t), \qquad t \geq t_0, t_0 \ \text{ fixed}, \tag{4.4}$$

where $\mathbf{x}(t) \in \mathbb{R}^n$ is the vector of the internal states and $\mathbf{u}(t) \in \mathbb{R}^m$ is the vector of the control variables (which we wish to choose optimally), $\mathbf{f}$ is an n-dimensional vector function and t is the time. The initial time and the initial state of the system are supposed to be known:

$$\mathbf{x}(t_0) = \mathbf{x}_0. \tag{4.5}$$

The system is subject to a set of generally nonlinear constraints, also called the algebraic path constraints:

$$\gamma(\mathbf{x}, \mathbf{u}) \geq 0, \tag{4.6}$$

$$\chi(\mathbf{x}, \mathbf{u}) = 0, \tag{4.7}$$

where γ and χ are constraint functions. These constraints must be taken into account while searching for the problem solution. The control variables $\mathbf{u} = [u_i], i = 1, \ldots, m$ should be in the admissible range, defined by the lower and upper bounds:

$$u_{\min_i} \leq u_i \leq u_{\max_i} \qquad i = 1, \ldots, m. \tag{4.8}$$

The system can also be subject to a terminal constraint that describes the target set:

$$\psi(\mathbf{x}(t_f), t_f) \leq 0. \tag{4.9}$$

The performance objectives may be achieved by minimising the performance index or the cost function described with Eq. (4.10) from the initial time t_0 until the final time t_f:

$$J(\mathbf{x}, \mathbf{u}) = \Phi(\mathbf{x}(t_f), t_f) + \int_{t_0}^{t_f} L(\mathbf{x}, \mathbf{u}, t)dt. \tag{4.10}$$

The final-state weighting function $\Phi(\mathbf{x}(t_f), t_f)$ and the weighting function $L(\mathbf{x}, \mathbf{u}, t)$ are selected according to the performance objectives.

The optimal control problem [1, 18] is to find the time-varying control input $\mathbf{u}(t)$ for the system described with Eq. (4.4) that minimises the performance index described with Eq. (4.10), while satisfying the constraints described with Eqs. (4.6)–(4.9).

The necessary conditions for optimality according to Pontryagin's maximum principle [1, 19] are the following. In order for $\mathbf{u}(t)$ to be optimal in the sense that it minimises the performance index described with Eq. (4.10), while satisfying the system equation described with Eq. (4.4) and the constraints described with Eq. (4.8), it is necessary that the condition

$$0 = \frac{\partial H}{\partial \mathbf{u}} = \frac{\partial L}{\partial \mathbf{u}} + \frac{\partial \mathbf{f}^{\mathrm{T}}}{\partial \mathbf{u}} \boldsymbol{\lambda}. \tag{4.11}$$

holds for the unconstrained portion of the path and the Hamiltonian function:

$$H(\mathbf{x}, \mathbf{u}, t) = L(\mathbf{x}, \mathbf{u}, t) + \boldsymbol{\lambda}^{\mathrm{T}} \mathbf{f}(\mathbf{x}, \mathbf{u}, t). \tag{4.12}$$

is minimised along the constrained portions of the control trajectory. Here, $\boldsymbol{\lambda}$ is an n-dimensional vector of time-dependent Lagrange multipliers, which are defined by the equation:

$$-\dot{\boldsymbol{\lambda}} = \frac{\partial H}{\partial \mathbf{x}} = \frac{\partial \mathbf{f}^{\mathrm{T}}}{\partial \mathbf{x}} \boldsymbol{\lambda} + \frac{\partial L}{\partial \mathbf{x}}, \qquad t \leq t_f. \tag{4.13}$$

The expression for the performance index described with Eq. (4.10) is sufficiently general to allow for the treatment of a wide class of practical problems, among which are:

- the minimum-time control, i.e. minimising the time to transfer the system from the initial state to the desired final state,
- the maximum productivity problem, i.e. maximising the amount of desired product at the final time.

Optimal control in discrete time is when the behaviour of a continuous-time system is not considered at all the instants of time t, but only for a sequence of instants k. The discrete-time systems are then described by the state-difference equation:

$$\mathbf{x}(k+1) = \mathbf{f}(\mathbf{x}(k), \mathbf{u}(k), k), \qquad k \geq 0, \tag{4.14}$$

where $\mathbf{x}(k)$ is the n-dimensional vector of the internal states, $\mathbf{u}(k)$ is the m-dimensional vector of the control variables that we wish to choose optimally, $\mathbf{f}$ is an n-dimensional vector function and k is the time instant.

The equations describing the continuous control problem have their discrete-time equivalents and the integral in the performance index is replaced by the summation of the discrete values over the finite N_h-step optimisation horizon.

$$J(\mathbf{x}, \mathbf{u}) = \Phi(\mathbf{x}(N_h), N_h) + \sum_{k=1}^{N_h-1} L(\mathbf{x}(k), \mathbf{u}(k), k). \tag{4.15}$$

There is a large variety of computational techniques for determining the optimal control for general nonlinear systems, but it is beyond the scope of this book. The reader is referred to other literature, e.g. [1, 18], for more information on this topic. From the presented problem, various special problems and solutions are derived.

For example, if the dynamics $\mathbf{f}$ is linear in $\mathbf{x}$ and $\mathbf{u}$, the cost is quadratic, and the noise is Gaussian, the resulting problem transfers to the so-called Linear Quadratic Gaussian and Linear Quadratic Regulator problems, which are convex and can be solved analytically [1].

The generalisation of the optimal control problem to stochastic systems is rather tricky. In the presence of noise, the dynamics is described as a stochastic differential equation

$$d\mathbf{x} = \mathbf{f}(\mathbf{x}, \mathbf{u}, t) + \mathbf{f}_C(\mathbf{x}, \mathbf{u}, t)d\boldsymbol{\nu}, \qquad t \geq t_0, t_0 \ \text{fixed}, \tag{4.16}$$

where $d\boldsymbol{\nu}$ is assumed to be Brownian-motion noise, which is transformed by a possibly state and control-dependent matrix $\mathbf{f}_C(\mathbf{x}, \mathbf{u}, t)$. The performance index or cost function is

$$J(t_0) = E\left(\Phi(\mathbf{x}(t_f), t_f) + \int_{t_0}^{t_f} L(\mathbf{x}, \mathbf{u}, t)dt\right). \tag{4.17}$$

Stochastic optimal control theory approaches the control problem by first specifying a cost function that is composed of some evaluation $\Phi(\mathbf{x}(t_f), t_f)$ of the final state, usually penalising deviations from the desired state $\mathbf{x}_r$ and the accumulated cost $L(\mathbf{x}, \mathbf{u}, t)$ of sending a control signal $\mathbf{u}$ at time t in the state $\mathbf{x}$, typically penalising excessive values of $\mathbf{u}$.

Finding an optimal control policy for a nonlinear system is a challenge. In theory, a global solution can be found by applying a dynamic programming method. *Dynamic programming* [1] in the context of optimal control is a recursive method for obtaining the optimal control as a function of the system's states. It is described with the Hamilton-Jacobi-Bellman equation for the continuous-time optimal control problem or with its discrete-time equivalent the Bellman equation for the discrete-time optimal control problem [1].

Another method that appears when finding solutions of the optimal control problem is *reinforcement learning* [20], which is an area of machine learning, concerned with finding actions based on past experience in a certain environment to maximise some cumulative discounted reward. *Approximate dynamic programming*, which is briefly addressed in Sect. 4.7, is a field where reinforcement learning has been studied in connection with the theory of optimal control.

The following sections describe the control methods emerging from the problem descriptions in this section that are usually variations or simplifications of the optimal control. Using the Gaussian process model does not necessarily mean that the stochastic control problem is dealt with. It could also be used in deterministic control problems.

4.3 Model Predictive Control

Model predictive control (MPC) is the name used to describe computer control algorithms that use an explicit process model to predict the future response of a controlled plant. According to the prediction made in a particular time frame, also known as the prediction horizon, the MPC algorithm optimises the manipulated variable over a chosen length, also known as the control horizon, to obtain the optimal future response of a plant. The first input value of the optimal control input sequence is sent to the plant and then the entire optimisation sequence is repeated during the next time period.

The popularity of MPC algorithms is, to a large extent, due to their ability to deal with the constraints that are frequently met in control practice and are often not well addressed by other approaches. MPC algorithms can handle the hard state and rate constraints on the inputs and states that are occasionally incorporated into the algorithms via an optimisation method.

Linear MPC approaches [21] started to appear in the early 1980s and are now well established in control practice (see [22] for an overview). Nonlinear model predictive control (NMPC) approaches [23] started to appear about 10 years later and have also found their way into control practice (e.g. [24]) though their popularity cannot be compared to that of linear MPC. This fact is connected with the difficulties associated with nonlinear model construction and with the lack of necessary confidence in the model. There were a number of contributions in the field of NMPC dealing with issues like stability, efficient computation, optimisation, constraints and others. Some contributions in this field can be found in [23, 25–27]. NMPC algorithms

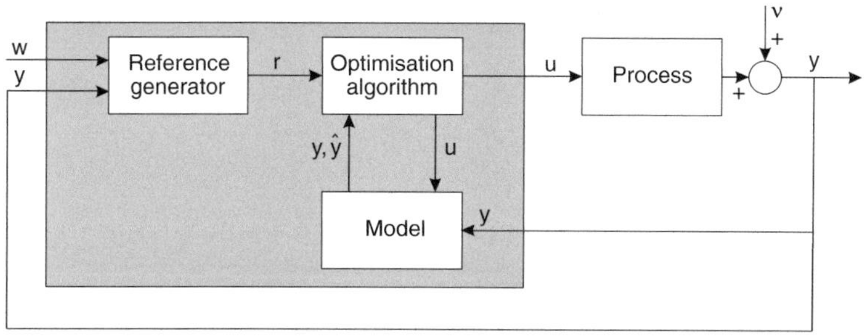

Fig. 4.7 Block diagram of an MPC system that illustrates the main principle where the control signal is obtained based on the difference between the process and model responses. This difference is then used for calculating the control signal over a selected horizon using the optimisation algorithm and process model embedded in the MPC

are based on various nonlinear models. Often, these models are developed as first-principles models, but other approaches, like black-box identification approaches, are also popular. Various NMPC algorithms are based on a neural-network model, e.g. [12], fuzzy models or local model networks, e.g. [28]. The quality of the control depends on the quality of the model. New developments in NMPC approaches are coming from resolving various issues: from faster optimisation methods to different process models.

In this section we focus on an NMPC principle with a GP model, but will start the description with a general description of MPC.

In general, MPC can be described with the block diagram in Fig. 4.7. The model used here is fixed, identified offline, which means that the control algorithm being used is not an adaptive one. The structure of the entire control loop is therefore less complex than the structure with a time-varying model.

The following items describe the basic idea of MPC, also illustrated in Fig. 4.8:

- The predictions of the system's output signal y values are calculated for each discrete sample k for a given horizon in the future ($j = N_1, \ldots, N_2$). The predictions are denoted as $\hat{y}(k + j)$ and represent a j-step-ahead prediction, given the information at time instant k, while N_1 and N_2 determine the lower and upper bounds of the prediction horizon. Commonly, N_1 is set to 0 and N_2 is then denoted as N_h. The lower and upper bounds of the output signal prediction horizon determine the coincidence horizon, within which a match between the output and the reference signal is expected. The output signal prediction is calculated from the process model. These predictions are also dependent on the control scenario in the future $u(k + j)$, $j = 0, \ldots, N_u - 1$, which will be applied from the moment k onwards.
- The reference trajectory is denoted by $r(k + j)$, $j = 0, \ldots, N_h$, which represents the reference response from the present value $y(k)$ to the set-point trajectory $w(k)$. The reference trajectory and the set-point trajectory may be the same in some MPC algorithms.

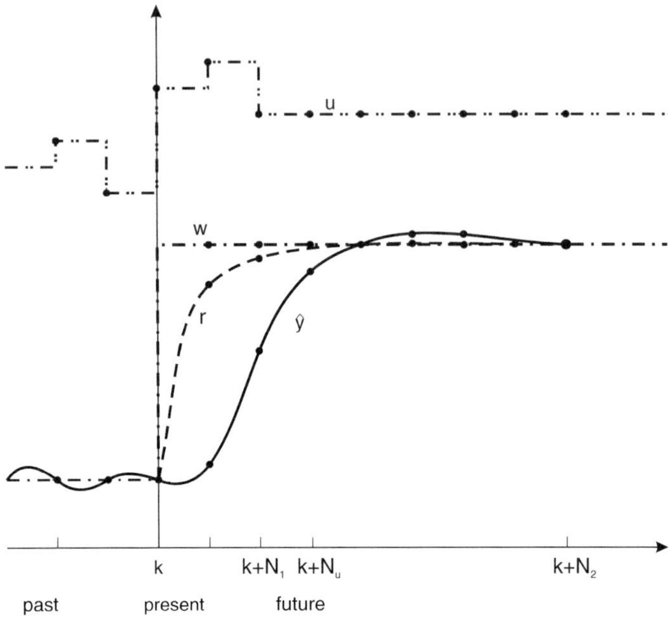

Fig. 4.8 The basic principle of MPC

- The future control sequence **u** containing $u(k+j)$, $j=0,\ldots,N_u-1$ is calculated by minimising the cost function (also called the objective function) such that the predicted error between $r(k+j)$ and $\hat{y}(k+j)$, $j=0,\ldots,N_h$ is a minimum. The structuring of the future control sequence can be used in some approaches.
- Only the first element $u(k)$ of the optimal control sequence $u(k+j)$, $j=0,\ldots,N_u-1$ is applied.

In the next sample, another measured output sample is available and the entire procedure is repeated. This principle is called the receding-horizon strategy [21].

The control objective is achieved by minimising the cost function, which penalises the deviations of the predicted controlled output value $\hat{y}(k+j)$ from a reference trajectory $r(k+j)$. This reference trajectory may depend on the measurements made up to time k. Its initial value may be the output measurement $y(k)$, but also a fixed set-point, or some predetermined trajectory. The minimisation of the cost function, in which the future control sequence described by the vector **u** is calculated, may be subject to various constraints (e.g. input, state, rates).

MPC [21] solves a constrained control problem and can generally be written as follows. The single-input, single-output case is elaborated here, but it can be generalised to the multiple-input, multiple-output case. A nonlinear, discrete-time system can be described in the input-output form:

$$y(k+1) = h\left(y(k)\ldots y(k-n), u(k)\ldots u(k-m)\right) + \nu_0(k) \tag{4.18}$$

or in the state space form:

$$\mathbf{x}(k+1) = f\left(\mathbf{x}(k), u(k)\right) + \boldsymbol{\nu}_1(k) \tag{4.19}$$

$$y(k) = g\left(\mathbf{x}(k), u(k)\right) + \nu_2(k) \tag{4.20}$$

where $\mathbf{x}(k) \in \mathbb{R}^n$, $u(k) \in \mathbb{R}$ and $y(k) \in \mathbb{R}$ are the state, input and output variables, $\boldsymbol{\nu}_i(k);\ i = 0, 1, 2$ are the Gaussian random variables representing disturbances, and h, f, g are the nonlinear continuous functions.

The associated input and state constraints of the general form are:

$$u(k) \in \mathcal{U}, \tag{4.21}$$

$$\mathbf{x}(k) \in \mathcal{X}, \tag{4.22}$$

where $\mathcal{U}$ is the set of constrained inputs and $\mathcal{X}$ is the set of constrained states. The optimisation problem is

$$V^*(k) = \min_{\mathbf{u}} J(\mathbf{u}, \mathbf{x}(k), r(k), u(k-1)), \tag{4.23}$$

where the cost function of a general form is

$$J(\mathbf{u}, \mathbf{x}(k), r(k), u(k-1)) = E\left(\sum_{j=0}^{N_h-1} l\left(\hat{\mathbf{x}}(k+j), u(k+j)\right)\right) \tag{4.24}$$

where l is a function called the stage cost function, and it is assumed that the cost falls to zero once the state has entered the set of optimal states $\mathcal{X}_0$, i.e. $l(\mathbf{x}, u) = 0$ if $\mathbf{x} \in \mathcal{X}_0$. The following terminal constraint is imposed:

$$\hat{\mathbf{x}}(k + N_h) \in \mathcal{X}_0. \tag{4.25}$$

The optimal solution to the optimisation problem described with Eq. (4.23), subject to input, state and terminal constraints, is the control sequence

$$\mathbf{u}_o = \left[u_o(k), u_o(k+1), \ldots, u_o(k+N_h-1)\right]. \tag{4.26}$$

Because only the first control input $u_o(k)$ in the optimal sequence $\mathbf{u}_o$ is applied to the system at time k, an implicit MPC law

$$\kappa_N(\mathbf{x}) = u_o(k), \tag{4.27}$$

that is time invariant and where the index N reminds us that it is based on the optimisation over the finite horizon N_h.

This is a general form and MPC formulations vary with the various models, cost functions, and parameters, for example, the length of the horizon.

A frequently used cost function is:

$$J(k) = ||E(\hat{y}(k + N_h)) - r(k + N_h)||^2_{\mathbf{P}} + \sum_{j=1}^{N_h-1} \left[||E(\hat{y}(k + j)) - r(k + j)||^2_{\mathbf{Q}} + ||u(k + j) - u(k + j - 1)||^2_{\mathbf{R}}\right]. \tag{4.28}$$

For $\mathbf{x} \in \mathbb{R}^n$, the Euclidean norm is $\|\mathbf{x}\| = \sqrt{\mathbf{x}^{\mathrm{T}}\mathbf{x}}$ and the weighted norm is defined for some symmetric and positive definite matrix $\mathbf{A}$ as $\|\mathbf{x}\|_{\mathbf{A}} = \sqrt{\mathbf{x}^{\mathrm{T}}\mathbf{A}\mathbf{x}}$. N_h is a finite horizon and $\mathbf{P}$, $\mathbf{Q}$ and $\mathbf{R}$ are positive definite matrices.

The MPC method based on online optimisation differs from dynamic programming or explicit MPC in terms of its implementation [29]. With the MPC using real-time optimisation, the optimal control action $\kappa_N(\mathbf{x}, k)$ for each time sample is determined by solving an optimisation problem online, while with the explicit MPC all the computations related to the optimisation are performed offline for a given set of initial states.

In our case, the process model for the calculation of the predicted outputs is the GP model, which predicts a normally distributed random variable with the mean value $\mu_{\hat{y}(k+j)} = E(\hat{y}(k + j))$ and the variance $\mathrm{var}(\hat{y}(k + j)) = E(\hat{y}(k + j)^2) - E(\hat{y}(k + j))^2$. Note that with long-term predictions, the predicted random variable is approximated in one of the ways described in Sect. 2.6.

There are many alternative ways in which an NMPC with GP models can be implemented.

Cost function. The cost function described with Eq. (4.24) is a general one and various special cost functions can be derived from it. It is well known that the selection of the cost function has a major impact on the amount of computation.

Optimisation problem for $\Delta\mathbf{u}$ *instead of* $\mathbf{u}$. This is not just a change of formalism; it also enables forms of MPC containing an integral action.

Process model. The process model can be determined offline and fixed for the time of operation or determined online during the operation of the controller.

Soft constraints. The use of constrained optimisation algorithms is very demanding for computation and soft constraints, i.e. the weights on the constrained variables in the cost function can be used to decrease the amount of computation. More on this topic can be found in, e.g. [21].

Linear MPC. It is worth noting that even though this is a constrained NMPC problem, it can be used in its specialised form as a robust linear MPC.

There are several issues of interest for applied NMPC.
The *stability* of the closed-loop system is one of most important. Some of the published overviews relating to closed-loop stability with MPC are in, e.g. [29, 30]. In general, the stability of a closed-loop system containing MPC can be ensured with:

- the terminal equality constraint, where the controller steers the state to the optimal state;
- the terminal cost function l_T employs the terminal penalty in Eq. (4.24) in addition to the stage cost function $l(\cdot)$;
- the terminal constraint set, where the controller steers the state to the set of optimal states $\mathcal{X}_0$ in a finite time, and a local stabilising controller is employed inside $\mathcal{X}_0$;
- the terminal cost and constraint set are employed.

The sufficient conditions [29] for closed-loop asymptotic (exponential) stability for deterministic systems are as follows:

1. the state constraint is satisfied in $\mathcal{X}_0$ ($\mathcal{X}_0 \subset \mathcal{X}$, $\mathcal{X}_0$ closed $0 \in \mathcal{X}_0$).
2. the input constraint is satisfied in $\mathcal{X}_0$ ($\kappa(\mathbf{x}) \subset \mathcal{U}, \forall \mathbf{x} \in \mathcal{X}_0$).
3. the set $\mathcal{X}_0$ is positively invariant under $\kappa(\cdot)$ ($f(\mathbf{x}, \kappa(\mathbf{x})) \in \mathcal{X}_0, \forall \mathbf{x} \in \mathcal{X}_0$).
4. the terminal cost function Δl_T is the change of a local Lyapunov function ($[l + \Delta l_T](\mathbf{x}, \kappa(\mathbf{x})) \leq 0, \forall \mathbf{x} \in \mathcal{X}_0$).

It is very convenient to have a terminal cost $l_T(\cdot)$ as close to the optimal value of the cost function in Eq. (4.24) with an infinite horizon. In the case of equality the benefits of infinite-horizon optimal control are achieved. Since the optimal value of a cost function with an infinite horizon is, in general, not known for nonlinear systems, at least a good approximation of $l_T(\cdot)$ with the optimal value of the cost function in the neighbourhood $\mathcal{X}_0$ of the target state is required. A useful result for the satisfaction of the sufficient conditions 1.–4. is that the finite-horizon optimal control problem described with Eq. (4.23) is equivalent to a modified infinite-horizon optimal problem in which $l(\cdot)$ is replaced with its upper bound. This makes it possible for the benefits of infinite-horizon optimal control to be achieved, even though the horizon is finite, when the listed sufficient conditions are satisfied [29].

Before the stability of MPC systems was addressed theoretically, MPCs were still applied in industrial practice, with control designers taking care of a few issues related to implementation that enabled stability as if an infinite horizon was applied. These issues are listed in [29]: they have restricted the design to stable plants and choosing a horizon that is large compared with the settling time of the plant.

Some other issues of interest with respect to the implementation of NMPC are briefly described as follows.

Efficient numerical implementation. A nonlinear programming optimisation algorithm is computationally demanding. Various approximations, e.g. using the simplifications of nonlinear functions, and other approaches, e.g. an approximate explicit solution, exist to decrease the computational load, often for special cases, like for special cost functions.

Robustness. This issue has a major impact on the applicability of the algorithm in practice. An overview of NMPC robustness, in general, can be found in [29]. The properties of a GP model in particular may influence the GP-NMPC contained closed-loop robustness. The fact that the GP model of the process contains information about the model's confidence allows the controller to optimise the manipulative variable in order to 'avoid' regions where confidence in the model is not high enough. This possibility itself makes the controller robust, if it is applied properly.

Various MPC methods can be applied with GP models, depending on the designer's choice and the imposed constraints. Using GP models does not impose any particular constraint on the cost function, the optimisation method, or any other element of choice for the MPC design.

NMPC can be treated as a deterministic or stochastic problem. Stochastic NMPC problems are formulated in applications where the system to be controlled is described by a stochastic model. Stochastic problems in general, like state estimation, have been studied for a long time, but the stochastic NMPC problem is just a small subset of the stochastic problems. The best known stochastic MPC approaches are based on parametric probabilistic models. Alternatively, stochastic systems can be modelled with nonparametric models, which can offer a significant advantage over parametric ones. This is related to the fact that the nonparametric probabilistic models, like GP models, provide information about the prediction uncertainties, which might be difficult to evaluate appropriately with parametric models.

An application of MPC with a GP model using the general cost function described by Eq. (4.28) can be found in, e.g. [31–34]. The MPC containing a GP multimodel, where GP models describing different operating regions are switched, can be found in [35]. MPC in the context of fault-tolerant control is described in [36]. Until recently, most of the applications or investigations of MPC based on GP models were one of three special forms. These three algorithms are *internal model control* (IMC), *predictive functional control* (PFC) and *approximate explicit control*, which are described in subsequent sections. Other algorithms can be found, like the one in [37], where the control is based on an estimation and multistep-ahead prediction of the system output response in combination with fuzzy models.

Internal Model Control

In this strategy, the controller is chosen to be an inverse of the plant model. Internal model control (IMC) is one of the most commonly used, model-based techniques for the control of nonlinear systems. It is closely related to MPC, and can be considered as a special case of MPC [38]. IMC with a GP model is elaborated in [39–42]. The description of IMC with a GP model is subsequently adopted from these references.

The general structure is shown in Fig. 4.9. Three advantageous properties of IMC listed in [42, 43] are:

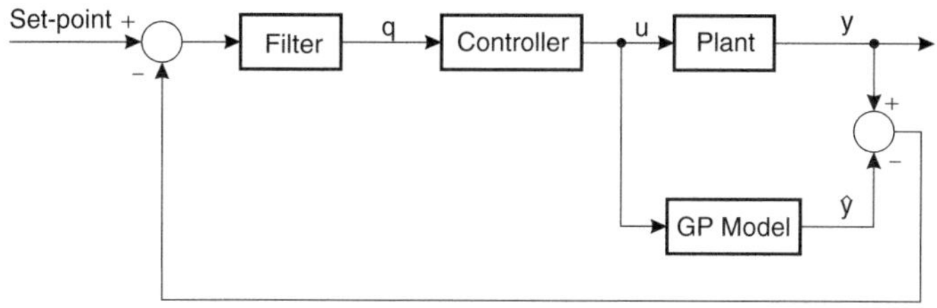

Fig. 4.9 General IMC structure

Stability Assuming the model of the plant is perfect, then if the controller and the plant are input-output stable, the closed-loop system in Fig. 4.9 is stable.

Perfect control Assuming that the controller is the exact model inverse and the closed-loop system in Fig. 4.9 is stable, then the control will be perfect, i.e. the output signal y equals the set-point for all disturbances.

Zero offset Assuming that the steady-state gain of the controller is equal to the inverse of the model gain and the closed-loop system in Fig. 4.9 is stable, then for asymptotically constant set-points and disturbances, there will be no offset, i.e. the output signal y asymptotically approaches to the set-point.

The filter in the loop shapes the desired closed-loop behaviour of the closed-loop system, reduces the gain of the feedback system at high frequencies and introduces a degree of robustness to the model mismatch.

The main difference between the various IMC approaches lies in the choice of the internal model and its inverse. It was shown in [44] that a GP model based on a squared exponential covariance function is not analytically invertible. Instead of calculating the exact inverse, a numerical approach, such as a successive approximation or the Newton–Raphson optimisation method, can be used to find the control effort that solves the following equation:

$$h(u(k)\dots u(k-m), y(k)\dots y(k-n)) - q(k) = 0, \tag{4.29}$$

where

$$h(u(k)\dots u(k-m), y(k)\dots y(k-n)) = \hat{y}(k+1) \tag{4.30}$$

and q is the controller input signal.

The GP model of h is trained as a one-step-ahead prediction model with the regressors $u(k)\dots u(k-m), y(k)\dots y(k-n)$ at different time instances and $y(k+1)$ at the corresponding time instances as the targets. The IMC strategy requires the use of the parallel model of Eq. (4.30). This GP model is then included in the IMC structure and the numerical inverse of Eq. (4.29) is found for each sample. The IMC works well when the control input and output values of the system are in the region where the model was trained. However, as soon as the system moves away from the well-modelled region, the control input value $u(k)$ on the left-hand side of Eq. (4.29) cannot be found to force the right-hand side of Eq. (4.29) to zero. The sub-optimal

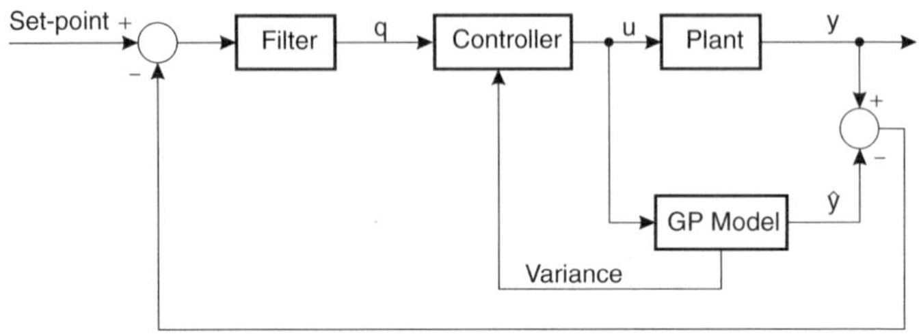

Fig. 4.10 Variance–constrained IMC structure

value of u that drives Eq. (4.29) closest to zero is applied instead and this can cause sluggish and, in certain cases, also unstable closed-loop system behaviour.

Since poor closed-loop performance is the result of the model being driven outside its trained region, the naive approach would be to constrain the control input. When the system is driven into this untrained portion of the operating space, the model will no longer represent the system well and the initial constraints of the control signal might not be relevant. However, the increase of the predicted variance will indicate a reduced confidence in the prediction. This increase of the variance can be used as a constraint in the optimisation algorithm utilised to solve Eq. (4.29). The concept is shown in Fig. 4.10.

The basic idea of the algorithm is to optimise the control effort so that the variance does not increase above its predefined limit. The variance constraint should be defined by the designer and, in general, it can be a function of some scheduling variable. The simple approach is for the variance constraint to be set at a constant value. Constrained nonlinear programming with random restarts might be used to optimise the control effort. The variance is predicted one-step ahead. The present information is not fed back to the input of the model, and it does not effect the variance prediction in the next step. The predicted variance, as a measure of the model's uncertainty, can be used as a constraint in the inversion algorithm, to improve the closed-loop response. The control effort can then be optimised so that the variance does not increase above a predefined limit. Since the GP model is not analytically invertible and numerical approaches have to be utilised to find the inverse of the model for each sample time, the associated computation load increases rapidly with the number of training-data points [42]. This is the main drawback of the GP modelling approach for IMC.

Predictive Functional Control

Predictive functional control (PFC) is, in principle, no different to the general MPC. Its distinct features are a relatively small number of coincidence points, the use of a reference trajectory, which always differs from the set-point trajectory, and that the future input signal is assumed to be a linear combination of a few simple basis functions. Coincidence points are points where the closed-loop response and the reference trajectory should coincide and are a simplified form of the coincidence horizon. More details can be found in, e.g. [21].

In the following description, the PFC with a single coincidence point and a constant controller output signal is used. Variants of this kind of NMPC with a GP model are described in [45–50]. NMPC based on a GP model was first introduced in [45].

A moving-horizon minimisation problem of the form

$$V^* = \min_{\mathbf{u}} \left[r(k+P) - E\left(\hat{y}(k+P)\right)\right]^2 \tag{4.31}$$

subject to

$$\operatorname{var}\left(\hat{y}(k+P)\right) \leq k_v, \tag{4.32}$$
$$||\mathbf{u}|| \leq k_{ih}, \tag{4.33}$$
$$||\Delta\mathbf{u}|| \leq k_{ir}, \tag{4.34}$$
$$||\mathbf{x}(k)|| \leq k_{sh}, \tag{4.35}$$
$$||\Delta\mathbf{x}(k)|| \leq k_{sr}, \tag{4.36}$$

is applied as the first presented choice, where $\mathbf{u}$ is the sequence of input-signal samples and contains $u(k+j)$, $j = 1 \ldots P$, $\Delta\mathbf{u}$ is the input rate and contains $u(k+j) - u(k+j-1)$; $j = 1, \ldots, P$, $\mathbf{x}$ is the state vector, $\Delta\mathbf{x} = \mathbf{x}(k) - \mathbf{x}(k-1)$ is the state rate, P is the coincidence point, i.e. the point where a match between the output and the reference value is expected, and inequalities from Eqs. (4.32) to (4.36) represent the constraint on the output variance k_v, the hard input constraint k_{ih}, the input rate constraint k_{ir}, the hard state constraint k_{sh} and the state rate constraint k_{sr}, respectively. These constraints are, in general, functions of some scheduling variable in the general form, but are frequently set to be constant values. The process model is a GP model.

The constrained nonlinear optimisation problem is solved for each sample time over a prediction horizon of length P. If the constraints from the minimisation problem described with Eq. (4.31) are omitted we obtain minimum-variance control for set-point tracking.

The *minimum-variance controller* [51] looks for a control signal u in the time instant k that will minimise the following performance objective:

$$J_{MV} = E\left(\left[r(k+P) - \hat{y}(k+P)\right]^2\right). \tag{4.37}$$

Taking the expected value of a variable squared gives the variance of that variable assuming the variable mean is zero. In this case, J_{MV} refers to the variance of the error between the reference value $r(k+P)$ and the controlled output value P-time steps in the future, $\hat{y}(k+P)$. The desired controller is thus the one that minimises this variance, and hence the name minimum-variance control.

The cost function described with Eq. (4.37) can be expanded with a penalty term λ on the control effort:

$$J_{MV_2} = E\left(\left[r(k+P) - \hat{y}(k+P)\right]^2\right) + \lambda u^2(k). \tag{4.38}$$

The term λ can be used for 'tuning' the performance of the closed-loop system. As mentioned in [52], this cost function can be written as:

$$J_{MV_2} = [r(k+P) - E(y(k+P))]^2 + \mathrm{var}\left(\hat{y}(k+P)\right) + \lambda u^2(k), \tag{4.39}$$

where the second term represents the model's uncertainty, which is available from the GP model prediction and can be used in the optimal control signal minimisation. Note that most conventional work has ignored it, or has added extra terms to the cost function, or has pursued other sub-optimal solutions.

The cost function described with Eq. (4.38) can be further expanded with other forms of penalty leading to a generalised minimum-variance control. A possible alternative to the cost function described with Eq. (4.39) is

$$J_{GMV} = Q\left(q^{-1}\right)[r(k+P) - E(\hat{y}(k+P))]^2 + \mathrm{var}\left(\hat{y}(k+P)\right) + R\left(q^{-1}\right)u^2(k), \tag{4.40}$$

where the polynomials $Q(q^{-1})$ and $R(q^{-1})$ are defined as:

$$Q\left(q^{-1}\right) = Q_0 + Q_1 q^{-1} + \cdots + Q_{n_q} q^{n_q}, \tag{4.41}$$

$$R\left(q^{-1}\right) = R_0 + R_1 q^{-1} + \cdots + R_{n_r} q^{n_r}, \tag{4.42}$$

where q^{-1} is a unit backward shift operator. The polynomial coefficients can be used as the tuning parameters. A similar cost function is used in [53].

Yet another possibility, used in [54], is an enhanced version of the generalised minimum-variance controller, known as the generalised linearising controller [55]

$$J_{GLC} = (1-\eta)E\left([r(k+P) - \hat{y}(k+P)]^2\right) + \eta\left[R\left(q^{-1}\right)u(k) - u_r\right]^2, \tag{4.43}$$

where u_r is the input signal associated with r, and the coefficients of $R(q^{-1})$ and η are the tuning parameters. When $\eta = 0$, then the cost function corresponds to the minimum variance cost function, and when $\eta = 1$ then the minimum of the cost function corresponds to a simple feedforward control. On the other hand, if the open-loop system is stable, but with an unstable inverse, then $\eta = 1$ provides a stable controller [54].

The optimal control signal u_{opt} can be obtained by minimising the cost functions:

$$V^* = \min_{\mathbf{u}} (J) , \tag{4.44}$$

where J is the cost function.

The minimisation can be made analytically by finding the extremum $\partial J / \partial u = 0$, but also numerically by using any appropriate optimisation method.

The references [52, 54] describe the control of an affine nonlinear system of the form,

$$y(k+1) = f(\mathbf{x}(k)) + g(\mathbf{x}(k))u(k) + \eta(k+1), \tag{4.45}$$

which allows a combination of the squared exponential and linear covariance function for the GP model and the combination with Minimum Variance control. This application is generalised to the multiple-input multiple-output case in [56].

Let us return to the cost function from the optimisation problem Eq. (4.31) of the form:

$$J(k) = E\left(\left[r(k+P) - \hat{y}(k+P)\right]^2\right). \tag{4.46}$$

Using the fact that $\mathrm{var}(\hat{y}) = E(\hat{y}^2) - E(\hat{y})^2$, the cost function can be written as [48]

$$J(k) = \left[r(k+P) - E\left(\hat{y}(k+P)\right)\right]^2 + \mathrm{var}\left(\hat{y}(k+P)\right). \tag{4.47}$$

The control strategy with the cost function described with Eq. (4.47) is 'to avoid' going into regions with a higher variance. The term 'higher variance' does not specify any particular value. In the case that the controller does not seem to be 'cautious' enough, a 'quick-and-dirty' option is to weight the variance term with a constant λ_{var} to enable the shaping of the closed-loop response according to the variance information [48]

$$J(k) = \left[r(k+P) - E\left(\hat{y}(k+P)\right)\right]^2 + \lambda_{\mathrm{var}}\mathrm{var}\left(\hat{y}(k+P)\right). \tag{4.48}$$

Besides the difference in the optimisation algorithm, the presented options also give a design choice as to how 'safe' the control algorithm is. In the case when it is very undesirable to go into 'unknown' regions, the constrained version might be a better option.

The application of the described predictive controller is given in the following example.

Example 4.2 Model predictive control of a pH neutralisation process

This example is adapted from [48]. A simplified schematic diagram of the pH neutralisation process taken from [57] is given in Fig. 4.11. The process consists of an acid stream (Q_1), a buffer stream (Q_2) and a base stream (Q_3) that are mixed in a tank T_1. Prior to mixing, the acid stream enters the tank T_2, which introduces additional flow dynamics. The acid and base flow rates are controlled with

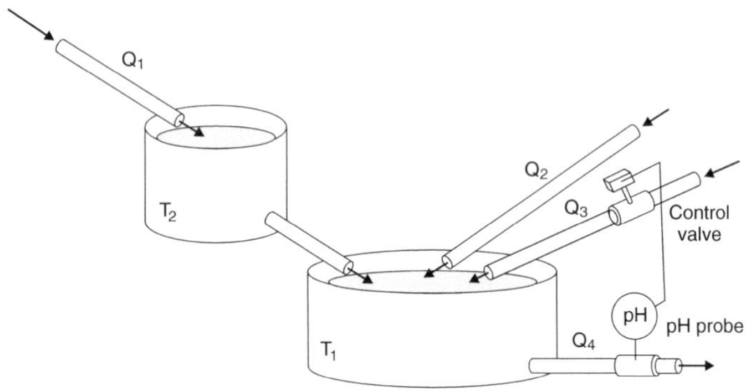

Fig. 4.11 The pH neutralisation system scheme

flow-control valves, while the buffer flow rate is controlled manually with a rotameter. The effluent pH (pH) is the measured variable. Since the pH probe is located downstream from the tank T_1, a time delay (T_d) is introduced into the pH measurement. In this study, the pH value is controlled by manipulating the base flow rate. A more detailed description of the process with a mathematical model and the necessary parameters is presented in [57].

The dynamic model of the pH neutralisation system shown in Fig. 4.11 is derived using the conservation equations and equilibrium relations. The model also includes the valve and transmitter dynamics as well as the hydraulic relationships for the tank outlet flows. The modelling assumptions include the perfect mixing, constant density, and complete solubility of the ions involved. The simulation model of the pH process that was used for the necessary data generation, therefore, contains various nonlinear elements.

The sampling time of 25 s was selected based on some responses and an iterative cut-and-try procedure.

An identification signal with a length of 400 samples was generated as a random signal with a uniform distribution and a sampling rate of 50 s. The GP model with the squared exponential covariance function (Eq. 2.14) for the functional part and the covariance function (Eq. 2.11) for the noise part are used for modelling the system. After selecting the regressors in an iterative manner, a third-order dynamic model is selected. The vector of eight hyperparameters $\boldsymbol{\theta}$ for the third-order GP model is obtained with marginal likelihood optimisation, where the hyperparameters denote the weights for each regressor representing the delayed input and output values, the noise variance and the variance of the vertical scale of the variation.

The GP model is validated with a simulation that propagates uncertainty with an exact matching of the statistical moments (Sect. 2.7.4). Such a validation is important for providing reassurance that the model behaves well in a multistep-ahead prediction. The response of the GP model to the identification signal is shown in Fig. 4.12. A very good fit can be observed for the identification input signal that was used for the optimisation. However, the obtained model contains information mainly in the region

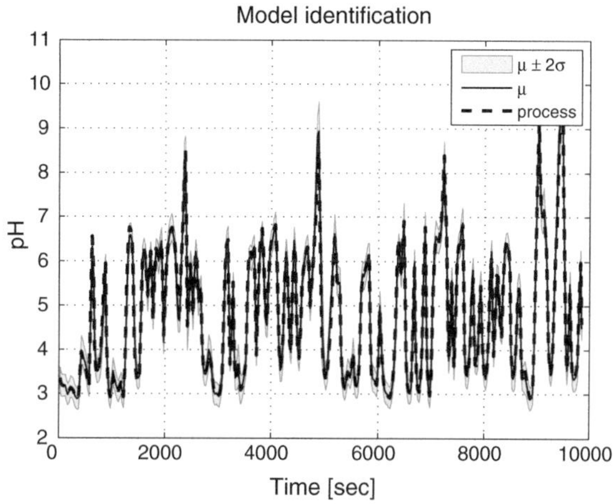

Fig. 4.12 Simulation response of the GP model to the excitation signal used for the identification

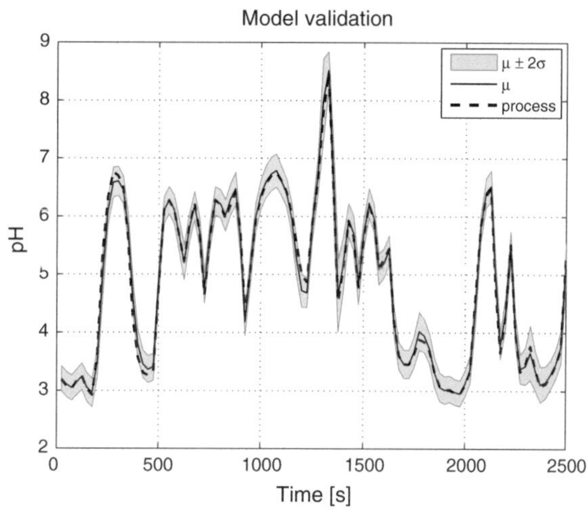

Fig. 4.13 Simulation response of the GP model to the excitation signal used for the validation

where $pH < 7$. The validation signal covers the same region as the identification signal. The identification and validation signals are obtained with a generator of random noise that has different initial values. The simulation response of the model to the validation signal and the comparison with the process response are depicted in Fig. 4.13. The values of the validation performance measures of the identification and validation data are given in Table 4.1. It is clear that the validation performance measures are better than the identification ones. This is not usual, but it my happen.

After the model is validated, it is utilised for the control design. See [58] for more issues relating to pH process modelling with GP models.

The described PFC algorithm is used for the pH process and tested with a simulation of the closed-loop response. The reference trajectory r is defined so that it approaches the set-point exponentially from the current output value.

Table 4.1 Values of the validation performance measures of the identification and validation data

Identification data			Validation data	
ℓ	SMSE	MSLL	SMSE	MSLL
222	0.0064	−2.4230	0.0058	−2.4435

ℓ is obtained at training and evaluate prediction, while other measures are performance measures of simulation results

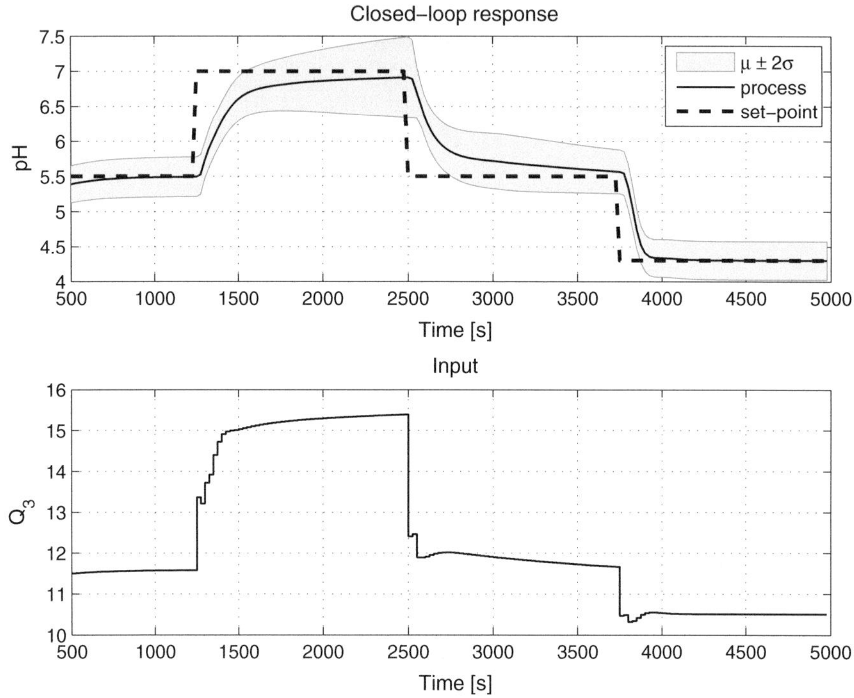

Fig. 4.14 Unconstrained case: response of the GP model-based control (*upper figure*) and the control signal (*bottom figure*)

The coincidence point is chosen to be 8 samples. The control signal, which is represented by a combination of a few basis functions, is a property of the PFC. In our case, the control signal is a constant, which is equivalent to MPC with a control horizon equal to 1. As we focus on the variance constraint, the rest of the constraints are not taken into account. The results for the unconstrained control are given in Figs. 4.14 and 4.15.

It is clear from the different set-point responses that the model differs from the process in different regions. It is clear that the variance increases as the output signal approaches regions that were not populated with sufficient identification data. It should be noted, however, that the predicted variance is the sum of the variance that can be interpreted as information about the degree of confidence in the model's

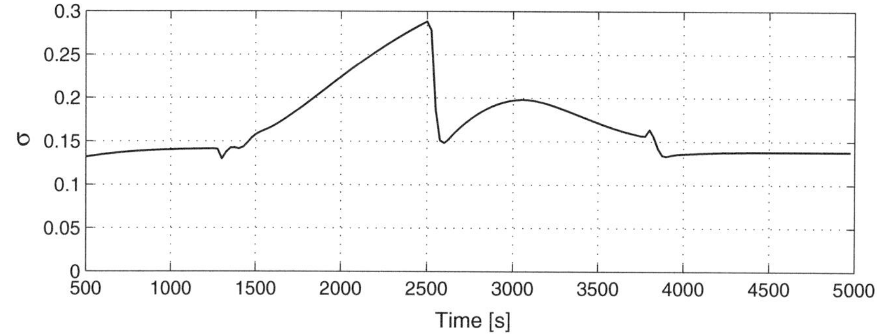

Fig. 4.15 Unconstrained case: Standard deviation corresponding to Fig. 4.14

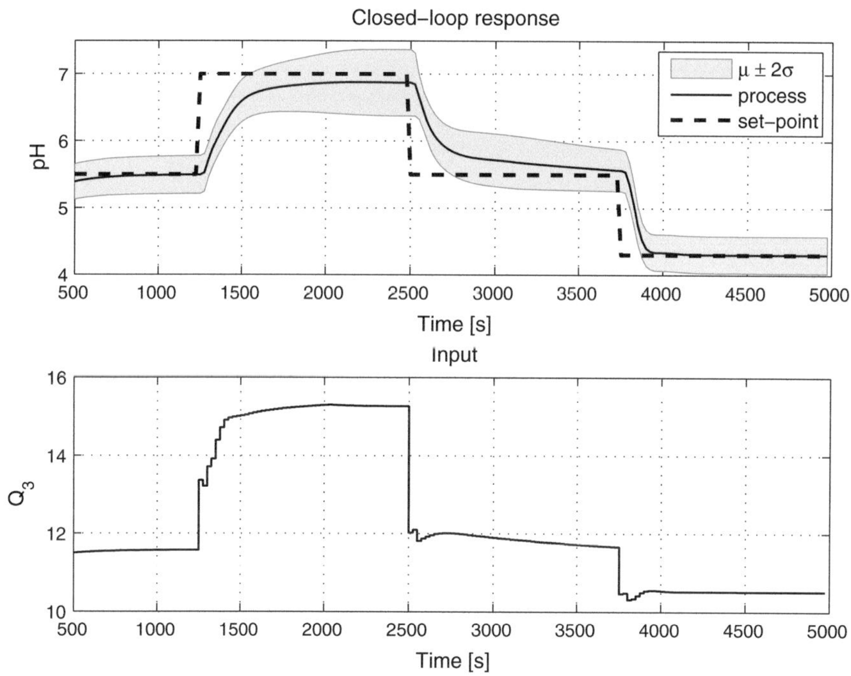

Fig. 4.16 Constrained case ($\sigma_{\max} = 0.25$): response of GP model-based control (*upper figure*) and control signal (*bottom figure*)

accuracy, i.e. depending upon the local density of the available identification data, and of the output response variance. When the variance increases too much, one design option is to optimise the response with constrained control. The results can be seen in Figs. 4.16 and 4.17.

It is clear from Figs. 4.16 and 4.17 that the closed-loop system response now avoids the region with large variance, at the cost of an increase in the steady-state

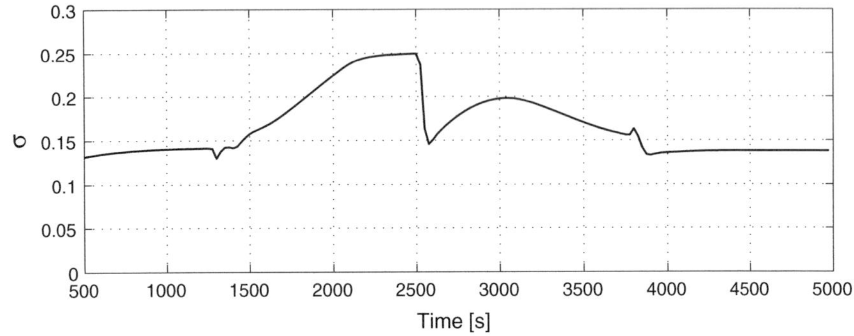

Fig. 4.17 Constrained case ($\sigma_{\max} = 0.25$): standard deviation corresponding to Fig. 4.16

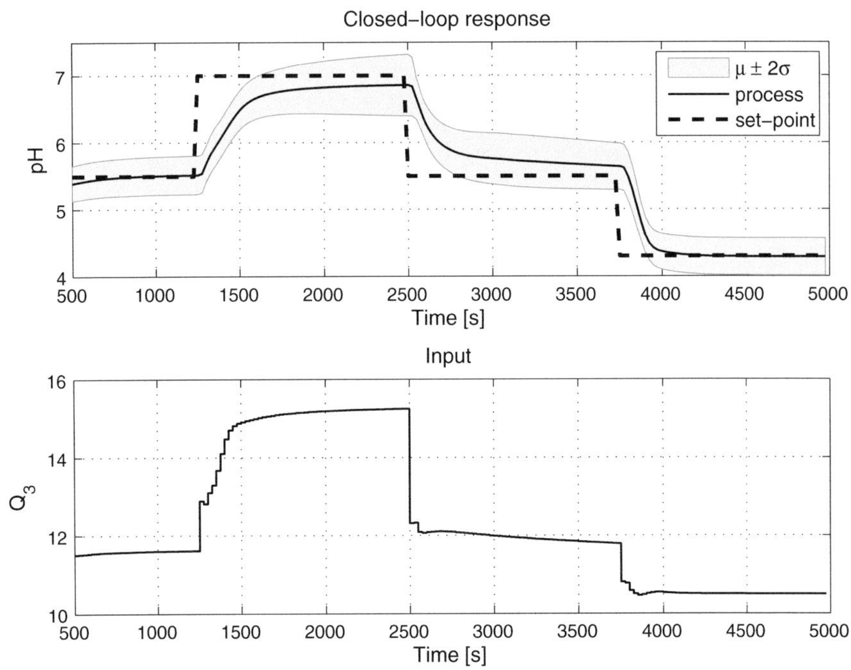

Fig. 4.18 Response of GP model-based control with 'soft constraints' (*upper figure*) and control signal (*bottom figure*)

error. This could also be interpreted as a trade-off between the designed performance and the safety.

The results with an alternative cost function incorporating a soft constraint described with Eq. (4.47) are presented in Figs. 4.18 and 4.19.

It is again clear from Figs. 4.18 and 4.19 that the closed-loop system response avoids the region with large variance at the cost of a steady-state error, as was the

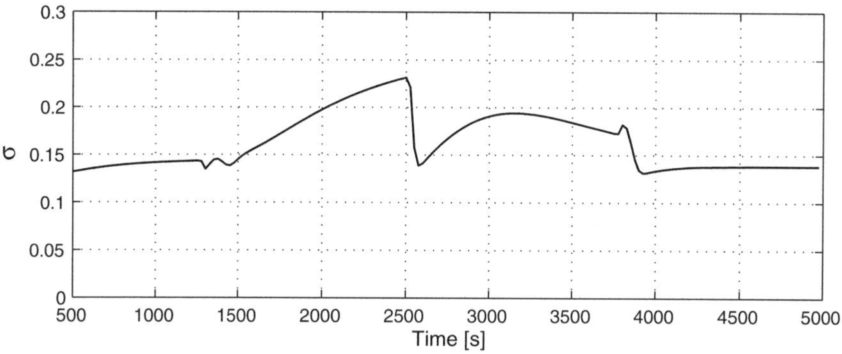

Fig. 4.19 Standard deviation corresponding to Fig. 4.18

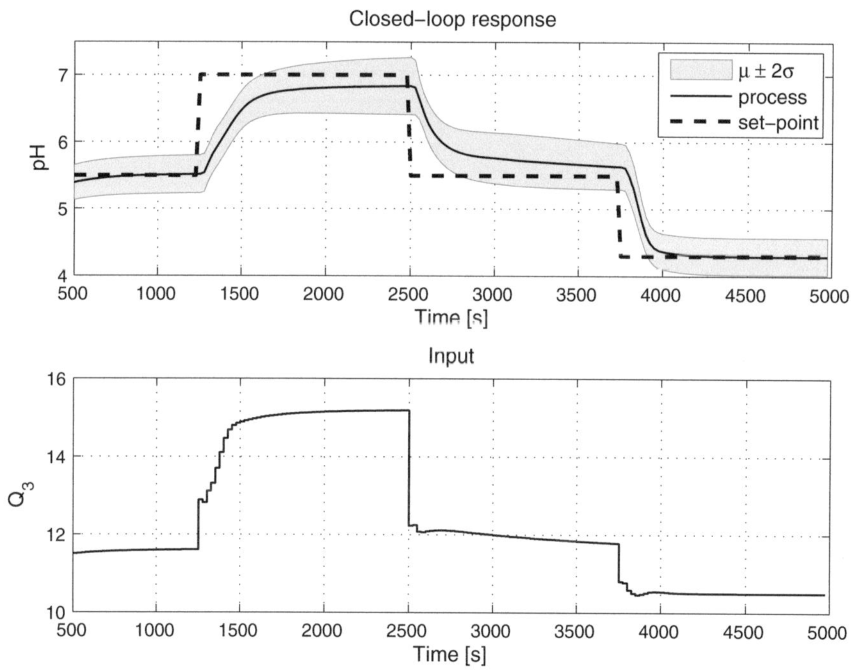

Fig. 4.20 Response of GP model-based control with 'soft constraints' (*upper figure*) and control signal (*bottom figure*)

case with the constrained control, but with a smaller computational burden than the constrained control case.

The NMPC with the cost function described by Eq. (4.47) with the weight on the variance $\lambda_{var} = 2$, using unconstrained optimisation, gives the results shown in Figs. 4.20 and 4.21, showing a reduction in the standard deviation of the predictions

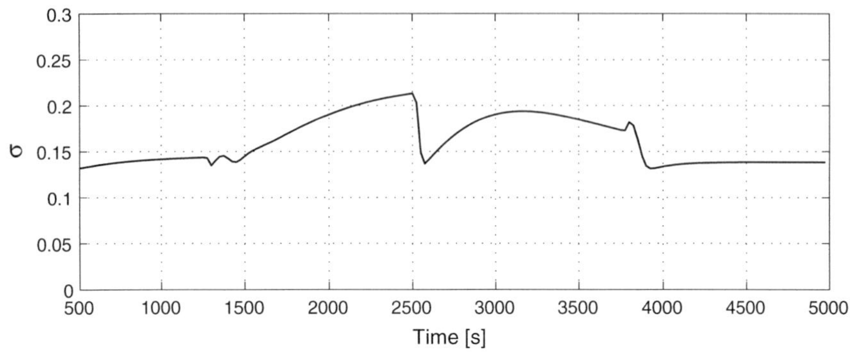

Fig. 4.21 Standard deviation corresponding to Fig. 4.20

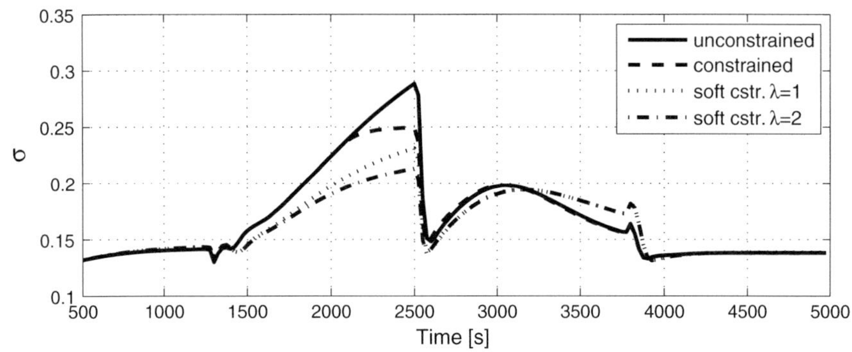

Fig. 4.22 The comparison of standard deviations corresponding to all four described cases

for the closed-loop response, compared to Fig. 4.21, and minor changes in the mean behaviour.

While differences in the mean values of the closed-loop responses in Figs. 4.14, 4.15, 4.16, 4.17, 4.18, 4.19, and 4.20 are not apparent, the comparison of standard deviations for the model predictions in Fig. 4.22 clearly shows the differences among the responses in all four cases.

Approximate Explicit Nonlinear Predictive Control

The MPC formulation described so far provides the control action $u(k)$ as a function of the states $\mathbf{x}(k)$ defined *implicitly* by the cost function and the constraints. In the past 10 years, several methods for an *explicit* solution to MPC problems have been suggested (see for example [25, 59, 60]). The main motivation behind *explicit* MPC is that an explicit state feedback law avoids the need for executing a numerical optimisation algorithm in real time, and is, therefore, potentially useful for applications where MPC has not traditionally been used, e.g. electromechanical systems requiring a fast response. By treating $\mathbf{x}(k)$ as a vector of parameters, the goal of the explicit methods is to solve the MPC problem offline with respect to all the values of $\mathbf{x}(k)$ of interest and to make the dependence of the control input values on the state values explicit. It has been shown in [61] that the feedback solution to MPC problems for constrained linear systems has an explicit representation as a piecewise linear (PWL) state feedback defined on a polyhedral partition of the state space. The benefits of an explicit solution, in addition to the efficient online computations, also include the verifiability of the implementation and the possibility to design embedded control systems. For the nonlinear MPC the benefits of explicit solutions are even greater than for linear MPC, since the computational efficiency and verifiability are even more important. In [25], approaches to an offline computation of explicit sub-optimal piecewise predictive controllers for general nonlinear systems with state and input constraints are presented, based on the multiparametric Nonlinear Programming (mp-NLP) ideas [62].

In [63], an approach to explicit stochastic NMPC based on parametric probabilistic models is published. An approximate mp-NLP approach to the offline computation of an explicit sub-optimal NMPC controller for constrained nonlinear systems based on a GP model (abbreviated as GP-NMPC) is proposed in [64]. The approach represents an extension of the approximate methods in [65] and [66]. The approximate explicit GP-NMPC approach was elaborated in [67]. When we subsequently refer to the NMPC problem, it is the NMPC problem based on a GP model, as it has been thus far.

Let us see what is the formulation of the NMPC problem as an mp-NLP problem.

Consider a nonlinear discrete-time system described with Eq. (4.19):

$$\mathbf{x}(k+1) = f(\mathbf{x}(k), u(k)) + \boldsymbol{\nu}_1(k)$$

where $\mathbf{x}(k) \in \mathbb{R}^n$ and $u(k) \in \mathbb{R}$ are the state and input variables, $\boldsymbol{\nu}_1(k) \in \mathbb{R}^n$ are the Gaussian disturbances, and $f : \mathbb{R}^n \times \mathbb{R} \to \mathbb{R}^n$ is a nonlinear continuous function. The case with more than one input is elaborated in the references above. Suppose the initial state $\mathbf{x}(k)$ and the control input values $u(k+j)$, $j = 0, 1, \ldots, N_h - 1$, are given. Then, the approximated probability distribution of the predicted states $\hat{\mathbf{x}}(k+j+1)$, $j = 0, 1, \ldots, N_h - 1$, which correspond to the given initial state $\mathbf{x}(k)$ and the control input values $u(k+j)$, $j = 0, 1, \ldots, N_h - 1$, can be approximated [68] with, e.g. one of methods described in Sect. 2.6

$$\begin{aligned} &p\left(\hat{\mathbf{x}}(k+j+1)\,|\hat{\mathbf{x}}(k+j), u(k+j)\right) \\ &\approx \mathcal{N}\left(E\left(\hat{\mathbf{x}}(k+j+1)\right), \mathrm{var}\left(\hat{\mathbf{x}}(k+j+1)\right)\right), \\ &j = 0, 1, \ldots, N_h - 1. \end{aligned} \tag{4.49}$$

A more general stochastic MPC problem is formulated in [69–72], where a probabilistic formulation of the cost also includes the probabilistic bounds of the predicted variable. The MPC problem considered here is of a more special form since the cost function described with Eq. (4.49) includes the mean value of the random variable. Nevertheless, the approximate approach to the explicit solution can be extended to the more general case of a stochastic MPC problem formulation that is beyond the scope of this book, but which can be found in [25].

The disturbance-rejection NMPC problem based on a GP model is considered in [73], where the goal is to steer the state vector $\mathbf{x}(k)$ to the origin. The reference-tracking NMPC problem based on a GP model is considered in [25, 64, 67], where the goal is to have the state vector $\mathbf{x}(k)$ track the reference signal with $\mathbf{r}(k) \in \mathbb{R}^n$.

Here, we describe the reference-tracking NMPC problem based on a GP model, but the principle of disturbance-rejection NMPC is based on the same idea. In the problem formulation, the cost function is like the one used in [61]. Suppose that a full measurement of the state $\mathbf{x}(k)$ is available at the current time sample k. For the current $\mathbf{x}(k)$, the reference-tracking NMPC solves the following optimisation problem [25, 64]:

$$V^*(\mathbf{x}(k), \mathbf{r}(k), u(k-1)) = \min_{\mathbf{u}} J(\mathbf{u}, \mathbf{x}(k), \mathbf{r}(k), u(k-1)) \tag{4.50}$$

subject to $\mathbf{x}(k)$ and:

$$E\left(\hat{\mathbf{x}}(k+j)\right) - 2\sigma_{\hat{\mathbf{x}}(k+j)} \geq \mathbf{x}_{\min} \quad j = 1, \ldots, N_h, \tag{4.51}$$

$$E\left(\hat{\mathbf{x}}(k+j)\right) + 2\sigma_{\hat{\mathbf{x}}(k+j)} \leq \mathbf{x}_{\max} \quad j = 1, \ldots, N_h, \tag{4.52}$$

$$u_{\min} \leq u(k+j) \leq u_{\max} \quad j = 0, 1, \ldots, N_h - 1, \tag{4.53}$$

$$\Delta u_{\min} \leq \Delta u(k+j) \leq \Delta u_{\max} \quad j = 0, 1, \ldots, N_h - 1, \tag{4.54}$$

$$\begin{aligned} &\max\{\left\| E(\hat{\mathbf{x}}(k+N_h)) - 2\sigma_{\hat{\mathbf{x}}(k+N_h)} - \mathbf{r}(k)\right\|, \\ &\quad \left\| E(\hat{\mathbf{x}}(k+N_h)) + 2\sigma_{\hat{\mathbf{x}}(k+N_h)} - \mathbf{r}(k)\right\|\} \leq \delta, \end{aligned} \tag{4.55}$$

$$\Delta u(k+j) = u(k+j) - u(k+j-1), \quad j = 0, 1, \ldots, N_h - 1, \tag{4.56}$$

$$\begin{aligned} &p\left(\hat{\mathbf{x}}(k+j+1)\,|\hat{\mathbf{x}}(k+j),\, u(k+j)\right) \\ &\approx \mathcal{N}\left(E\left(\hat{\mathbf{x}}(k+j+1)\right), \operatorname{var}\left(\hat{\mathbf{x}}(k+j+1)\right)\right), \\ &j = 0, 1, \ldots, N_h - 1, \end{aligned} \tag{4.57}$$

where σ is the standard deviation, $\mathbf{u} = [u(k), u(k+1), \ldots, u(k+N_h-1)]$ and the cost function is:

$$\begin{aligned} J(\mathbf{u}, \mathbf{x}(k), \mathbf{r}(k), u(k-1)) &= \left\| E(\hat{\mathbf{x}}(k+N_h)) - \mathbf{r}(k)\right\|_{\mathbf{P}}^2 \\ &+ \sum_{j=0}^{N_h-1} \left[\left\| E(\hat{\mathbf{x}}(k+j)) - \mathbf{r}(k)\right\|_{\mathbf{Q}}^2 + \|\Delta u(k+j)\|_{\mathbf{R}}^2\right]. \end{aligned} \tag{4.58}$$

Here, N_h is a finite horizon and $\mathbf{P}$, $\mathbf{Q}$, $\mathbf{R}$ are squared and positive definite matrices. From a stability point of view it is desirable to choose δ in the terminal constraint described with Eq. (4.55) to be sufficiently small in accordance with the stability conditions [29] explained at the beginning of this subsection. If the prediction horizon N_h is large and the GP model has a small prediction uncertainty, then it is more likely that the choice of a small δ will be possible.

The MPC problem considered here has a more special form since the cost function described with Eq. (4.58) includes the mean value of the random variable. However, the approximate approach to the explicit solution of the problem, presented in the continuation, can be easily extended to the more general case of a MPC problem formulation where the optimisation is performed on the expected value of the cost function.

We introduce an extended state vector:

$$\tilde{\mathbf{x}}(k) = [\mathbf{x}^{\mathrm{T}}(k),\ \mathbf{r}^{\mathrm{T}}(k),\ u(k-1)]^{\mathrm{T}} \in \mathbb{R}^{\tilde{n}}, \quad \tilde{n} = 2n + 1. \tag{4.59}$$

Let $\tilde{\mathbf{x}}$ be the value of the extended state at the current time sample k. Then, the optimisation problem described with Eqs. (4.50)–(4.58) can be formulated in a compact form as follows [25, 64]:

$$V^*(\tilde{\mathbf{x}}) = \min_{\mathbf{u}} J(\mathbf{u}, \tilde{\mathbf{x}}) \quad \text{subject to} \quad \gamma(\mathbf{u}, \tilde{\mathbf{x}}) \leq 0, \tag{4.60}$$

where $\gamma(\mathbf{u}, \tilde{\mathbf{x}}) \leq 0$ represents the nonlinear constraints generally.

The NMPC problem defines an mp-NLP, since it is NLP in $\mathbf{u}$ parameterised by $\tilde{\mathbf{x}}$. An optimal solution to this problem is denoted by $\mathbf{u}_o = [u_o(k),\ u_o(k+1),\ \ldots,\ u_o(k+N_h-1)]$, and the control input value is chosen according to the receding-horizon policy $u(k) = u_o(k)$. Define the set of N_h-step feasible initial states as follows:

$$\mathcal{X}_f = \{\tilde{\mathbf{x}} \in \mathbb{R}^{\tilde{n}} |\ \gamma(\mathbf{u}, \tilde{\mathbf{x}}) \leq 0 \text{ for some } \mathbf{u} \in \mathbb{R}^{N_h}\}. \tag{4.61}$$

If δ in Eq. (4.55) is chosen such that at least one solution to the optimisation problem described with Eqs. (4.50)–(4.58) exists, then $\mathcal{X}_f$ is a nonempty set.

In parametric programming problems, we seek the solution $\mathbf{u}_0(\tilde{\mathbf{x}})$ as an explicit function of the parameters $\tilde{\mathbf{x}}$ in some set $\mathcal{X} \subseteq \mathcal{X}_f \subseteq \mathbb{R}^{\tilde{n}}$. The explicit solution allows us to replace the computationally expensive real-time optimisation with a simple function evaluation.

In general, an exact solution to the mp-NLP problem described with Eq. (4.60) cannot be found. However, there are some approximate approaches that can be used to solve it [25]. In this section, the principle of the computational method [64] for constructing a PWL approximate solution of the reference-tracking NMPC problem is described.

In general, the problem described with Eq. (4.60) can be nonconvex with multiple local minima. Therefore, it would be necessary to apply a good initialisation of the problem described with Eq. (4.60) so as to find a close-to-global solution. One possible way to obtain this is to find a close-to-global solution at a point $\mathbf{v}_0 \in \mathcal{X}_g$, where $\mathcal{X}_g \subset \mathcal{X}$ is a hyper-rectangular region. The close-to-global solution at the point $\mathbf{v}_0$ can be found by comparing the local minima corresponding to several initial guesses and then to use this solution as an initial guess for the neighbouring points $\mathbf{v}_i \in \mathcal{X}_g$, $i = 1, 2, \ldots, N_1$, i.e. to propagate the solution.

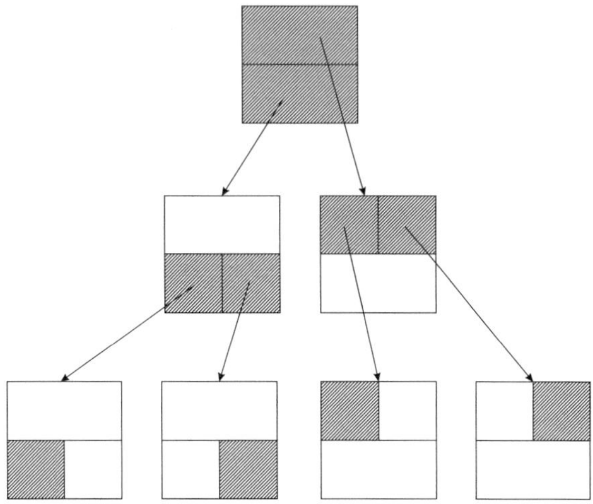

Fig. 4.23 $k-d$ tree partition of the state space used in the approximation

We restrict our attention to the hyper-rectangle $\mathcal{X} \subset \mathbb{R}^{\tilde{n}}$ where we seek to approximate the optimal solution $\mathbf{u}_o(\tilde{\mathbf{x}})$ to the problem described with Eq. (4.60). We require that the state space partition is orthogonal and can be represented as a so-called $k-d$ tree [18, 74] (Fig. 4.23). The main idea of the approximate mp-NLP approach is to construct a feasible PWL approximation $\hat{\mathbf{u}}(\tilde{\mathbf{x}})$ to $\mathbf{u}_o(\tilde{\mathbf{x}})$ on $\mathcal{X}$, where the constituent affine functions are defined on hyper-rectangles covering $\mathcal{X}$. In the case of convexity, it suffices to compute the solution of Eq. (4.60) at the $2^{\tilde{n}}$ vertices of a considered hyper-rectangle $\mathcal{X}_g$ by solving up to $2^{\tilde{n}}$ NLPs. In the case of non-convexity, it would not be sufficient to impose the constraints only at the vertices of the hyper-rectangle $\mathcal{X}_g$. One approach to resolving this problem is to include some interior points in addition to the set of vertices of $\mathcal{X}_g$ [65]. These additional points can represent the vertices and the facet centres of one or more hyper-rectangles contained in the interior of $\mathcal{X}_g$. Based on the solutions at all the points, a feasible local linear approximation $\hat{\mathbf{u}}_{ll}(\tilde{\mathbf{x}}) = \mathbf{K}_{ll}\tilde{\mathbf{x}} + \mathbf{g}_{ll}$ to the optimal solution $\mathbf{u}_o(\tilde{\mathbf{x}})$, valid for the whole hyper-rectangle $\mathcal{X}_g$, is determined. This is done by solving

$$\min_{\mathbf{K}_{ll},\, \mathbf{g}_{ll}} \sum_{i=0}^{N_1} \left(J(\mathbf{K}_{ll}\mathbf{v}_i + \mathbf{g}_{ll},\ \mathbf{v}_i) - V^*(\mathbf{v}_i) + \quad \alpha \left\|\mathbf{K}_{ll}\mathbf{v}_i + \mathbf{g}_{ll} - \mathbf{u}_o(\mathbf{v}_i)\right\|_2^2\right) \tag{4.62}$$

subject to

$$\gamma\,(\mathbf{K}_{ll}\mathbf{v}_i + \mathbf{g}_{ll},\ \mathbf{v}_i) \leq 0\ ,\ \ \forall \mathbf{v}_i \in \mathcal{V}^0, \tag{4.63}$$

where $\mathcal{V}^0$ is the finite set of points that represent the vertices and the facet centres of one or more hyper-rectangles contained in the interior of $\mathcal{X}_g$.

In Eq. (4.62), the parameter $\alpha > 0$ is a weighting coefficient.

Suppose that a state feedback $\hat{\mathbf{u}}_{ll}(\tilde{\mathbf{x}})$ which is feasible on $\mathcal{V}^0 \subseteq \mathcal{X}_g$ has been determined. Then, for the cost function approximation error in $\mathcal{X}_g$ we have:

$$\epsilon(\tilde{\mathbf{x}}) = \hat{V}(\tilde{\mathbf{x}}) - V^*(\tilde{\mathbf{x}}) \leq \epsilon_0 \ , \quad \tilde{\mathbf{x}} \in \mathcal{X}_g, \tag{4.64}$$

where $\hat{V}(\tilde{x}) = J(\hat{\mathbf{u}}_{ll}(\tilde{\mathbf{x}}), \tilde{\mathbf{x}})$ is the sub-optimal cost and $V^*(\tilde{\mathbf{x}})$ denotes the cost corresponding to the close-to-global solution $\mathbf{u}_o(\tilde{\mathbf{x}})$, i.e. $V^*(\tilde{\mathbf{x}}) = J(\mathbf{u}_o(\tilde{\mathbf{x}}), \tilde{\mathbf{x}})$. The reader is referred to [25] for more details.

The following iterative procedure that terminates a PWL approximate solution of the mp-NLP problem described with Eq. (4.60) is proposed in [65]:

1. Initialise the partition to the whole hyper-rectangle and mark the hyper-rectangle as unexplored.
2. Select one of the unexplored hyper-rectangles. If no such hyper-rectangle exists, terminate the design procedure.
3. Compute a solution to the optimisation problem of Eq. (4.60) at the centre point of the present hyper-rectangle. If the optimisation problem has a feasible solution, go to the next step. Otherwise, split the present hyper-rectangle into two hyper-rectangles by applying the heuristic rules described in [65], mark the new hyper-rectangles as unexplored, and return to the previous step.
4. Define a finite set of points including the vertices and some facet points of the present hyper-rectangle. Compute a solution to the optimisation problem described by Eq. (4.60) for each of the points. If the optimisation problem has a feasible solution at all these points, go to step 6. Otherwise, go to the next step.
5. Compute the size of the present hyper-rectangle using some metric. If it is smaller than some given tolerance, mark the hyper-rectangle as infeasible and explored and return to step 2. Otherwise, split the hyper-rectangle by applying the heuristic rules described in [65], mark the new hyper-rectangles as unexplored, and return to step 2.
6. Compute an affine state feedback as an approximation to be used in the present hyper-rectangle. If no feasible solution is found, split the hyper-rectangle into two hyper-rectangles by applying the heuristic rules described in [65]. Mark the new hyper-rectangles as unexplored and return to step 2.
7. Compute an estimate of the error bound in the present hyper-rectangle. If the error bound is less than some prescribed tolerance, mark the hyper-rectangle as explored and feasible, and return to step 2. Otherwise, split the hyper-rectangle by applying the heuristic rules described in [65], mark the new hyper-rectangles as unexplored and return to step 2.

The following algorithm represents compactly the described procedure for explicit NMPC design [25]:

Algorithm: EXPLICITNMPC($\mathcal{X}$)

---

procedure SPLIT($\mathcal{X}_g$)
split the current hyper-rectangle into two and mark both as unexplored
go to 2.

main
1.
initialise the partition to the whole solution hyper-rectangle
and mark it as unexplored
2.
select one of the unexplored hyper-rectangles
if no such hyper-rectangle exists
  then end the algorithm
3.
compute a solution at the centre
if not feasible
  then SPLIT($\mathcal{X}_g$)
4.
define a finite set of points
if solutions in the points are feasible
  then go to 6.
5.
compute the size of the solutions according to a metric
if size > tolerance
  then SPLIT($\mathcal{X}_g$)
mark the set as infeasible and explored and **go to** 2.
6.
compute an affine state feedback as an approximation
if not feasible
  then SPLIT($\mathcal{X}_g$)
7.
compute an estimate of the error bound
if error bound > tolerance
  then mark region as feasible and explored
  else SPLIT($\mathcal{X}_g$)

The presented approximate mp-NLP approach is a practical computational method to handle non-convex mp-NLP problems. It does not necessarily lead to guaranteed properties, like feasibility and closed-loop stability, but when combined with verification and analysis methods it gives a practical tool for the development and implementation of explicit NMPC [65]. It should also be noted that in contrast to the conventional MPC based on real-time optimisation, the explicit MPC makes the rigorous verification and validation of the controller performance much easier [66]. Hence, problems due to a lack of convexity and numerical difficulties can be addressed during the design and implementation.

While the explicit solution allows us to replace the computationally expensive real-time optimisation with a simple function evaluation that enables fast online operation, it requires more effort during the offline computation. The computational complexity of the algorithm for the design of the explicit NMPC increases with the input and state dimensions and the 'curse of dimensionality' applies. Nevertheless, it is important to note that it is the online computational complexity that counts and not the offline, for which enough time is usually available. In any case, there are methods for complexity reduction of the approximate explicit NMPC solution, e.g. [75].

Example 4.3 This example is adopted from [64]. Consider the system described by the following nonlinear state space model:

$$x(k+1) = x(k) - 0.5\tanh(x(k) + u^3(k)) + \nu_1(k) \tag{4.65}$$

where ν_1 is a white noise with variance 0.0025 and zero mean.

This dynamic system was identified with a GP model using the regressors $x(k)$ and $u(k)$ and the output value $x(k+1)$.

The described mp-NLP approach is applied to design an explicit reference-tracking GP-NMPC controller for the system described with Eq. (4.65) based on the obtained GP model. In the GP-NMPC problem formulation (Eq. (4.50)), the predicted state $\hat{x}(k+j+1)$ of the system described with Eq. (4.65) is used. This prediction is obtained in the following way. First, we obtain the prediction of $\hat{y}(k+j+1)$ from the GP model of the system described with Eq. (4.65):

$$\begin{aligned}&\hat{y}(k+j+1)|\hat{y}(k+j), u(k+j) \sim \\ &\mathcal{N}(E(\hat{y}(k+j+1)), \mathrm{var}(\hat{y}(k+j+1))); \\ &j = 0, 1, \ldots, N_h - 1.\end{aligned} \tag{4.66}$$

Then, the predicted $\hat{x}(k+j+1)$ is:

$$\hat{x}(k+j+1) = \hat{y}(k+j+1) + m(x), \tag{4.67}$$

where $m(x)$ is the mean value of the state of the system described with Eq. (4.65) obtained for the generated control signals, i.e. $m(x) = \frac{1}{M}\sum_1^M x$. The iterative, multistep-ahead prediction is made by feeding back, at each time step, the predictive mean only. The following control input and rate constraints are imposed on the system:

$$-1 \le u \le 1; -0.5 \le \Delta u \le 0.5 \tag{4.68}$$

The prediction horizon is $N_h = 8$ and the terminal constraint is:

$$\max\left(||E(\hat{x}(k+N_h)) - 2\sigma_{\hat{x}(k+N_h)} - r(k)||,\right.$$
$$\left.||E(\hat{x}(k+N_h)) + 2\sigma_{\hat{x}(k+N_h)} - r(k)||\right) \leq \delta \tag{4.69}$$

where $\delta = 0.015$. The weighting matrices in the cost function described with Eq. (4.58) are $Q = 10$, $R = 1$, $P = 10$. The GP-NMPC minimises the cost function described with Eq. (4.58) subject to the GP model of Eqs. (4.66)–(4.67) and the constraints described with Eqs. (4.68) and (4.69). The formulated GP-NMPC problem results in the optimisation problem described with Eq. (4.60) with 8 optimisation variables and 33 constraints. One internal region $\mathcal{X}_g^1 \subset \mathcal{X}_g$ is used. This results in the problem described with Eq. (4.62), which has 32 optimisation variables and 285 constraints. In Eq. (4.62) the value $\alpha = 10$ is chosen. The approximation tolerance is selected in the following way:

$$\epsilon = \max(\epsilon_a, \epsilon_r \min_{\tilde{\mathbf{x}} \in \mathcal{X}_g} V^*(\tilde{\mathbf{x}})) \tag{4.70}$$

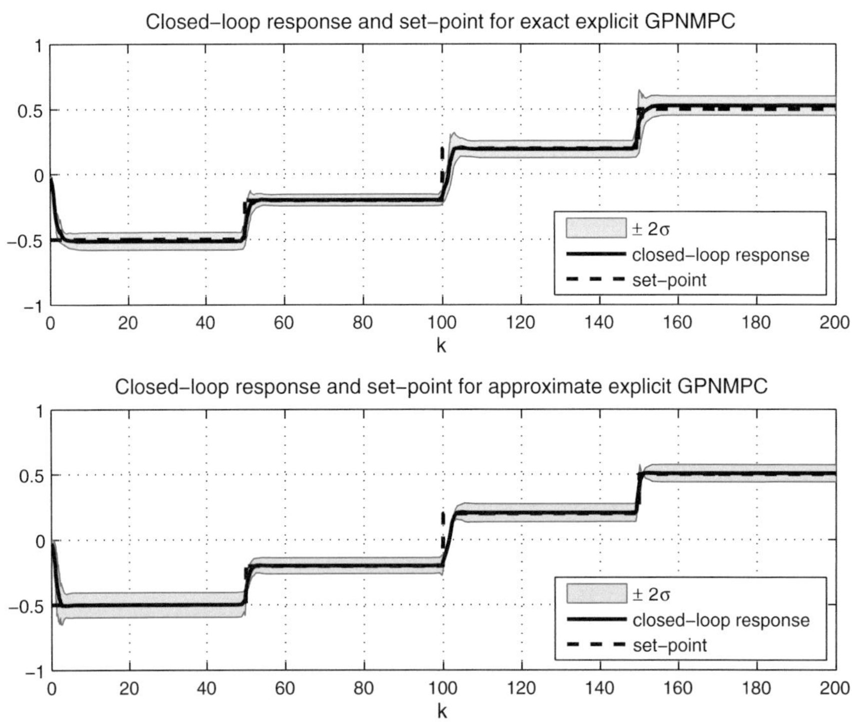

Fig. 4.24 The closed-loop response on the set-point tracking with the 95 % confidence interval of the state variable predicted with the GP model. The closed-loop response with the exact explicit GP-NMPC is shown in the *upper graph*, the closed-loop response of the explicit GP-NMPC is shown in the *lower graph*

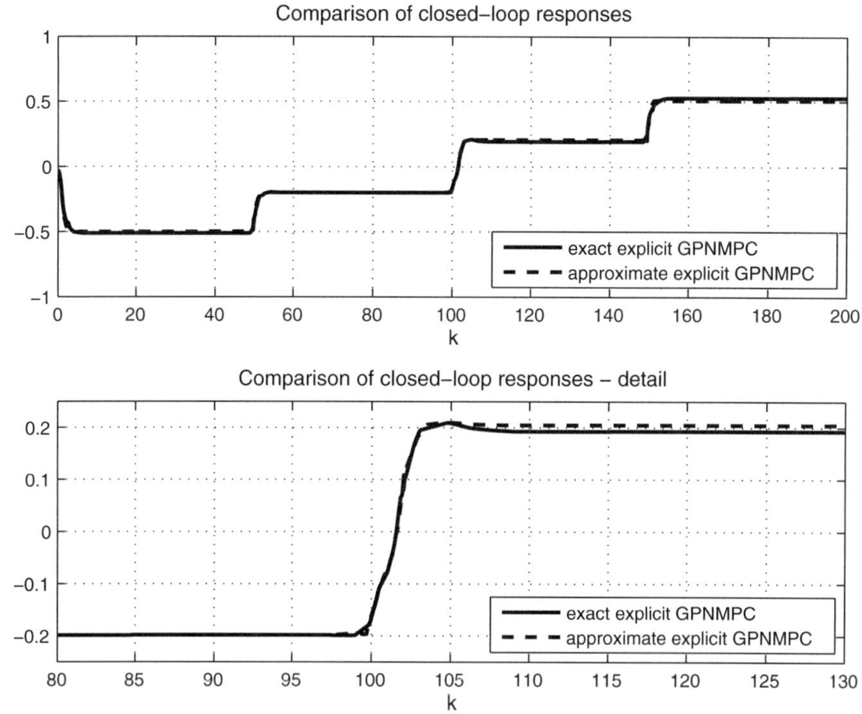

Fig. 4.25 The comparison of closed-loop responses with the exact and approximate explicit GP-NMPC. The *dashed line* is with the approximate explicit GP-NMPC, the *full line* is with the exact GP-NMPC. The detail of comparison is given in the *lower graph*, while the entire response is given in the *upper graph*

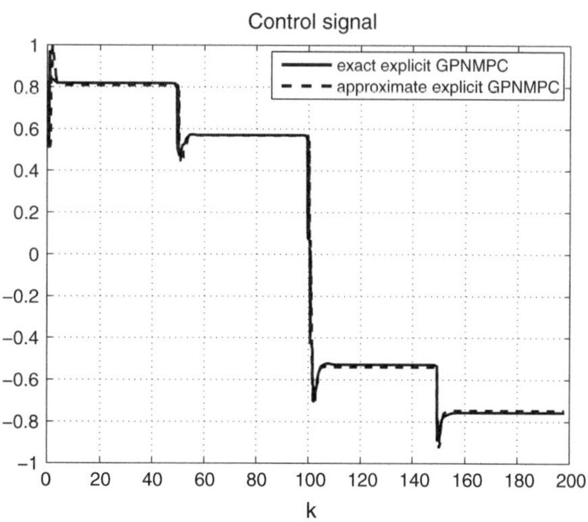

Fig. 4.26 The control input. The *dashed line* is with the approximate explicit GP-NMPC and the *full line* is with the exact GP-NMPC

where $\epsilon_a = 0.005$ and $\epsilon_r = 0.05$ are the absolute and the relative tolerances, respectively. The extended state vector is $\tilde{\mathbf{x}}(k) = [x(k), r(k), u(k-1)]^{\mathrm{T}} \in \mathbb{R}^3$, which leads to a three-dimensional state space to be partitioned. The latter is defined by $\mathcal{X} = \{[-1.2, 1.2] \times [-0.7, 0.7] \times [-1, 1]\}$. The partition has 1419 regions and 18 levels of search in the $k-d$ tree. In total, 24 arithmetic operations are needed in real time to compute the control input value (18 comparisons, 3 multiplications and 3 additions). The performance of the closed-loop system was simulated for the following set-point and reference change ($r(k) = w(k)$):

$$r(k) = -0.5, k \in [0; 50]; r(k) = -0.2, k \in [51; 100]$$
$$r(k) = 0.2, k \in [101; 150]; r(k) = 0.5, k \in [151; 200]$$

and the initial conditions for the state and control variable $x(0) = 0$ and $u(0) = 0$, respectively. The resulting closed-loop response is shown in Figs. 4.24, 4.25 and 4.26.

The results show that the exact and the approximate solutions are very similar.

4.4 Adaptive Control

An adaptive controller [51] is one that continuously adapts to what is considered as a changing process. Such adaptive controllers emerged in the early 1960s. At first, these controllers were mainly adapting themselves on the basis of linear models with changing parameters. Since then several authors have proposed the use of nonlinear models as a basis on which to build nonlinear adaptive controllers. These are meant for the control of time-varying, nonlinear systems or of time-invariant nonlinear systems that are modelled as a varying, simplified, nonlinear model.

Various divisions of adaptive control structures are possible. One possible division [51] is into open-loop and closed-loop adaptive systems.

Open-loop adaptive systems, also *feedforward adaptive systems*, are the gain-scheduling or parameter-scheduling controllers that are described in Sect. 4.5.

Closed-loop adaptive systems, also *feedback adaptive systems*, may be further divided into dual and non-dual adaptive systems.

Dual-adaptive systems are those where the optimisation of the information collection and the control action are pursued at the same time. The control signal should ensure that the system output signal cautiously tracks the desired reference value and at the same time excites the plant sufficiently to accelerate the identification process. Cautious control means that in the case of an uncertain model of a stable process, the control signal should be smaller, i.e. more cautious, than the control signal with a certain model and after adaptation. Adaptive control systems with these two properties of cautious control and excitation are referred to as *adaptive dual control* [76, 77]. Since the specifications of the closed-loop system are such that the output signal should normally vary as little as possible, these two properties are in conflict.

The solution to the dual-control problem is based on dynamic programming and the resulting functional equation is often the Hamilton-Jacobi-Bellman equation or its discrete-time counterpart the Bellman equation, which were already mentioned in relation to the optimal control problem in Sect. 4.2. This is a difficult task and only a small number of such controllers have been developed [51].

A similar manifestation of model adaptation and control optimisation that are pursued at the same time can also be observed in the method of reinforcement learning when applied in optimal control, where the exploration/exploitation tradeoff is met [20].

The difficulties in finding the optimal solution for adaptive dual control led to suboptimal dual adaptive controllers [76, 77] obtained either by various approximations or by reformulating the problem, e.g. modifications of the loss function. Such a reformulated, adaptive dual-control problem is when a special cost function is considered, which consists of two added parts: control losses and an uncertainty measure. This is appealing for applications with a GP model that provides measures of uncertainty.

The way in which the identified system in adaptive dual control becomes exited is a reason for concern, due to safety concerns relating to control implementations in practice.

A possible adaptive dual-control principle using GP models is described in [78].

In general, many adaptive controllers are based on the separation principle [77] that implies a separate estimation of the system model, i.e. system parameters, and the application of this model for control design. The separation principle holds, for instance, in the case of Gaussian noise, when the process is linear and the cost function is a quadratic function. When the identified model is presumed to be the same as the true system for control design and adaptation, then an adaptive controller of this kind is said to be based on the certainty-equivalence principle, and such an adaptive controller is referred to as *non-dual adaptive controller*. The control actions of non-dual adaptive controller do not take any active actions that would influence the uncertainty.

Non-dual-adaptive systems are divided into:

- model-reference adaptive systems, where the controller adapts based on the difference between the output responses of the reference model and the process;
- model identification adaptive systems, where the certainty-equivalence principle is used;
- iterative learning for control, where the control works in a repetitive mode;
- other adaptive systems.

The following sections describe those adaptive controllers that use a GP model of dynamic systems or a GP model represents a significant part of the control system: gain scheduling is described in Sect. 4.5, model identification adaptive control is described in Sect. 4.6 and iterative learning control is described in Sect. 4.7.

4.5 Gain Scheduling

The *gain-scheduling* method is probably the most widespread nonlinear control design method. It has been successfully applied in fields ranging from process control to aerospace engineering. The basic idea behind the approach is called the divide-and-conquer method, where a nonlinear system is divided into local subsystems that are modelled as linear dynamic systems. A linear control problem is then solved for each of these subsystems. The global control solution, called gain-scheduling control, is afterwards put together from partial local solutions. Overviews of the gain-scheduling method and its applications can be found in [79–81].

Gain scheduling is an effective and economical method of industrial control, whereby changes in the operating point lead to corresponding variations in the parameters of the linearised models of the plant about these operating states. Its frequent use in industrial practice has made it indispensable to the control-design engineer.

Gain scheduling as a control-design method is very closely related to the divide-and-conquer modelling techniques [28, 82] called multimodel systems, local model networks, linear parameter-varying systems, Takagi-Sugeno fuzzy models, etc. The version of gain scheduling with a finite number of local controllers, known as blended multimodel control systems [28, 83–85] which is a commonly used type of multimodel control system, is closely related to the concepts of Takagi-Sugeno fuzzy controllers as well as to the concept of controller switching.

The divide-and-conquer method is based on a series-expansion linearisation of a nonlinear system to be used for the control design about a single trajectory or equilibrium point. Consider the nonlinear system,

$$\begin{aligned}\dot{\mathbf{x}}(t) &= f_t(\mathbf{x}(t), u(t)),\\ y(t) &= g_t(\mathbf{x}(t), u(t))\end{aligned} \tag{4.71}$$

where $\mathbf{x} \in \mathbb{R}^n$, $u \in \mathbb{R}$, $y \in \mathbb{R}$ are the vectors of the states, the input and the output signals, respectively. Let $(\bar{\mathbf{x}}(t), \bar{u}(t), \bar{y}(t))$ denote a specific trajectory of the nonlinear system, which could also simply be an equilibrium operating point, in which case $\bar{\mathbf{x}}$ is constant. Neglecting higher order terms, it follows from the series-expansion theory that the nonlinear system, Eq. (4.71), may be approximated, locally to the trajectory, $(\bar{\mathbf{x}}(t), \bar{u}(t), \bar{y}(t))$, by the linear time-varying system

$$\delta\dot{\mathbf{x}} = \nabla_{\mathbf{x}} f_t(\bar{\mathbf{x}}, \bar{u})\delta\mathbf{x} + \nabla_u f_t(\bar{\mathbf{x}}, \bar{u})\delta u, \tag{4.72}$$

$$\delta y = \nabla_{\mathbf{x}} g_t(\bar{\mathbf{x}}, \bar{u})\delta\mathbf{x} + \nabla_u g_t(\bar{\mathbf{x}}, \bar{u})\delta u, \tag{4.73}$$

where

$$\begin{aligned}\delta\mathbf{x} &= \mathbf{x} - \bar{\mathbf{x}},\\ \delta u &= u - \bar{u},\\ y &= \delta y + \bar{y}.\end{aligned} \tag{4.74}$$

The series-expansion linearisations associated with a family of equilibrium points $(\bar{\mathbf{x}}_e, \bar{u}_e, \bar{y}_e)$ form a series-expansion linearisation family describing a nonlinear system to be used for control design in the family of equilibrium points.

A linear, time-invariant controller is then designed to ensure the appropriate closed-loop performance when employed with the process linearisation. This procedure is repeated for the family of equilibrium operating points, covering the envelope of operation, whilst ensuring that the linear-controller designs have compatible structures; for example, when a smoothly gain-scheduled controller is required, the linear-controller designs are selected to permit smooth interpolation, in some appropriate manner, between the designs. In addition to the synthesis of a family of linear controllers, the gain-scheduling design approach requires the determination of a suitable nonlinear controller realised from the family of linear controllers.

The traditional gain-scheduled controller, which is adjusted with reference to an externally measured vector of variables, $\varrho(\mathbf{x}(t), u(t))$, has the form

$$\begin{aligned} \dot{\mathbf{x}} &= \mathbf{A}_c(\varrho(t))\mathbf{x} + \mathbf{b}_c(\varrho(t))u, \\ y &= \mathbf{c}_c(\varrho(t))\mathbf{x} + d_c(\varrho(t))u, \end{aligned} \tag{4.75}$$

where $\mathbf{A}_c, \mathbf{b}_c, \mathbf{c}_c$ and d_c are the controller system matrix, the input, the output and the input-output vectors, respectively. The dynamic properties change with the so-called scheduling vector $\varrho(\mathbf{x}(t), u(t))$. But, provided [86] that the rate of change is not too rapid, then the dynamic properties of the time-varying controller described with Eq. (4.75) are similar to those of the linear controllers obtained by 'freezing' the value of $\varrho(t)$. This means that the nonlinear controller inherits the dynamic properties of the family of linear controllers.

A more thorough explanation of gain scheduling can be found in the literature [4, 80, 81, 86]. A block scheme showing the general principle of gain-scheduling control is given in Fig. 4.27.

A number of nonlinear identification methods, including conventional nonparametric GP-model identification methods, provide models that can only be used with model-based predictive control. The FSGP model is, on the other hand, a parametric model with probabilistic parameters that can be used for a wider range of control-design methods, but those utilising black-box GP models.

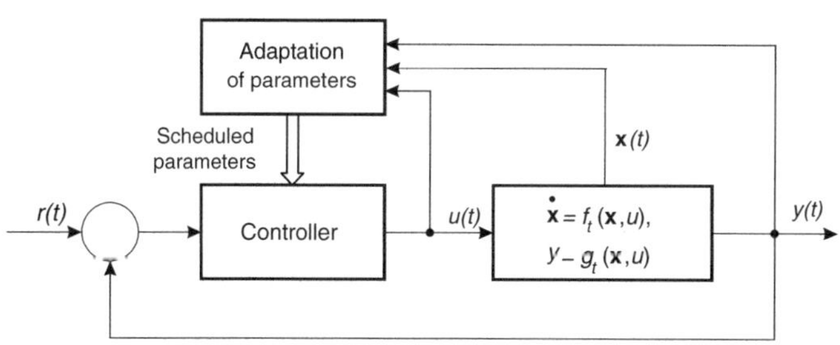

Fig. 4.27 General block scheme of the closed-loop system with a gain-scheduling controller

One possible control-design approach is gain-scheduling control design. In this case the local controllers are designed for a selected local model of the process. Gain-scheduling control based on a GP model, i.e. on the FSGP model introduced in Chap. 3, is described in [87, 88].

The selection of local process models for control design depends on the region where the closed-loop dynamics is expected and is, in general, not the same as the set of local models used for the process modelling. It is sensible to keep the system where its model is good, i.e. where the variances of the local models' parameters are small. The variances of the GP models contained in the FSGP model provide this information. The parameters of the local controllers depend on the same scheduling variables as the associated local model parameters of the process.

The *stability* of the closed-loop nonlinear system can be analysed based on the theory from [89] in the framework of VBL, explained in Sect. 3.3.2, and can be summarised as follows [81].

The relationship between the solution to a nonlinear system and the solutions to the members of the associated velocity-based linearisation family can be used to derive conditions relating to the stability of a nonlinear system to the stability of its velocity-based linearisations. General stability analysis methods, such as the small-gain theory and the Lyapunov theory [4], can be applied to derive velocity-based stability conditions for equilibrium points. Furthermore, the bounded-input bounded-output (BIBO) stability of the nonlinear system described with Eq. (4.71) is guaranteed, provided the members of its velocity-based linearisation family are uniformly stable. The unboundedness of the state $\mathbf{x}$ implies that its derivation $\dot{\mathbf{x}}$ is unbounded, under the assumption that the input signal u is bounded, and the class of inputs and initial conditions is restricted to limit the rate of evolution of the nonlinear system to be sufficiently slow [89].

Provided that the rate of evolution is sufficiently slow, the nonlinear system inherits the stability robustness of the members of the velocity-based linearisation family [89]. This velocity-based result involves no restriction on near-equilibrium operation other than that implicit in the slow variation requirement; for example, for some systems where the slow variation condition is automatically satisfied, the class of allowable input-signal values and initial conditions is unrestricted and the stability analysis is global.

The basic idea of modelling and control design is illustrated in the next simple example.

Example 4.4 The gain-scheduling control design is illustrated with the second-order discrete nonlinear system used in Example 3.5 and described with Eq. (3.73):

$$\begin{aligned} y(k+1) &= f\,(y(k), y(k-1), u(k)) \\ &= k_1 y(k) + k_2 y^2(k) + k_3 y(k-1) + k_3 u(k) y(k) + k_4 u(k), \end{aligned}$$

with the constants $k_1 = 0.893$, $k_2 = 0.0371$, $k_3 = -0.05$, $k_4 = 0.157$ and the sampling time $T_s = 1$ s.

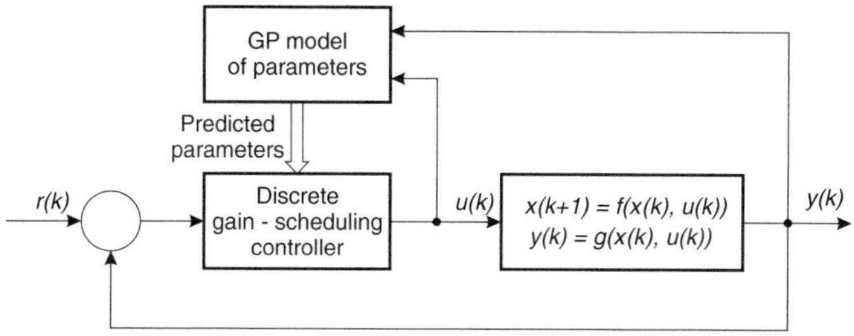

Fig. 4.28 Block scheme of the closed-loop system with a gain-scheduling controller, with the design based on the FSGP model

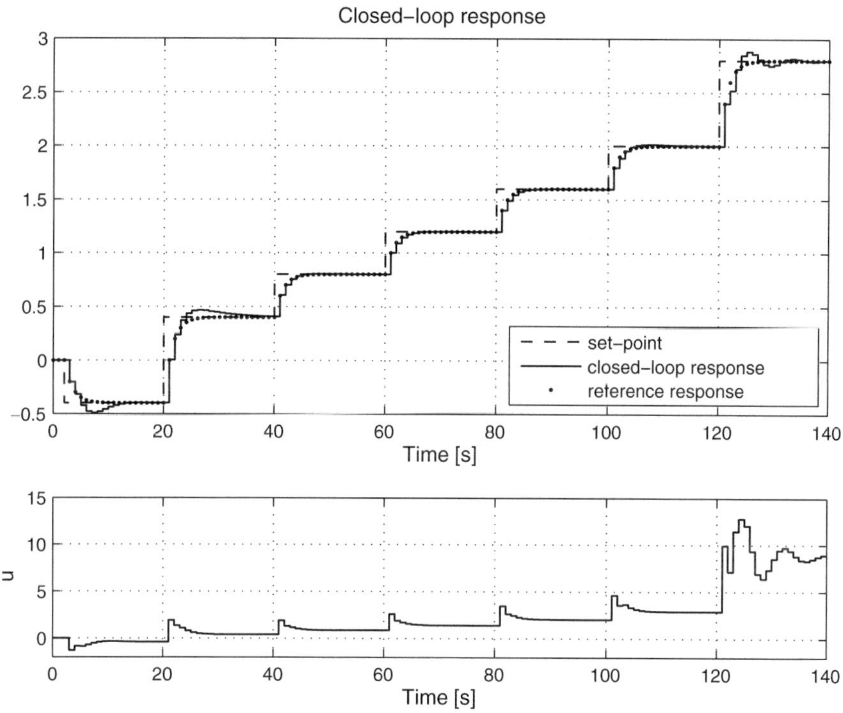

Fig. 4.29 Closed-loop response to the set-point changes after zero-order-hold on the output and a comparison with the reference response. The region between 60 and 100 s is the region with low uncertainty of the used model from Example 3.5

The development of a fixed-structure GP model for the process described with Eq. (3.73) is described in Example 3.5, where the GP models are identified to model the variable parameters of the selected linear structure.

A possible application of the developed model is control design. A controller that is suitable for LPV models is a gain-scheduling controller or a local controller network. GP models with varying parameters mean that local controllers can also be designed at different points to the points where the local models were identified, or all of the varying-parameters models can be incorporated into the controller, resulting in a gain-scheduled controller. GP models become a scheduling or adapting mechanism for varying the parameters of the scheduled controller.

The block scheme of the closed-loop system applied in our case is shown in Fig. 4.28.

The response of a closed-loop system with a second-order compensatory controller realised using VBL that can also be interpreted as a discrete gain-scheduled PID controller is given in Fig. 4.29, and compared with a reference response, which is defined as the first-order response with a unity gain and a time constant of 1.5 s.

The operation of the closed-loop system is first pursued in the region where the FSGP model used for the control design has a higher uncertainty. Afterwards it

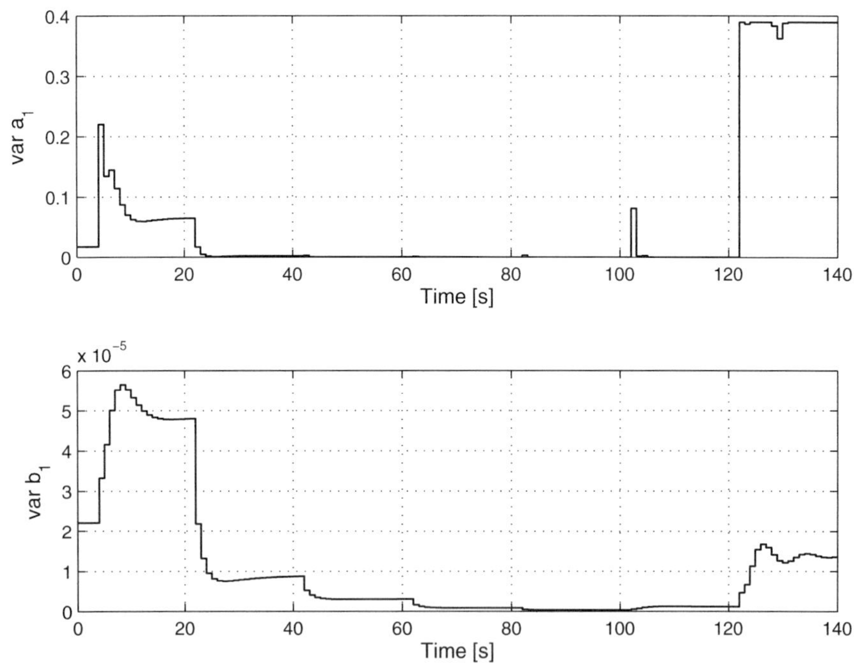

Fig. 4.30 Predicted variance of the parameters a_1 (*top*) and b_1 (*bottom*) during a simulation of the closed-loop system. The region between 60 and 100 s is the region of the used model with low uncertainty

operates in the region with low uncertainty of the used model and towards the end it again operates in the region with a high uncertainty of the used model. It is clear from Fig. 4.29 that the closed-loop response at the beginning and at the end does not match the reference response very well, as it does later on. Due to the integral action of the gain-scheduling controller, the mismatch between the closed-loop response (with the zero-order-hold) and the reference response is relatively minor. Nevertheless, the mismatch would increase when moving away from the region with high trust in the used model.

The online calculated variances of the FSGP model parameters, which are shown in Fig. 4.30, indicate when the closed-loop system operates in the region with the confident model and when it is out of the region.

4.6 Model Identification Adaptive Control

A block scheme showing the general principle of model identification adaptive control is given in Fig. 4.31.

When using the GP model for adaptive control, the GP model is identified online and this model is used in the control algorithm. It is a good idea for the advantages of GP models to be considered in control design, which relates GP model-based adaptive control at least to the suboptimal adaptive dual-control principles.

The uncertainty of the model predictions obtained with the GP model are dependent, among other factors, on the local training-data density, and the model complexity is automatically related to the amount and distribution of the available data—more complex models need more evidence to make them more trustworthy. Both aspects are very useful in sparsely populated transient regimes. Moreover, since weaker prior assumptions are typically applied in a nonparametric model, the bias is typically smaller than in parametric models.

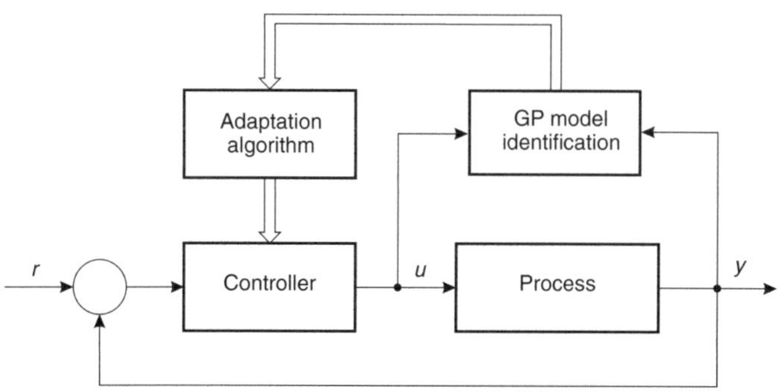

Fig. 4.31 General block scheme of the closed-loop system with adaptive controller

The above ideas are indeed related to the work carried out on adaptive dual control, where the main effort has focused on the analysis and design of adaptive controllers based on the use of the uncertainty associated with the parameters of models with a fixed structure [53, 76].

The major differences in the so far published adaptive systems based on GP models are in the way that the online model identification is pursued.

Increasing the size of the covariance matrix, i.e. the 'blow-up model', with in-streaming data and repeating model optimisation is used in [52–54, 56, 90], where more attention is devoted to control algorithms and their benefits based on the information gained from the GP model and not so much on the model identification itself.

Another adaptive-control-algorithm implementation is control with feedback for cancelling the nonlinearities, already described in Sect. 4.1 with online training of an inverse model. This type of adaptive control with the covariance matrix increasing with in-streaming data is described in [15, 91]. Two types of online training for the mentioned feedforward contained control are described in [92, 93]. The first type is with the moving-window strategy, where the old data is dropped from the online trained model, while the new data is accommodated. The second type of training accommodates only new data with sufficient information gain.

Similar adaptive control strategies applied in robot control can also be found in [94, 95], while adaptive control based on local GP models is described in [96]. The combination of GP model identification with conventional control algorithms is reported in [97].

In contrast to all the referenced adaptive controllers, the adaptive-control-system principle described in [98] is based on the evolving GP model described in Sect. 2.5.3.

The basic idea of control based on the evolving-system model [98] is that the process's GP model evolves with in-streaming data and the information about the system from the model is then used for its control. One option is that the information can be in the form of a GP model prediction for one or several steps ahead, which is then used to calculate the optimal control input sequence for the controlled system, e.g. for MPC control.

Different possibilities exist for the evolving GP model, depending on the level of changes we accommodate in the evolving-system model, as described in Sect. 2.5.3. On the other hand, various control algorithms can also be used, depending on the GP model or the closed-loop requirements.

Closed-loop *stability* is also a very important issue in the case of adaptive control. GP modelling is a computational-intelligence-based method. The GP model is probabilistic and nonparametric, and the conventional analytical methods for closed-loop analysis do not apply directly, except in special cases, e.g. a GP model with a linear covariance function under some assumptions. Due to the lack of published analytical results for closed-loop stability when GP models are involved, computer simulations present the main general-purpose analysis tool for closed-loop systems with GP model-based adaptive control.

A lot of GP model-based adaptive control algorithms from the referenced publications are based on the adaptive minimum-variance controller. One of the reasons for this is that the adaptive minimum-variance controller explores the variance that is readily available with GP model prediction. An example of such an adaptive control application, though with increasing dimensions of the covariance matrix with in-streaming data, is given in [90] and can be considered as an adaptive model predictive control.

An illustrative example that shows the operation of the minimum-variance control based on an evolving GP model is given next.

Example 4.5 Adaptive control with an evolving GP model

Consider the nonlinear dynamic system [17] used in Example 4.1 and described by

$$y(k+1) = \frac{y(k)}{1 + y^2(k)} + u^3(k) + \nu, \tag{4.76}$$

where u is the system's input signal, y is the system's output signal, ν is the white noise of a normal distribution with a standard deviation of 0.1 that contaminates the system's response and the sampling time is one second. The nonlinearity in the region of interest for the benchmark system is indicated by the grey mesh in Fig. 4.32.

The requirement of closed-loop control is that it follows the set-point, depicted in Fig. 4.33, as closely as possible. We start off with the empty active set of a GP model and with some default hyperparameter values $\ln \boldsymbol{\theta} = [0; 0; 1; -1]$, which are quite different compared to the optimal ones. The set-point signal is a combination of periodic pulses in three different regions. The first region is between 0.5 and 1.5, the

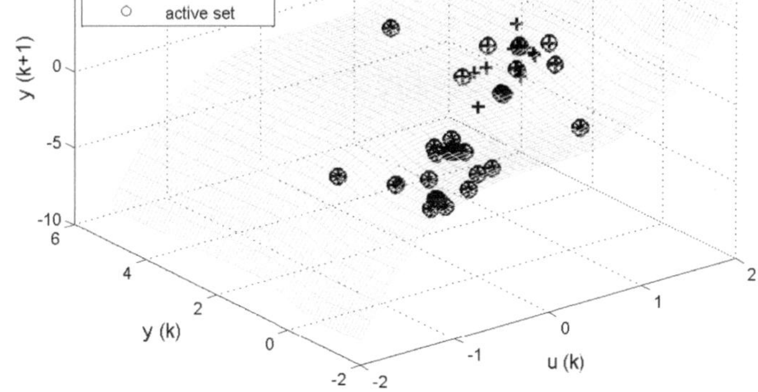

Fig. 4.32 Observed data and the most informative data—active set shown on the surface of a selected nonlinear system. The *gray mesh* denotes the nonlinear mapping of the system to be controlled, the *pluses*, *crosses* and *stars* denote first, second and third regions, respectively, while *circles* denote the active set

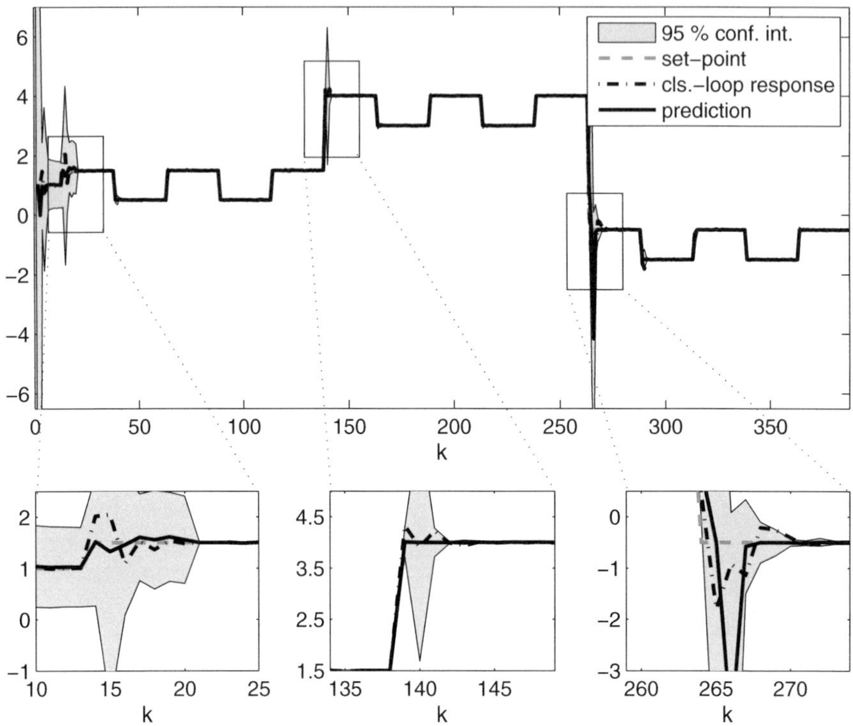

Fig. 4.33 Simulation of a controller based on an evolving GP model. The *grey dashed line* denotes the set-point signal and the *dash-dot line* denotes the output signal of the system, while the *solid line* denotes the mean value of the prediction and the *gray band* denotes the 95 % prediction confidence interval based on the current GP model

second one between 3 and 4, and the last one between −0.5 and −1.5. The priority for such a signal is to show that the proposed approach for a control system based on evolving GP models is able to learn from scratch, without any prior model, and to update with respect to the changes in the dynamics. The data stream contains only 388 data points, shown in Fig. 4.33, which serves for the demonstration requirements. We pre-set the maximal active size to 50 data points. The used control cost function is a variation of the minimum-variance cost function from Eq. (4.37)

$$J(k) = [r(k) - [y(k-1) - E(\hat{y}(k-1))] - E(\hat{y}(k))]^2. \qquad (4.77)$$

The term $y(k-1) - E(\hat{y}(k-1))$ is used to make the control algorithm insensitive to errors in the steady-state gain by subtracting the discrepancy between the latest plant output value and the latest, most likely output value of the model.

As the controller has no prior knowledge about the system, the system's output signal oscillates at the beginning, Fig. 4.33. Nevertheless, the controller observes enough data to successfully, but with some overshoot, follow the first step. Afterwards, the controller easily follows the set-point signal, even in the second and the third regions, where the nonlinearity is locally different. The complete nonlinearity, including all three regions, can be seen in Fig. 4.32, where pluses, crosses and stars denote the first, second and third regions, respectively. However, at the beginning of these regions some overshoots also appear, but the output signal quickly settles down.

The final active set is shown in Fig. 4.32. The most informative data points, selected from in-streaming data with the proposed evolving method, are denoted with circles. It is clear that the selected data points are evenly distributed through all the nonlinear space, which indicates that the proposed method successfully adapts the GP model according to the operating regions. The times when the GP model is updated are depicted in Fig. 4.34 as dashed light-grey lines. It is clear that most updates occur, as expected, during changes of the set-point signal, especially during changes in the dynamics. The situation is similar for the hyperparameter values, whose traces through the process are also shown in Fig. 4.34, denoted as solid lines. It is clear the hyperparameter values are mostly changing in a region of the first three steps when most of the new information about the system is obtained. Once the near-optimal values are reached, the hyperparameter values change on a much smaller scale. Fig. 4.35 shows the corresponding optimal control signal u.

The main purpose of this implementation of closed-loop control is the adaptation of the model according to the system's dynamics. Therefore, once enough data about the system is obtained, the controller can easily follow the set-point signal and adapts the GP model. But the controller can be further improved to somehow explore an unknown space, especially at the beginning of the process or for any other cases when it is not possible to follow the set-point signal due to a lack of information about the dynamics. With such an improvement, the controller is able to follow almost arbitrary changes in the process dynamics.

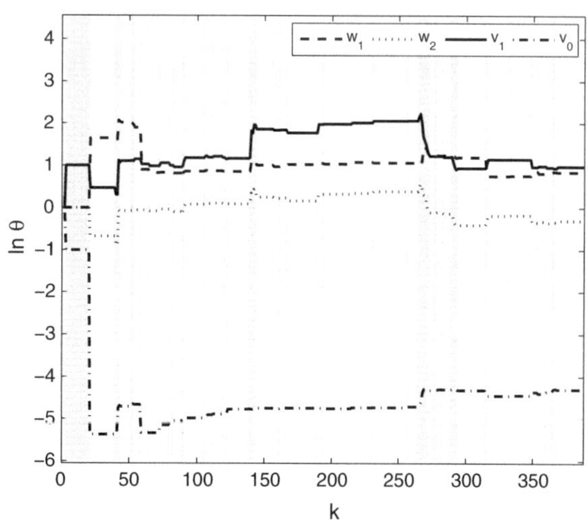

Fig. 4.34 Traces of the hyperparameter values changing over time and the active set updates. The different, *black lines* denote the hyperparameter values and the *gray lines* denote the time instants when the GP model was updated with new data

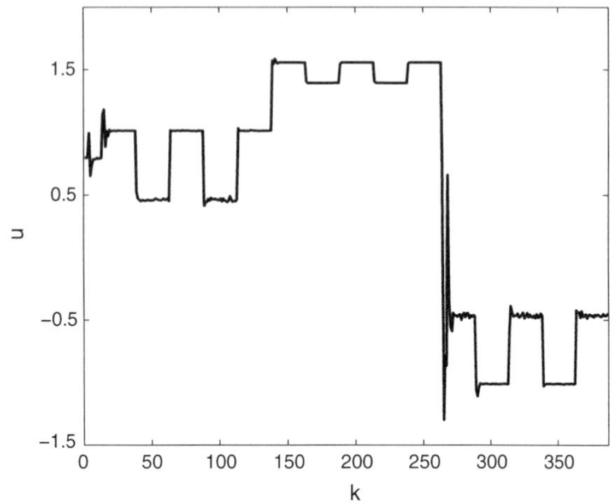

Fig. 4.35 Optimal control signal u based on the minimum-variance controller using the GP model's predictions

4.7 Control Using Iterative Learning

The iterative learning of a control signal, control sequence or control law comprises a family of methods used to improve the transient response and the performance or to establish a stable response of systems, processes or devices. They can also be named methods for controller auto-tuning. Such concepts in control theory are, for example, iterative learning control [99] and iterative feedback tuning [100]. Both methods are intended to improve the performance of the controlled system based on episodes or trials of the system's operation. The iterative learning control method generates a feedforward input signal to achieve a given desired trajectory by the iteration of experiments. The iterative feedback tuning method, on the other hand, adjusts the design parameter of the feedback controller via the experiments.

The concept of learning in control is often found in the field of machine learning as *reinforcement learning* used for control [20]. Reinforcement learning is concerned with how software agents map situations to actions in an environment so as to maximise some notion of reward. It belongs to the field called approximate dynamic programming.

Reinforcement learning searches for a policy that is good, based on past experience. If it is assumed that the problem being studied is episodic, and that no matter what course of actions the agent takes termination is inevitable, then the expectation of the total reward is well-defined, for any policy and any initial distribution over the states. Here, a policy refers to a mapping that, in general, assigns some probability distribution over the actions to all possible histories. In control engineering this mapping is called control law. The problem then is to specify an algorithm that can be used to find a policy, i.e. a control law, with the maximum expected return. The variety of policies may be extremely large.

To circumvent the problem of a large number of policies, some structure can be assumed and also samples generated from one policy to influence the estimates made for another are allowed. There are many approaches for achieving this. Dynamic programming achieves a more efficient search of a policy through the concept of a value function. The value function is the best possible value of the objective, written as a function of the system's state.

The utility of reinforcement learning goes far beyond control law learning and is used in game theory, information theory, etc. Nevertheless, if we focus on control systems then reinforcement learning controllers are stochastic search engines and can be considered as iterative learning algorithms. They evaluate the outcome of an action after the action and its effect are over. An early example of using a GP model for control using reinforcement learning appeared in, e.g. [101].

If we look at the reinforcement learning algorithm as a type of controller auto-tuning, the actions as a control signal [102] and the value function as the optimal values of the cost function, then together with the already mentioned control law, instead of a policy we get a vocabulary that is more familiar to the control community.

Since a considerable portion of the research regarding GP models comes from the machine-learning community, a significant part of the control-learning algorithms also refers to the reinforcement learning concept. One of the first such concepts is the method named *Gaussian Process Dynamic Programming* (GPDP), the details of which are described in [102, 103]. The following description is summarised from [102, 104]. The evolution of the method can be followed through time with the publications [102–107].

GPDP is an approximate dynamic programming method for solving the optimal control problem, where the performance indices, the so-called value functions in the dynamic programming recursion, are online modelled by GPs. The value functions are usually referred to as the state-value function and state-action-value function, because they evaluate the performance based on the state and the state and control sequence, respectively. A brief description of the method is as follows.

The discrete-time system with one input signal described with Eq. (4.78) is considered throughout the method

$$\mathbf{x}(k+1) = f(\mathbf{x}(k), u(k)) + \boldsymbol{\nu}(k), \tag{4.78}$$

where $\mathbf{x}$ is a vector of states, u is a control signal and $\boldsymbol{\nu} \sim \mathcal{N}(\mathbf{0}, \boldsymbol{\Sigma}_n)$ is a Gaussian-distributed-noise random variable, where $\boldsymbol{\Sigma}_n$ is the diagonal. The transition or system function f mapping a pair $(\mathbf{x}(k), u(k))$ to the successor state $\mathbf{x}(k+1)$ is assumed to evolve smoothly over time and be time-invariant [102].

The control law in a GPDP is a deterministic mapping from a state vector into a control input value that assigns a value of the control signal to each state, i.e. a nonlinear state controller.

For an initial state $\mathbf{x}_0$ and the selected control sequence, the expected cumulative cost of the finite N_h-step optimisation horizon is:

$$J(\mathbf{x}_0) = E\left(\Phi(\mathbf{x}(N_h)) + \sum_{k=1}^{N_h-1} L(\mathbf{x}(k), u(k))\right). \tag{4.79}$$

The function $\Phi(\mathbf{x}(N_h))$ is a control-independent terminal cost that is incurred during the last time step of the optimisation horizon N_h. The immediate cost is denoted as $L(\mathbf{x}(k), u(k))$. An optimal control sequence $\mathbf{u}$ for the N_h-step problem minimises Eq. (4.79) for any initial state $\mathbf{x}_0$. The associated minimised cost, the so-called state-value function V^*, satisfies Bellman's equation [1, 102]:

$$V^*(\mathbf{x}) = \min_{\mathbf{u}} J(\mathbf{x}, u) = \min_{\mathbf{u}}[L(\mathbf{x}, u) + \gamma E\left(V^*(\mathbf{x}')|(\mathbf{x}, u)\right)], \tag{4.80}$$

for any state vector $\mathbf{x}$. $J(\mathbf{x}, u)$ is the state-action value function and the factor $\gamma \in (0, 1]$ weights the cost. The successor state for a given state-action pair $(\mathbf{x}, u)$ is denoted by $\mathbf{x}'$. Assuming a time-additive cost and Markovian transitions, the minimal expected cumulative cost can be calculated using dynamic programming.

The GPDP method assumes in advanced a known, i.e. offline learned, model of the system dynamics and describes the performance indices $J(\mathbf{x}, u)$ and $V^*(\mathbf{x})$ with GP models. To find an optimal control input sequence, guiding the system from an initial state to the goal state, dynamic programming is used.

A GPDP that is enhanced with active learning is called an ALGPDP. Simultaneously with the building of GP models for performance indices, i.e. value functions, the GP model of the process dynamics is also built online in ALGPDP.

In the case that there are more system outputs, there are as many GP sub-models of the controlled process developed as there are outputs. For a stochastic system, the noise term $\boldsymbol{\nu}$ in the system Eq. (4.78) is the process noise. The obtained dynamic GP model of the underlying stochastic transition function f contains two sources of uncertainty. First, is the uncertainty about the underlying system function itself, and, second, is the uncertainty induced by the process noise. With an increasing amount of absorbed information the first source of uncertainty tends to zero, whereas the stochasticity due to the process noise $\boldsymbol{\nu}$ is always present. Therefore, only the uncertainty about the model vanishes with the time of operation.

The online algorithm for dynamics model building exploits the information that is already computed within the GPDP to a large extent. Bayesian active learning of the process model is incorporated into the GPDP, such that only a relevant part of the state space is explored. The optimisation part regarding control law is not affected with this enhancement.

Only the best candidate data should be used for learning. The objective function, called the utility function, that rates the quality of the candidate data and therefore the relevance of the state space is:

$$J_U = \rho E(V^*(k)) + \frac{\beta}{2} \ln \mathrm{var}(V^*(k)) \tag{4.81}$$

where V^* is the value function, modelled by the GP model, that satisfies Bellman's equation for all the states and with weighting factors ρ and β.

The ALGPDP starts from a small initial set $\mathcal{X}(N_h)$ of state vectors, where N_h is the length of the optimisation horizon. Using Bayesian active learning, new state vectors are added to the current set $\mathcal{X}(k)$, at any time step k. The set $\mathcal{X}(k)$ contains the same training input data for both the GP model describing the system and the GP models describing the performance indices. During each time step the GP models are updated to incorporate the most recent information. The ALGPDP is a GP-model-based learning control algorithm with a proximity to adaptive dual control.

The general ALGPDP algorithm is as follows [102]:

Algorithm: ALGPDPCONTROL

---

initialise process-dynamics model
compute terminal cost $\Phi(\mathbf{x}(N_h))$
train GP model for terminal cost $\Phi(\mathbf{x}(N_h))$
repeat
  comment: dynamic programming recursion
  train GP model of transition function, i.e. process-dynamics model
  for each state $\mathbf{x}_i \in \mathcal{X}$
    do {
      for each control action $u_j \in \mathcal{U}$
        do {compute state-action value function J
      train GP model for state-action value function J
      select the best control input sequence for the current state
      update state-value function with optimal state-action value function
  train GP model for state-value function V^*
until *end*

The reader is referred to [102] for details and a demonstration of the method. Unfortunately, according to the method's authors [102], ALGPDP faces difficulties when directly applied to a dynamic system because it is often not possible to experience arbitrary state transitions. Moreover the GPDP method does not scale that well to high dimensions. The application of ALGPDP to control blood glucose is reported in [108].

A more promising method for engineering control applications, the *Probabilistic Inference and Learning for COntrol* (PILCO) method, is described in [109–111].

PILCO is a *policy search method*, i.e. a control law search method, and no value function is used, in contrast to the GPDP method.

The general idea of the method is to identify the system-dynamics model online in episodes with reinforcement learning. The learned model is used to simulate a closed-loop system internally and to search for the optimal closed-loop control law, taking into account the probabilistic model of the process.

Like in the case of GPDP, the discrete-time system described with Eq. (4.78) is considered throughout the method and the same considerations apply.

The control law in PILCO is a deterministic mapping from a state vector into a control vector that assigns a value of the control signal to each state, i.e. a nonlinear state controller.

The objective of the method is to find a control law, i.e. a policy, that minimises the expected long-term cost, i.e. the value function described with Eq. (4.79).

The PILCO algorithm can be divided into three layers: a top level for the controller adaptation, an intermediate layer for the approximate inference for long-term predictions and a bottom layer for the identification of the dynamics model. The intermediate layer serves to simulate the closed-loop system using the selected control law. The bottom layer, for the identification of the dynamics model, is implemented as an online identification of the model. A start state $\mathbf{x}_0$ is required by the algorithm in the beginning.

The high-level description of the method starts with an initialisation of the controller with arbitrary values. The possible control law realisations are a linear state controller in the case of a linear control law or a radial-basis-function network for a nonlinear control law. The method involves adaptation in two stages. First, when interacting with the process, information about the system, e.g. experience, is collected and the internal probabilistic dynamics model is updated, based on both historical and novel observations. Second, the control law is refined in the light of the updated dynamics model and applied in the simulated closed-loop system. The controller's parameters are refined with gradient-based optimisation techniques. The model-optimised control law is then applied to the real system to gather novel information about the closed-loop response. The subsequent model update accounts for possible discrepancies between the predicted and the actually encountered state trajectory.

The general algorithm is as follows [110]:

Algorithm: PILCOControl

---

initialise controller to random values
apply an episode of an initial control signal to the process
record the episode data from the closed-loop system
repeat
  learn probabilistic model of the process *bottom layer*
  model-based policy search
  repeat
    simulate closed-loop system *intermediate layer*
    compute expected long-term cost J
    compute gradient-based improvement of control law
    update simulated controller parameters
  until *convergence*
  update parameters of controller realisation *top layer*
  apply the episode of the control signal to the process
  record the episode data from the closed-loop system
until *end*

The PILCO method was applied to real systems, mainly robotic systems as the most appropriate for the iterative learning of control. Applications are reported in, e.g. [110–114]. The use of PILCO in a continuous-time setting is reported in [115, 116].

Some other applications of reinforcement learning using GP models for robotic or electromechanical systems are reported in, e.g. [117, 118]. In general, control law or policy search methods are very suitable for applications in robotics. An overview of the policy search methods in robotics control is given in [119]. Active research and development can be detected in this field, see, e.g. [120, 121], and more results are envisaged in the future.

References

1. Levine, W.S. (ed.): The Control Handbook. CRC/IEEE Press, Boca Raton, FL (1996)
2. Singh, M.G. (ed.): Systems and Control Encyclopedia: Theory, Technology, Applications, vol. 1–8. Pergamon press, Oxford (1987)
3. Wang, H.: Bounded Dynamic Stochastic Systems, Modelling and Control. Advances in Industrial Control. Springer, London (2000)
4. Khalil, H.K.: Nonlinear Systems, 3rd edn. Prentice-Hall, Upper Saddle River, NJ (2002)
5. Slotine, J.J.E., Li, W.: Applied Nonlinear Control. Prentice Hall, Englewood Cliffs, NJ (1991)
6. Chowdhary, G., Kingravi, H., How, J., Vela, P.: Bayesian nonparametric adaptive control of time-varying systems using Gaussian processes. In: Proceedings of the American Control Conference, pp. 2655–2661 (2013)
7. Grande, R., Chowdhary, G., How, J.: Nonparametric adaptive control using Gaussian processes with online hyperparameter estimation. In: Proceedings of 2013 IEEE 52nd Annual Conference on Decision and Control (CDC), pp. 861–867 (2013). doi:10.1109/CDC.2013.6759990
8. Kocijan, J., Grancharova, A.: Application of Gaussian processes to the modelling and control in process engineering. In: Balas, V.E., Koprinkova-Hristova, P., Jain, L.C. (eds.) Innovations in Intelligent Machines-5. Studies in Computational Intelligence, vol. 561, pp. 155–190. Springer, Heidelberg (2014). doi:10.1007/978-3-662-43370-6_6
9. Hansen, J., Murray-Smith, R., Johansen, T.A.: Nonparametric identification of linearizations and uncertainty using Gaussian process models–application to robust wheel slip control. In: Proceedings of the Joint 44th IEEE Conference on Decision and Control and European Control Conference (CDC-ECC), pp. 7994–7999. Sevilla (2005)
10. Chowdhary, G., Kingravi, H., How, J., Vela, P.: Bayesian nonparameteric model reference adaptive control using Gaussian processes. In: Proceedings of the AIAA Guidance, Navigation, and Control (GNC) Conference, pp. 874–879 (2013)
11. Abusnina, A., Kudenko, D., Roth, R.: Gaussian process-based inferential control system. In: de la Puerta, J.G., Ferreira, I.G., Bringas, P.G., Klett, F., Abraham, A., de Carvalho, A.C., Herrero, A., Baruque, B., Quintián, H., Corchado, E., (eds.) International Joint Conference SOCO'14-CISIS'14-ICEUTE'14. Advances in Intelligent Systems and Computing, vol. 299, pp. 115–124. Springer International Publishing (2014). doi:10.1007/978-3-319-07995-0_12
12. Norgaard, M., Ravn, O., Poulsen, N.K., Hansen, L.K.: Neural Networks for Modelling and Control of Dynamic Systems: A Practitioner's Handbook. Advanced Textbooks in Control and Signal Processing. Springer, New York, NY (2000)
13. Craig, J.J.: Introduction to Robotics: Mechanics and Control. Pearson Education International, Upper Saddle River, NJ (2005)
14. Nguyen-Tuong, D., Seeger, M., Peters, J.: Computed torque control with nonparametric regression models. In: Proceedings of the 2008 American Control Conference, ACC 2008, pp. 1–6. Seattle, WA (2008)
15. Nguyen-Tuong, D., Seeger, M., Peters, J.: Model learning with local Gaussian process regression. Advanced Robotics **23**(15), 2015–2034 (2009)

16. Nguyen-Tuong, D., Peters, J., Seeger, M., Schölkopf, B.: Learning inverse dynamics: a comparison. In: Proceedings of the European Symposium on Artificial Neural Networks (ESANN), pp. 13–18. Bruges (2008)
17. Narendra, K.S., Parthasarathy, K.: Identification and control of dynamical systems using neural networks. IEEE Trans. Neural Netw. **1**(1), 4–27 (1990)
18. Grancharova, A., Johansen, T.A.: Survey of explicit approaches to constrained optimal control. In: Murray-Smith, R., Shorten, R. (eds.) Switching and Learning in Feedback Systems. Lecture Notes in Computer Science, vol. 3355, pp. 47–97. Springer, Berlin (2005)
19. Ray, W.H.: Advanced Process Control. McGraw-Hill Book Company, New York, NY (1981)
20. Lewis, F.L., Vrabie, D., Vamvoudakis, K.G.: Reinforcement learning and feedback control: using natural decision methods to design optimal adaptive controllers. IEEE Control Syst. **32**(6), 76–105 (2012)
21. Maciejowski, J.M.: Predictive Control with Constraints. Pearson Education Limited, Harlow (2002)
22. Qin, S.J., Badgwell, T.A.: An overview of industrial model predictive control technology. In: Kantor, J., Garcia, C.E., Carnahan, B. (eds.) Fifth International Conference on Chemical Process Control. AIChE Symposium Series, pp. 232–256. American Institute of Chemical Engineers, New York, NY (1997)
23. Allgöwer, F., Zheng, A. (eds.): Nonlinear Model Predictive Control. Progress in System and Control Theory, vol. 26. Birkhäuser Verlag, Basel (2000)
24. Qin, S.J., Badgwell, T.A.: Nonlinear Model Predictive Control. Progress in System and Control Theory. Birkhäuser, Basel (2000)
25. Grancharova, A., Johansen, T.A.: Explicit Nonlinear Model Predictive Control: Theory and Applications. Lecture Notes in Control and Information Sciences. Springer, New York, NY (2012)
26. Grüne, L., Pannek, J.: Nonlinear Model Predictive Control: Theory and Algorithms. Communications and Control Engineering, 1st edn. Springer, New York, NY (2011)
27. Kouvaritakis, B., Cannon, M. (eds.): Nonlinear Predictive Control, Theory and Practice. IEE Control Engineering Series, vol. 61. IEE, Stevenage (2001)
28. Murray-Smith, R., Johansen, T.A. (eds.): Multiple Model Approaches to Modelling and Control. Taylor and Francis, London (1997)
29. Mayne, D.Q., Rawlings, J.B., Rao, C.V., Scokaert, P.O.M.: Constrained model predictive control: stability and optimality. Automatica **36**, 789–814 (2000)
30. Kwon, W.H., Han, S., Ahn, C.K.: Advances in nonlinear predictive control: a survey on stability and optimality. Int. J. Control, Autom., Syst. **2**(1), 15–22 (2004)
31. Grancharova, A., Kocijan, J.: Stochastic predictive control of a thermoelectric power plant. In: Proceedings of the International Conference Automatics and Informatics 07, pp. I–13–I–16. Sofia (2007)
32. Burkhart, M., Heo, Y., Zavala, V.: Measurement and verification of building systems under uncertain data: a Gaussian process modeling approach. Energy Build. **75**, 189–198 (2014)
33. Lourenço, J., Lemos, J., Marques, J.: Control of neuromuscular blockade with Gaussian process models. Biomed. Signal Process. Control **8**(3), 244–254 (2013). doi:10.1016/j.bspc.2012.10.007
34. Shen, G., Cao, Y.: A Gaussian process based model predictive controller for nonlinear systems with uncertain input-output delay. Appl. Mech. Mater. **433–435**, 1015–1020 (2013)
35. Hu, M., Sun, Z.: Multimodel nonlinear predictive control with Gaussian process model. Lect. Notes Electr. Eng. **238**, 1895–1902 (2014)
36. Maciejowski, J., Yang, X.: Fault tolerant control using Gaussian processes and model predictive control. In: Conference on Control and Fault-Tolerant Systems (SysTol), Nice (2013)
37. Palm, R.: Multiple-step-ahead prediction in control systems with Gaussian process models and TS-fuzzy models. Eng. Appl. Artif. Intell. **20**(8), 1023–1035 (2007)
38. Lee, J.H., Morari, M., Garcia, C.E.: State-space interpretation of model predictive control. Automatica **30**(4), 707–717 (1994)

39. Gregorčič, G., Lightbody, G.: Gaussian processes for internal model control. In: Rakar, A. (ed.) Proceedings of 3rd International PhD Workshop on Advances in Supervision and Control Systems, A Young Generation Viewpoint, pp. 39–46. Strunjan (2002)
40. Gregorčič, G., Lightbody, G.: From multiple model networks to the Gaussian processes prior model. In: Proceedings of IFAC ICONS conference, pp. 149–154. Faro (2003)
41. Gregorčič, G., Lightbody, G.: Gaussian process approaches to nonlinear modelling for control. Intelligent Control Systems using Computational Intelligence Techniques, vol 70., IEE Control Series, IEE, London, pp 177–217 (2005)
42. Gregorčič, G., Lightbody, G.: Gaussian process internal model control. Int. J. Syst. Sci. **43**(11), 2079–2094 (2012)
43. Economou, C.G., Morari, M., Palsson, B.O.: Internal model control: extension to nonlinear systems. Ind. Eng. Chem. Process Des. Dev. **25**, 403411 (1986)
44. Gregorčič, G., Lightbody, G.: Internal model control based on Gaussian process prior model. In: Proceedings of the 2003 American Control Conference., ACC 2003 Denver, CO, pp 4981–4986 (2003)
45. Kocijan, J., Murray-Smith, R., Rasmussen, C.E., Likar, B.: Predictive control with Gaussian process models. In: Proceedings of IEEE Region 8 EUROCON 2003: Computer as a Tool, vol. A, pp. 352–356. Ljubljana (2003)
46. Kocijan, J., Leith, D.J.: Derivative observations used in predictive control. In: Proceedings of IEEE Melecon conference, vol. 1, pp. 379–382. Dubrovnik (2004)
47. Kocijan, J., Murray-Smith, R., Rasmussen, C.E., Girard, A.: Gaussian process model based predictive control. In: Proceedings of 4th American Control Conference (ACC 2004), pp. 2214–2218. Boston, MA (2004)
48. Kocijan, J., Murray-Smith, R.: Nonlinear predictive control with a Gaussian process model. In: Murray-Smith, R., Shorten, R. (eds.) Switching and Learning in Feedback Systems, Lecture Notes in Computer Science, vol. 3355, pp. 185–200. Springer, New York, NY (2005)
49. Likar, B., Kocijan, J.: Predictive control of a gas-liquid separation plant based on a Gaussian process model. Comput. Chem. Eng. **31**(3), 142–152 (2007)
50. Ažman, K., Kocijan, J.: Non-linear model predictive control for models with local information and uncertainties. Trans. Inst. Meas. Control **30**(5), 371–396 (2008)
51. Isermann, R., Lachman, K.H., Matko, D.: Adapt. Control Syst. International Systems and Control Engineering. Prentice Hall, Upper Saddle River, NJ (1992)
52. Murray-Smith, R., Sbarbaro, D.: Nonlinear adaptive control using nonparametric Gaussian process prior models. In: Proceedings of IFAC 15th World Congress. Barcelona (2002)
53. Sbarbaro, D., Murray-Smith, R.: Self-tuning control of nonlinear systems using Gaussian process prior models. In: Murray-Smith, R., Shorten, R. (eds.) Switching and Learning in Feedback Systems. Lecture Notes in Computer Science, vol. 3355, pp. 140–157. Springer, Heidelberg (2005)
54. Sbarbaro, D., Murray-Smith, R.: An adaptive nonparametric controller for a class of nonminimum phase non-linear system. In: Proceedings of IFAC 16th World Congress. Prague (2005)
55. Goodwin, G.C., Rojas, O., Takata, H.: Nonlinear control via generalized feedback linearization using neural networks. Asian J. Cont. **3**(2), 79–88 (2001)
56. Sbarbaro, D., Murray-Smith, R., Valdes, A.: Multivariable generalized minimum variance control based on artificial neural networks and Gaussian process models. In: Yin, F.L., Wang, J., Guo, C. (eds.) Advances in Neural Networks—ISNN 2004. Lecture Notes in Computer Science, vol. 3174, pp. 52–58. Springer, Berlin (2004)
57. Henson, M.A., Seborg, D.E.: Adaptive nonlinear control of a pH neutralization process. IEEE Trans. Control Syst. Technol. **2**, 169–183 (1994)
58. Kocijan, J., Banko, B., Likar, B., Girard, A., Murray-Smith, R., Rasmussen, C.E.: A case based comparison of identification with neural networks and Gaussian process models. In: Proceedings of IFAC ICONS conference, pp. 137–142. Faro (2003)
59. Alessio, A., Bemporad, A.: A Survey on Explicit Model Predictive Control. In: Magni, L., Raimondo, D.M., Allgöwer, F. (eds.) Nonlinear Model Predictive Control: Towards New

Challenging Applications. Lecture Notes in Control and Information Sciences, vol. 384, pp. 345–369. Springer, Berlin (2009)
60. Pistikopoulos, E.N., Georgiadis, M.C., Dua, V. (eds.): Multi-Parametric Model-Based Control. Wiley-VCH, Weinheim (2007)
61. Bemporad, A., Morari, M., Dua, V., Pistikopoulos, E.N.: The explicit linear quadratic regulator for constrained systems. Automatica **38**, 3–20 (2002)
62. Fiacco, A.V.: Introduction to Sensitivity and Stability Analysis in Nonlinear Programming. Academic Press, San Diego, CA (1983)
63. Grancharova, A., Johansen, T.A.: A computational approach to explicit feedback stochastic nonlinear model predictive control. In: Proceedings of the IEEE Conference on Decision and Control, pp. 6083–6088. Atlanta, GA (2010)
64. Grancharova, A., Kocijan, J., Johansen, T.A.: Explicit stochastic nonlinear predictive control based on Gaussian process models. In: Proceedings of European Control Conference (ECC), pp. 2340–2347. Kos (2007)
65. Grancharova, A., Johansen, T.A., Tøndel, P.: Computational aspects of approximate explicit nonlinear model predictive control. In: Findeisen, R., Allgöwer, F., Biegler, L.T. (eds.) Assessment and Future Directions of Nonlinear Model Predictive Control, Lecture Notes in Control and Information Sciences, vol. 358, pp. 181–190. Springer, Berlin (2007)
66. Johansen, T.A.: Approximate explicit receding horizon control of constrained nonlinear systems. Automatica **40**, 293–300 (2004)
67. Grancharova, A., Kocijan, J., Johansen, T.A.: Explicit stochastic predictive control of combustion plants based on Gaussian process models. Automatica **44**(4), 1621–1631 (2008)
68. Kocijan, J., Murray-Smith, R.: Gaussian processes: Prediction at a noisy input and application to iterative multiple-step ahead forecasting of time-series. In: Murray-Smith, R., Shorten, R. (eds.) Switching and Learning in Feedback Systems. Lecture Notes in Computer Science, vol. 3355, pp. 158–184. Springer (2005)
69. Kouvaritakis, B., Cannon, M., Couchman, P.: MPC as a tool for sustainable development integrated policy assessment. IEEE Trans. Autom. Control **51**, 145–149 (2006)
70. Couchman, P., Kouvaritakis, B., Cannon, M.: LTV models in MPC for sustainable development. Int. J. Control **79**, 63–73 (2006)
71. Couchman, P., Cannon, M., Kouvaritakis, B.: Stochastic MPC with inequality stability constraints. Automatica **42**, 2169–2174 (2006)
72. Cannon, M., Couchman, P., Kouvaritakis, B.: MPC for stochastic systems. In: Findeisen, R., Allgöwer, F., Biegler, L.T. (eds.) Assessment and future directions of nonlinear model predictive control. Lecture Notes in Control and Information Sciences, vol. 358, pp. 255–268. Springer, Berlin (2007)
73. Grancharova, A., Kocijan, J.: Explicit stochastic model predictive control of gas-liquid separator based on Gaussian process model. In: Proceedings of the International Conference on Automatics and Informatics, pp. B–85–B–88. Sofia (2011)
74. Bentley, J.L.: Multidimensional binary search trees used for associative searching. Commun. ACM **18**, 509–517 (1975)
75. Grancharova, A., Johansen, T.A.: Approaches to explicit nonlinear model predictive control with reduced partition complexity. In: Proceedings of European Control Conference, pp. 2414–2419. Budapest (2009)
76. Filatov, N., Unbehauen, H.: Survey of adaptive dual control methods. IEE Proc.—Control Theory Appl. **147**(1), 119–128 (2000)
77. Wittenmark, B.: Adaptive dual control. In: Control Systems, Robotics and Automation, Encyclopedia of Life Support Systems (EOLSS), Developed under the auspices of the UNESCO. Eolss Publishers, Oxford (2002)
78. Alpcan, T., Shames, I., Cantoni, M., Nair, G.: Learning and information for dual control. In: Proceedings of the Control Conference (ASCC), 2013 9th Asian, pp. 1–6 (2013). doi:10.1109/ASCC.2013.6606212
79. Rugh, W.J.: Analytic framework for gain-scheduling. IEEE Control Syst. Mag. **11**, 79–84 (1991)

80. Rugh, W.J., Shamma, J.S.: Research on gain scheduling. Automatica **36**, 1401–1425 (2000)
81. Leith, D.J., Leithead, W.E.: Survey of gain-scheduling analysis and design. Int. J. Control **73**, 1001–1025 (2000)
82. Nelles, O.: Nonlinear System Identification. Springer, New York, NY (2001)
83. Leith, D.J., Leithead, W.E.: Analytic framework for blended multiple model systems using linear local models. Int. J. Control **72**, 605–619 (1999)
84. Kocijan, J., Hvala, N., Strmčnik, S.: System and control: theory and applications, chap. Multimodel Control of Wastewater Treatment Reactor, pp. 49–54. World Scientific and Engineering Society, Singapore (2000)
85. Čokan, B., Kocijan, J.: Soft computing and industry: recent applications, chap. Some realisation issue of fuzzy gain-scheduling controllers: a robotic manipulator case study, pp. 191–199. Springer, New York, NY (2002)
86. Leith, D.J., Leithead, W.E.: Appropriate realisation of MIMO gain-scheduled controllers. Int. J. Control **70**(1), 13–50 (1998)
87. Ažman, K., Kocijan, J.: An application of Gaussian process models for control design. In: Proceedings of the International Control Conference ICC 2006, p. 4. Glasgow (2006)
88. Ažman, K., Kocijan, J.: Fixed-structure Gaussian process model. Int. J. Syst. Sci. **40**(12), 1253–1262 (2009)
89. Leith, D.J., Leithead, W.E.: Gain scheduled and nonlinear systems: dynamic analysis by velocity-based linearisation families. Int. J. Control **70**(2), 289–317 (1998)
90. Murray-Smith, R., Sbarbaro, D., Rasmussen, C.E., Girard, A.: Adaptive, cautious, predictive control with Gaussian process priors. In: Proceedings of 13th IFAC Symposium on System Identification. Rotterdam (2003)
91. Nguyen-Tuong, D., Peters, J.: Learning robot dynamics for computed torque control using local Gaussian processes regression. In: Symposium on Learning and Adaptive Behaviors for Robotic Systems, pp. 59–64 (2008)
92. Nguyen-Tuong, D., Seeger, M., Peters, J.: Real-time local GP model learning, vol. 264, chap. From Motor Learning to Interaction Learning in Robots, pp. 193–207. Springer (2010)
93. Nguyen-Tuong, D., Peters, J.: Incremental online sparsification for model learning in real-time robot control. Neurocomputing **74**(11), 1859–1867 (2011)
94. Park, S., Mustafa, S., Shimada, K.: Learning based robot control with sequential Gaussian process. In: Proceedings of the 2013 IEEE Workshop on Robotic Intelligence in Informationally Structured Space (RiiSS), pp. 120–127 (2013). doi:10.1109/RiiSS.2013.6607939
95. Su, Y., Wu, Y., Soh, H., Du, Z., Demiris, Y.: Enhanced kinematic model for dexterous manipulation with an underactuated hand. In: Proceedings of the 2013 IEEE/RSJ International Conference on Intelligent Robots and Systems (IROS), Tokyo, pp. 2493–2499 (2013)
96. Park, S., Mustafa, S., Shimada, K.: Learning-based robot control with localized sparse online Gaussian process. In: Proceedings of the 2013 IEEE/RSJ International Conference on Intelligent Robots and Systems (IROS), pp. 1202–1207 (2013). doi:10.1109/IROS.2013.6696503
97. Calliess, J., Osborne, M., Roberts, S.: Nonlinear adaptive hybrid control by combining Gaussian process system identification with classical control laws. In: Proceedings of the Novel Methods for Learning and Optimization of Control Policies and Trajectories for Robotics, ICRA (2013)
98. Petelin, D., Kocijan, J.: Control system with evolving Gaussian process model. In: Proceedings of IEEE Symposium Series on Computational Intelligence, SSCI 2011. IEEE, Paris (2011)
99. Moore, K.L.: Iterative Learning Control for Deterministic Systems. Advances in Industrial Control. Springer, London (1993)
100. Hjalmarsson, H.: Iterative feedback tuning—an overview. Int. J. Adapt. Control Signal Process. **16**(5), 373–395 (2002)
101. Engel, Y., Szabo, P., Volkinshtein, D.: Learning to control an octopus arm with Gaussian process temporal difference methods. In: Weiss, Y., Schölkopf, B., Platt, J. (eds.) Advances in Neural Information Processing Systems, vol. 18, pp. 347–354. MIT Press, Cambridge, MA (2006)

102. Deisenroth, M.P., Rasmussen, C.E., Peters, J.: Gaussian process dynamic programming. Neurocomputing **72**(7–9), 1508–1524 (2009)
103. Deisenroth, M.P., Rasmussen, C.E., Peters, J.: Model-based reinforcement learning with continuous states and actions. In: Proceedings of the European Symposium on Artificial Neural Networks (ESANN), pp. 19–24. Bruges (2008)
104. Deisenroth, M.P., Rasmussen, C.E.: Bayesian inference for efficient learning in control. In: Proceedings of Multidisciplinary Symposium on Reinforcement Learning (MSRL). Montreal (2009)
105. Rasmussen, C.E., Kuss, M.: Gaussian processes in reinforcement learning. In: Thurn, S., Saul, L., Schölkopf, B. (eds.) Advances in Neural Information Processing Systems Conference, vol. 16, pp. 751–759. MIT Press, Cambridge, MA (2004)
106. Deisenroth, M.P., Peters, J., Rasmussen, C.E.: Approximate dynamic programming with Gaussian processes. In: Proceedings of American Control Conference (ACC), pp. 4480–4485. Seattle, WA (2008)
107. Rasmussen, C.E., Deisenroth, M.P.: Probabilistic inference for fast learning in control. In: Recent Advances in Reinforcement Learning. Lecture Notes on Computer Science, vol. 5323, pp. 229–242. Springer (2008)
108. De Paula, M., Martinez, E.: Probabilistic optimal control of blood glucose under uncertainty. Comput. Aided Chem. Eng. **30**, 1357–1361 (2012)
109. Deisenroth, M.P.: Efficient Reinforcement Learning using Gaussian Processes. Ph.D. Thesis, Karlsruhe Institute of Technology, Karlsruhe (2010)
110. Deisenroth, M.P., Rasmussen, C.E.: PILCO: A model-based and data-efficient approach to policy search. In: Proceedings of the 28th International Conference on Machine Learning (ICML 2011). Bellevue, WA (2011)
111. Deisenroth, M.P., Rasmussen, C.E., Fox, D.: Learning to control a low-cost manipulator using data-efficient reinforcement learning. In: Proceedings of the International Conference on Robotics: Science and Systems (R:SS 2011). Los Angeles, CA (2011)
112. Deisenroth, M.P., Fox, D.: Multiple-target reinforcement learning with a single policy. In: Proceedings of the 28th International Conference on Machine Learning (ICML 2011). Bellevue, WA (2011)
113. Deisenroth, M.P., Calandra, R., Seyfarth, A., Peters, J.: Toward fast policy search for learning legged locomotion. In: IROS'12, pp. 1787–1792 (2012)
114. Cliff, O., Sildomar, T., Monteiro: Evaluating techniques for learning a feedback controller for low-cost manipulators. In: IEEE International Conference on Intelligent Robots and Systems, pp. 704–709 (2013)
115. Hall, J., Rasmussen, C.E., Maciejowski, J.: Reinforcement learning with reference tracking control in continuous state spaces. In: Proceedings of the 50th International Conference on Decision and Control (2011)
116. Hall, J., Rasmussen, C., Maciejowski, J.: Modelling and control of nonlinear systems using Gaussian processes with partial model information. In: Conference on Decision and Control (CDC) (2012)
117. Jung, T., Stone, P.: Gaussian processes for sample efficient reinforcement learning with rmax-like exploration. Lect. Notes Comput. Sci. **6321**, 601–616 (2010)
118. Bischoff, B., Nguyen-Tuong, D., Markert, H., Knoll, A.: Learning control under uncertainty: A probabilistic value-iteration approach. In: ESANN 2013 proceedings, 21st European Symposium on Artificial Neural Networks, Computational Intelligence and Machine Learning, pp. 209–214 (2013)
119. Deisenroth, M., Neumann, G., Peters, J.: A survey on policy search for robotics. Found. Trends Robot. **2**, 1–142 (2013). doi:10.1561/2300000021
120. Kupcsik, A.G., Deisenroth, M.P., Peters, J., Neumann, G.: Data-efficient generalization of robot skills with contextual policy search. In: AAAI (2013)
121. Neumann, G., Daniel, C., Paraschos, A., Kupcsik, A., Peters, J.: Learning modular policies for robotics. Front. Comput. Neurosci. **8**(62), 1–13 (2014). doi:10.3389/fncom.2014.00062

Chapter 5
Trends, Challenges and Research Opportunities

Several research topics remain to be fully explored before we are able to say that the application of GP models for control is a mature technology, ready to use in everyday engineering practice. The fact that the potential of GP models for control design is still not fully explored in engineering practice can be seen by the small, but constantly increasing, number of practical, i.e. industrial, applications reported in the literature so far.

Research opportunities can be roughly divided into issues concerning dynamic-systems modelling with GP models related to control design, issues concerning control design itself and, thirdly, some general issues related to control design and applications. The given list of issues is subjective and heavily based on the on-going research activities throughout the world.

Modelling trends and issues. The computational burden that increases with the increasing amount of data contained in the model, caused mainly by the calculation of the inverse covariance matrix, directs researchers to find more efficient methods for an inverse-covariance-matrix calculation or input-data selection.

The issue of automatically polishing data and finding informative portions is reported as one of key issues in dynamic-systems identification in general [1] and remains one of the current challenges in GP modelling research. Efficient modelling, in general, facilitates control design. For example, solutions for issues like errors-in-variables, e.g. [2, 3] are always in demand.

The issue of online model identification is the one that is closely linked to adaptive control methods. In the machine-learning community it is known as online learning, which is not limited to sequentially in-streaming data. An efficient method for the online identification of the GP model remains to be found.

Methods for developing and utilising GP state-space models, e.g. [4, 5], are of great interest, especially for control design methods. Consequently, this is at present a dynamic research field.

J. Kocijan, *Modelling and Control of Dynamic Systems Using Gaussian Process Models*, Advances in Industrial Control,
DOI 10.1007/978-3-319-21021-6_5

Control design trends and issues. Some of the reported control methods have not addressed disturbance rejection, which is crucial in control systems, but have been more focused on set-point tracking. The complete assessment of control methods also requires a disturbance-rejection analysis, which in many cases still remains an unexplored issue.

Current research on control methods often deals with adaptive and predictive control. Some other results that have the potential for control applications include the modelling of switching systems, e.g. [6].

If control methods are meant to be used in engineering practice, more results on robust control design methodologies are necessary. GP models offer a lot of potential for robust control design and offer a lot of research and application opportunities.

Some general issues that need to be looked at are the benchmarking of control methods with the purpose being to assess different control methods properly and fairly. A giant step to bring research results closer to engineering practice is the integration of knowledge and a software upgrade from pieces scattered around, mainly on the internet, into user-friendly integrated software. Examples of software can be found at, e.g. [7], or the software description in, e.g. [8, 9].

Research opportunities also lie in developing methods and procedures for various kinds of analyses of system models and closed-loop systems. Model-simulation stability and closed-loop stability are only two, very different, but important and challenging problems among many to be addressed.

One of the perspective applications for the GP model that has not been addressed in this book is *fault detection and isolation* (FDI). It is related to modelling as well as to the control of systems. FDI is concerned with monitoring a system, identifying when a fault has occurred, and revealing the type of fault and its location. Fault detection can, generally speaking, be sorted into two categories. The first one contains signal-processing-based methods and the second one contains model-based methods.

Signal-processing-based FDI methods [10] are based on classifying a fault from sensor measurements, which indicate a fault, directly. *Model-based FDI methods* are based on the analysis of the discrepancy between the sensor readings and expected values, derived from some model. This model can be a first-principle model or a model obtained from measured data. In model-based FDI methods the question may be raised as to how to avoid false alarms provoked by the presence of modelling errors.

Some attempts to address FDI using GP models can already be found in the literature. Model-based methods using GP models that are mainly concerned with the detection of change-points, i.e. abrupt variations in the statistical parameters of a data sequence, in the context of modelling for prediction, are, e.g. [6, 11–14] or for performance monitoring, e.g. [15]. The early stages of the method using GP models to avoid false alarms due to modelling errors can be found in [16, 17].

GP signal-processing-based FDI methods may be found in [18, 19]. Both methods are based on a GP latent-variable model [20] that can be interpreted as a GP-based kernel PCA method for nonlinear systems.

Prognostics and health management methods [21], on the other hand, go a step further from FDI. They are focused on predicting the time at which a system or a component will no longer perform its intended function with certainty. GP models for prognostics are used in [22–24].

FDI as well as prognostics offer additional space for applications' development and the utilisation of GP models.

The number of research opportunities concerning GP models of dynamic systems is large and it seems that many issues need to be solved before the control design method based on GP models becomes an everyday tool for control design in engineering practice. Nevertheless, a lot of results are already available that confirm the potential and benefits of GP model-based control.

References

1. Ljung, L.: Perspectives on system identification. In: Proceedings of IFAC 17th World Congress 2008, pp. 1–6. Seoul (2008)
2. Frigola, R., Rasmussen, C.E.: Integrated pre-processing for Bayesian nonlinear system identification with Gaussian processes. In: 52nd IEEE Conference on Decision and Control (CDC) (2013)
3. McHutchon, A., Rasmussen, C.E.: In: Shawe-Taylor, J., Zemel, R., Bartlett, P., Pereira, F., Weinberger, K. (eds.) Advances in Neural Information Processing Systems. Gaussian process training with input noise, vol. 24, pp. 1341–1349 (2011)
4. Turner, R., Deisenroth, M.P., Rasmussen, C.E.: State-space inference and learning with Gaussian processes. In: Proceedings of 13th International Conference on Artificial Intelligence and Statistics, vol. 9, pp. 868–875. Sardinia (2010)
5. Frigola, R., Lindsten, F., Schön, T.B., Rasmussen, C.E.: Bayesian inference and learning in Gaussian process state-space models with particle MCMC. In: Bottou, L., Burges, C., Ghahramani, Z., Welling, M., Weinberger, K. (eds.) Advances in Neural Information Processing Systems 26, pp. 3156–3164 (2013)
6. Saatçi, Y., Turner, R., Rasmussen, C.E.: Gaussian process change point models. In: Proceedings of the 27th Annual International Conference on Machine Learning, pp. 927–934 (2010)
7. The Gaussian processes web site. http://www.gaussianprocess.org/#code
8. Rasmussen, C.E., Nickisch, H.: Gaussian Processes for Machine Learning (GPML) toolbox. J. Mach. Learn. Res. **11**, 3011–3015 (2010)
9. Vanhatalo, J., Riihimäki, J., Hartikainen, J., Jylänki, P., Tolvanen, V., Vehtari, A.: GPstuff: Bayesian modeling with Gaussian processes. J. Mach. Learn. Res. **14**, 1175–1179 (2013)
10. Venkatasubramanian, V., Rengaswamy, R., Yin, K., Kavuri, S.N.: A review of process fault detection and diagnosis: Part I: Quantitative model-based methods. Comput. Chem. Eng. **27**(3), 293–311 (2003)
11. Kocijan, J., Přikryl, J.: Soft sensor for faulty measurements detection and reconstruction in urban traffic. In: Proceedings 15th IEEE Mediterranean Electromechanical Conference (MELECON), pp. 172–177. Valletta (2010)
12. Osborne, M.A., Garnett, R., Roberts, S.J.: Active data selection for sensor networks with faults and changepoints. In: IEEE International Conference on Advanced Information Networking and Applications (2010)
13. Osborne, M.A., Garnett, R., Swersky, K., de Freitas, N.: Prediction and fault detection of environmental signals with uncharacterised faults. In: 26th AAAI Conference on Artificial Intelligence (AAAI-12). Toronto (2012)

14. Kullaa J (2013) Detection, identification, and quantification of sensor fault in a sensor network. Mechanical Systems and Signal Processing 40(1):208–221. doi:10.1016/j.ymssp.2013.05.007
15. Butler, S., Ringwood, J., O'Connor., F.: Exploiting SCADA system data for wind turbine performance monitoring. In: Conference on Control and Fault-Tolerant Systems (SysTol) October 9–11, Nice (2013)
16. Juričić, a., Ettler, P., Kocijan, J.: Fault detection based on Gaussian process models: An application to the rolling mill. In: ICINCO 2011—Proceedings of the 8th International Conference on Informatics in Control, Automation and Robotics, vol. 1, pp. 437–440 (2011)
17. Juričić, D., Kocijan, J.: Fault detection based on Gaussian process model. In: Troch, I., Breitenecker, F. (eds.) Proceedings of the 5th Vienna Symposium on Mathematical Modeling (MathMod). Vienna (2006)
18. Eciolaza, L., Alkarouri, M., Lawrence, N.D., Kadirkamanathan, V., Fleming, P.: Gaussian process latent variable models for fault detection. In: Proceedings of the 2007 IEEE Symposium on Computational Intelligence and Data Mining (CIDM 2007), pp. 287–292. Honolulu, HI (2007)
19. Serradilla, J., Shi, J.Q., Morris, J.A.: Fault detection based on Gaussian process latent variable models. Chem. Intel. Lab. Sys. **109**(1), 9–21 (2011)
20. Lawrence, N.: Gaussian process latent variable models for visualization of high dimensional data. In: Thrun, S., Saul, L., Schlkopf, B. (eds.) Advances in Neural Information Processing Systems, pp. 329–336. The MIT Press, Cambridge, MA (2004)
21. Vachtsevanos, G., Lewis, F.L., Roemer, M., Hess, A., Wu, B.: Intelligent Fault Diagnosis and Prognosis for Engineering Systems. Wiley, Hoboken (2006)
22. Kocijan, J., Tanko, V.: Prognosis of gear health using Gaussian process model. In: Proceedings IEEE Eurocon 2011 International Conference Computer as a tool. Lisbon (2011)
23. Liu, D., Pang, J., Zhou, J., Peng, Y., Pecht, M.: Prognostics for state of health estimation of lithium-ion batteries based on combination Gaussian process functional regression. Microelectron. Reliab. **53**(6), 832–839 (2013)
24. Mohanty, S., Das, S., Chattopadhyay, A., Peralta, P.: Gaussian process time series model for life prognosis of metallic structures. J. Intel. Mater. Syst. Struct. **20**(8), 887–896 (2009)

Chapter 6
Case Studies

This chapter will describe three examples of the use of GP models. Each example covers selected issues of GP model applications for dynamic systems modelling and control in practice.

The examples have been selected to demonstrate the utilisation potential of GP models and to highlight the various issues that have been mentioned in the book. MATLAB® has been used as the main computational tool, but other software has also been used when necessary for the implementation.

Gas–liquid separator (Sect. 6.1) This example demonstrates the modelling of a semi-industrial plant and the application of the developed model for model-predictive control. The plant is equipped with industrial sensors and actuators, and the control is implemented through industrially used hardware, which gives realistic industrial conditions.

Urban traffic system (Sect. 6.2) The detection of faults and the reconstruction of measurements on a selected region of the Prague traffic system show the application of GP models for fault detection and the use of model predictions. The acquired data, which can be considered as discrete-event data on the microlevel, is interpreted as stochastic data on the macrolevel and processed accordingly.

Ozone as an air pollutant (Sect. 6.3) This example from environmental sciences shows the online modelling and prediction of a highly complex environmental system. The used data was collected from the city of Burgas region.

J. Kocijan, *Modelling and Control of Dynamic Systems Using Gaussian Process Models*, Advances in Industrial Control,
DOI 10.1007/978-3-319-21021-6_6

6.1 Gas–Liquid Separator Modelling and Control

The following example shows a case study for system identification to illustrate the methods from Chap. 2 and the model-predictive control design from Chap. 4 for an industrial-like plant with some industrial application aspects encountered as it is presented in [1].

Gas–liquid separation plants are frequently encountered in the chemical process industry. Various sorts of separators exist. The semi-industrial process plant used for the case study is a unit for separating the gas from the liquid that forms part of a larger pilot plant. The role of this separation unit is to capture the flue gases under low pressure from the effluent channels by means of a water flow, to cool them down and then supply them under high enough pressure to other parts of the pilot plant. The pressure and level control is an inevitable part of the plant's operation.

Process and Problem Description

The scheme of the plant is shown in Fig. 6.1.

The flue gases coming from the effluent channels are absorbed by the water flow into the water-circulation pipe through the injector.

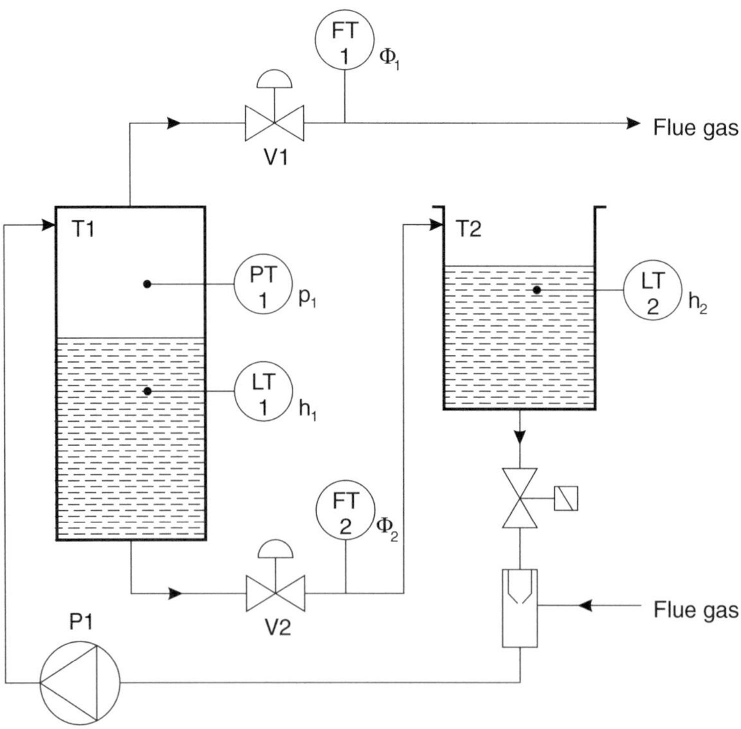

Fig. 6.1 The scheme of the gas–liquid separation plant

The water flow is generated by the water ring pump. The speed of the pump is kept constant. The pump feeds the mixture of water and gas into the tank, where the gas is separated from the water. Hence, the accumulated gas in the tank forms a sort of ‘gas cushion’ with an increased internal pressure. Owing to this pressure, the flue gas is blown out from the tank into the neutralisation unit which is omitted from Fig. 6.1. On the other hand, the ‘cushion’ forces the water to circulate back to the reservoir. The quantity of water in the circuit is constant.

In order to understand the basic relations among the variables and to illustrate the nonlinearity of the process, a theoretical model of the gas–liquid separation pressure sub-system of interest is introduced. The model, described in more detail in [2], was obtained as follows.

The equation for the isothermal gas change is used to obtain the differential equation for the air pressure in the separator.

$$\frac{pV}{m} = rT = \text{const}, \tag{6.1}$$

where p, V, m, r and T are the absolute air pressure, the gas volume, the mass of the air, the gas constant and the temperature inside the tank T1, respectively. The derivative of Eq. (6.1) with regard to time is equal to 0, which leads to the next expression:

$$mV\frac{dp}{dt} = pV\frac{dm}{dt} - mp\frac{dV}{dt}. \tag{6.2}$$

The mass of air m is

$$m = \rho V, \tag{6.3}$$

where ρ is the density of the air. The time derivative of the mass of gas is proportional to the difference between the input and output air flows

$$\frac{dm}{dt} = \rho_0(\Phi_{air} - \Phi_1), \tag{6.4}$$

where ρ_0 denotes the density of normal atmospheric air.

Substituting Eq. (6.4) for $\frac{dm}{dt}$ and Eq. (6.3) for m in Eq. (6.2), then $\frac{dp}{dt}$ can be written

$$\frac{dp}{dt} = \frac{p\rho_0}{\rho V}(\Phi_{air} - \Phi_1) + \frac{pS_1}{V}\frac{dh_1}{dt}, \tag{6.5}$$

where h_1 is the liquid level in the tank T1 and S_1 is the cross-sectional area of the tank T1.

Taking into account from Eq. (6.1) that $\frac{p}{p_0} = \frac{\rho}{\rho_0}$ and that $p = p_0 + p_1$, where p_0 denotes normal atmospheric pressure and p_1 is the relative air pressure in the tank T1, we obtain:

$$\frac{dp_1}{dt} = \frac{1}{V}\left[p_0(\Phi_{air} - \Phi_1) + (p_0 + p_1)S_1\frac{dh_1}{dt}\right]. \tag{6.6}$$

The volume of the gas V inside the tank T1 equals

$$V = S_1(h_{T1} - h_1), \tag{6.7}$$

where h_{T1} denotes the height of the tank T1.

The change of the liquid level in the tank T1 can be described as

$$\frac{dh_1}{dt} = \frac{1}{S_1}(\Phi_w - \Phi_2), \tag{6.8}$$

where Φ_w is the known constant water flow through the pump P1. The substitution of Eqs. (6.7) and (6.8) into Eq. (6.6) leads to

$$\frac{dp_1}{dt} = \frac{1}{S_1(h_{T1} - h_1)}\left[p_0(\Phi_{air} - \Phi_1) + (p_0 + p_1)(\Phi_w - \Phi_2)\right]. \tag{6.9}$$

The air flow through the valve V1 is

$$\Phi_1 = K_1\sqrt{p_1}, \tag{6.10}$$

and the water flow through the valve V2 is

$$\Phi_2 = K_2\sqrt{p_2 + k_w(h_1 - h_{T2})}. \tag{6.11}$$

$K_1 = k_1 R_1^{(u_1-1)}$ and $K_2 = k_2 R_2^{(u_2-1)}$ are the so called equal percentage valve characteristics with exponential relation between command signal and flow through the valve. Other variables are as follows: u_i is the command signal of the valve Vi, $i = 1, 2$, R_i is the ratio of the flows for the maximum and minimum opening of the valve Vi, $i = 1, 2$, k_i is the flow coefficient of the valve Vi, $i = 1, 2$, h_{T1} is the height of the tank T2 and k_w is the proportional factor between water level in meters and pressure in bars.

The air flow to the tank T1 is a nonlinear function of the pressure p_1. This functional dependency was obtained empirically and approximated by the function

$$\Phi_{air} = \alpha_0 + \alpha_1 p_1 + \alpha_2 p_1^2, \tag{6.12}$$

where α_i, $i = 1, 2, 3$, are constant parameters obtained by data fitting.

Equations (6.10)–(6.12) are substituted into Eq. (6.9), and Eq. (6.11) into (6.8). Consequently, the gas–liquid separation pressure sub-system of interest can be described by a set of two Eqs. (6.13) and (6.14) [3] as follows

$$\frac{dp_1}{dt} = \frac{1}{S_1(h_{T_1} - h_1)}(p_0(\alpha_0 + \alpha_1 p_1 + \alpha_2 p_1^2 \\ -k_1 R_1^{u_1-1}\sqrt{p_1}) + (p_0 + p_1)(\Phi_w - k_2 R_2^{u_2-1}\sqrt{p_1 + k_w(h_1 - h_{T_2})})), \tag{6.13}$$

$$\frac{dh_1}{dt} = \frac{1}{S_1}(\Phi_w - k_2 R_2^{u_2-1}\sqrt{p_1 + k_w(h_1 - h_{T_2})}). \tag{6.14}$$

From the model presented it is clear that the nonlinear process is of a multivariable nature (two inputs and two outputs with dynamic interactions between the channels). In our case, a liquid-level feedback control was implemented on the plant. Consequently, the dynamic system could be approached as a single-input single-output dynamic system with the command signal of the valve V_1 as the control input, the liquid level in the tank T1 as the measured disturbance input and the pressure in the tank T1 as the output. It can be seen from Eqs. (6.13) that the pressure is nonlinearly related to the level and the input flow, which results in different dynamic behaviour depending on the operating region.

The real-time experiments were pursued in the environment schematically shown in Fig. 6.2.

A user-friendly experimentation with the process plant is made possible by an interface with the Matlab/Simulink environment. This interface enables PLC access with the Matlab/Simulink using the DDE protocol via a Serial Communication Link RS-232 or TCP/IPv4 over the Ethernet IEEE-802.3. Control algorithms for the experimentation can be prepared in the Matlab code or as Simulink blocks and extended with functions/blocks that access the PLC. In our case, all the schemes for the data acquisition were put together as Simulink blocks.

Modelling

From Eq. (6.13) it can be seen that the pressure p_1, the liquid level h_1 and the valve-opening signal u_1 contribute a great part of information that could be used for the plant model identification.

Based on the response and an iterative cut-and-try procedure, a sampling time of 15 s was selected. The identification data consists of pseudo-random changes of the valve signal u_1 in regions with a different liquid level h_1, so that as wide region as possible was encompassed in 967 samples for each signal.

Selections of the covariance function and the regressors are the only remaining choices that have to be made. There are no other identification parameters to be chosen. Nevertheless, the selection of the data and, consequently, the information that is contained in the data are crucial for the quality of the model. The more information about the process dynamics that is contained in the data the better is the identified nonparametric model that will contain this data. The accuracy and the amount of data affect the predicted output distribution: the lower the data accuracy or the sparser the input data, the wider is the predicted output distribution. This means that the quality of the identification data can be recognised in the variance of the output distribution.

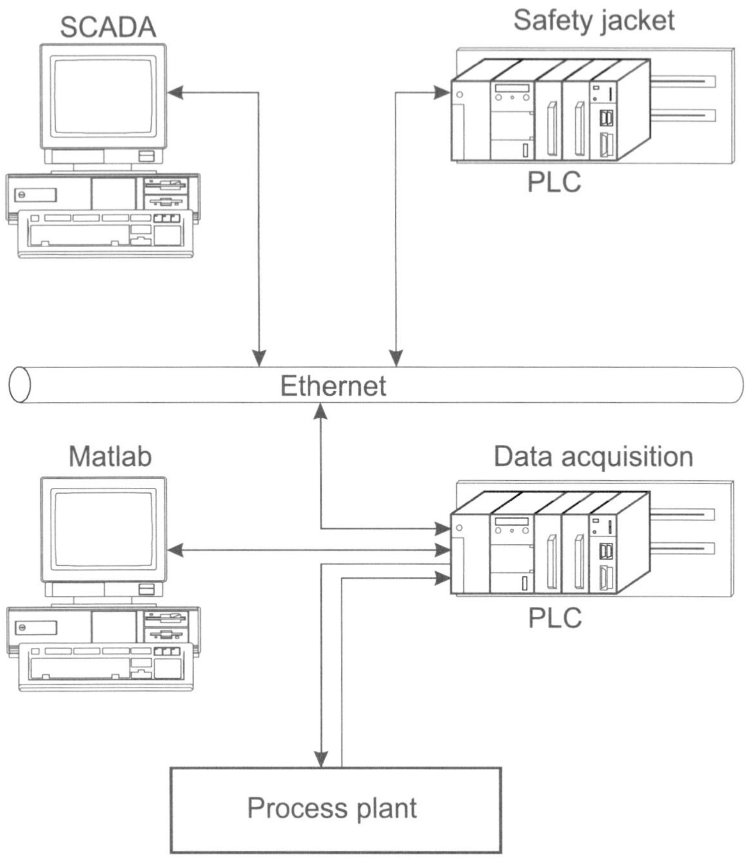

Fig. 6.2 Experimental set-up for the data acquisition and control algorithm testing

The nonlinearities describing the process model are presumed to be smooth. A squared exponential covariance function with one hyperparameter for each regressor described with Eq. (2.14) is selected and consequently no other prior knowledge about the relations between the data is used.

Input regressors are to be selected next. The following GP-NARX structure is selected for the model:

$$\begin{aligned} p_1(k+1) = f(&p_1(k), \ldots, p_1(k-n_p), h_1(k), \ldots, h_1(k-m_h), \\ &u_1(k), \ldots, u_1(k-m_u)) + \nu, \end{aligned} \tag{6.15}$$

where n_p, m_h and m_u denote the maximum delays for samples of pressure, liquid level and input signal, respectively, and ν is a white noise with unknown variance. The samples of the identification signals then form the $D \times N$ matrix of regressors composed of regressors:

$$\mathbf{z} = [p_1(k), \ldots, p_1(k - n_p), h_1(k), \ldots, h_1(k - m_h), u_1(k), \ldots, u_1(k - m_u)]^{\mathrm{T}} \tag{6.16}$$

and the $N \times 1$ vector of corresponding targets. N is the number of samples and $D = n_p + m_h + m_u$.

The optimal number of delayed variables is not known in advance so a method of subset selection based on iterative GP model training with different regressors and a different model order n_p, is pursued. The selection procedure started with straightforward models, for which $n_p = m_h = m_u = n$ for decreasing values of n from 3 downwards are considered:

Model M4:

$$\mathbf{z} = [p_1(k), p_1(k-1), p_1(k-2), p_1(k-3), h_1(k), h_1(k-1), h_1(k-2),\\ h_1(k-3), u_1(k), u_1(k-1), u_1(k-2), u_1(k-3)]^{\mathrm{T}};$$

Model M3:

$$\mathbf{z} = [p_1(k), p_1(k-1), p_1(k-2), h_1(k), h_1(k-1), h_1(k-2),\\ u_1(k), u_1(k-1), u_1(k-2)]^{\mathrm{T}};$$

Model M2:

$$\mathbf{z} = [p_1(k), p_1(k-1), h_1(k), h_1(k-1), u_1(k), u_1(k-1)]^{\mathrm{T}};$$

Model M1:

$$\mathbf{z} = [p_1(k), h_1(k), u_1(k)]^{\mathrm{T}}.$$

The models are compared according to their predictive performance on identification data and simulation performance on validation data. Table 6.1 shows the log marginal likelihood (ℓ) for the identification data and the standardised mean-squared error (SMSE, Eq. 2.54) and the mean standardised log loss (MSLL, Eq. 2.57) for the validation data for each model after training.

The model that has the highest probability for the identification data, i.e. the smallest negative log likelihood, is model M2. The model with the lowest MSLL for the simulation error on the validation data and the lowest SMSE for the simulation

Table 6.1 Values of the validation performance measures of the identification and validation data with the best scores in bold

Model	Id. data ℓ	Valid. data SMSE	Valid. data MSLL
M4	3594	0.0409	−1.27
M3	3586	0.0405	−1.38
M2	**3610**	0.0377	−1.52
M1	3551	**0.0341**	**−1.68**

error on the validation data is model M1. The differences in all the performance measures are small, which means that the models are quite similar. The values of the hyperparameters indicate that no regressors are to be dropped from the selection for the tested model orders. Consequently, model M1 is selected, which is in line with favouring simple models for the control design and in line with the known background of the system described with Eq. (6.13).

In [4], a slightly different way of choosing regressors is described that is based on the validation of the prediction data only, but it yields the same final regressor selection.

Consequently, the first-order model M1 of the form described with Eq. (6.17) is selected over the alternatives

$$p_1(k+1) = f(p_1(k), h_1(k), u_1(k)) + \nu. \tag{6.17}$$

Therefore, it is presumed that the process to be identified is characterised as predominantly a first-order system, which can be confirmed by looking at the step responses at various liquid levels in Fig. 6.3.

In our case, the pressure $p_1(k)$ is fed back as a distribution, and the predicted mean and variance are calculated with the analytical approximation of statistical moments with Taylor expansion as described in Chap. 2.

The vector of hyperparameters of the first-order GP model obtained by the optimisation [5] is:

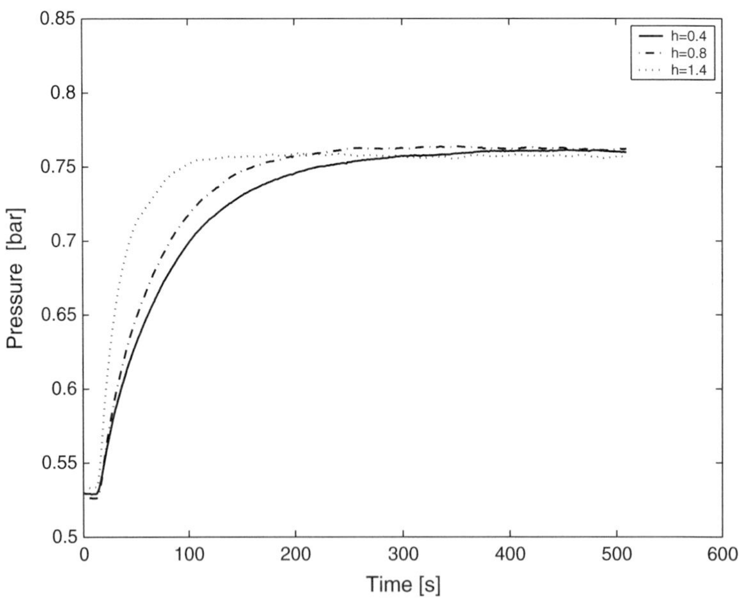

Fig. 6.3 Step responses of pressure for different values of liquid level

$$\theta = [w_1, w_2, w_3, v_0, v_1], \tag{6.18}$$

where the hyperparameter w_1 corresponds to the pressure signal p_1, w_2 corresponds to the valve signal u_1, w_3 corresponds to the level signal h_1, v_0 is the estimated white-noise variance, and v_1 is the estimate of the variance of the vertical scale of variation. Three hyperparameters w_1, w_2, and w_3 are inversely proportional to the scales of variation in the direction of each regressor that forms a three-dimensional space.

The validation signals are different from the identification signals that are given in Fig. 6.4, though they are of the same kind. All the signals were obtained from experiments on the plant using the experimentation environment described in the previous section. The valve opening for the identification signal is between 42 and 54 %, which is within the regular range in which the valve operates, i.e. between 33 and 66 %. An opening smaller than 33 % is not commonly used, because the pressure increases over the safety limits. An opening larger than 66 % is viable, but this causes very small pressure changes due to the valve's nonlinearity. Regardless of minimising the effect of the valve's nonlinearity, the system's response is still nonlinear, due to the changes in the liquid level.

The response of the model to a validation signal and a comparison with the process response obtained from experiments on the plant are given in Fig. 6.5 and response's double standard deviation in Fig. 6.6. While the valve-opening signal is noise free, the measured liquid-level and pressure signals are noisy and the noise is not white, due to the used closed-loop control of the level.

The results in Fig. 6.5 confirm the predominantly first-order dynamics, but also indicate some plant's response that is not captured by the first-order model.

Control Design

The main control objective is to achieve a uniform closed-loop performance for the pressure control in the entire liquid-level operating region. This means that the pressure dynamic response should be approximately equal, regardless of the liquid level. The way that this control objective is fulfilled using a technique different from the one proposed here can be found in [6].

The rationale for the selected case study is as follows.

- The used plant contains features of industrial processes so that the control implementation and commissioning would be a good test for other real-life applications.
- The dynamics of the plant appear relatively simple, but not all the features of the semi-industrial plant can be seen from the presented first-principle model. However, it is known that MPC algorithms can deal with complex dynamics, regardless of the model used, under the condition that the model describes the process well and reliably enough. The stress in our example is on demonstrating the feasibility of the implementation in real-time, regardless of issues such as the computational burden.

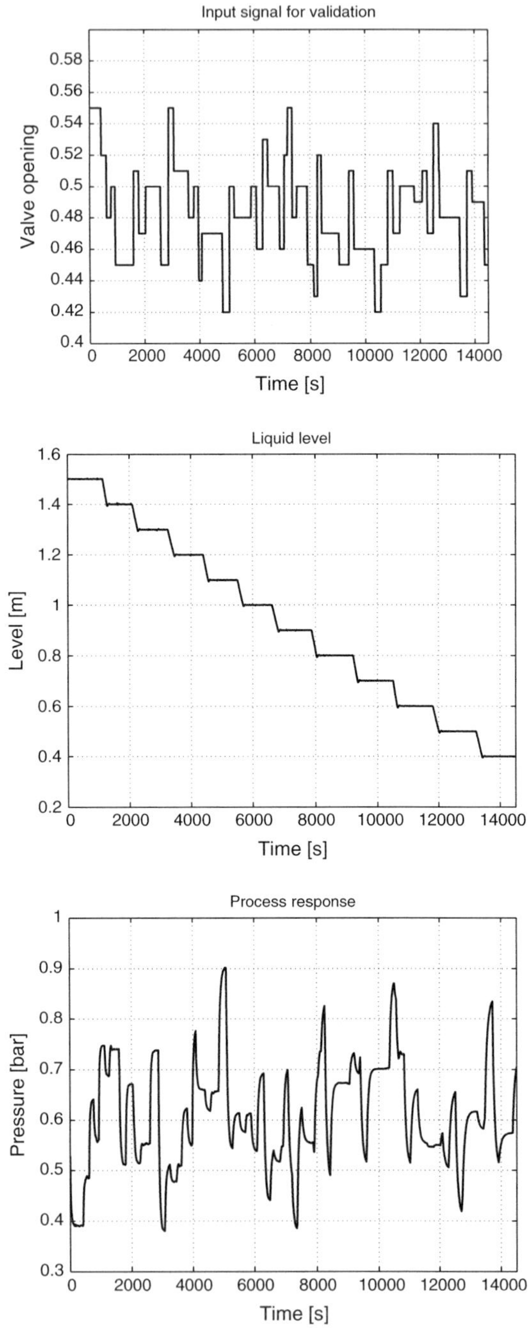

Fig. 6.4 Validation signals for the modelling of the gas–liquid separator. The valve-opening signal in the *top figure* is a noise-free pseudo-random sequence with discrete levels, the liquid level is increasing step-wise over the entire operating region and the pressure response is in the *bottom figure*

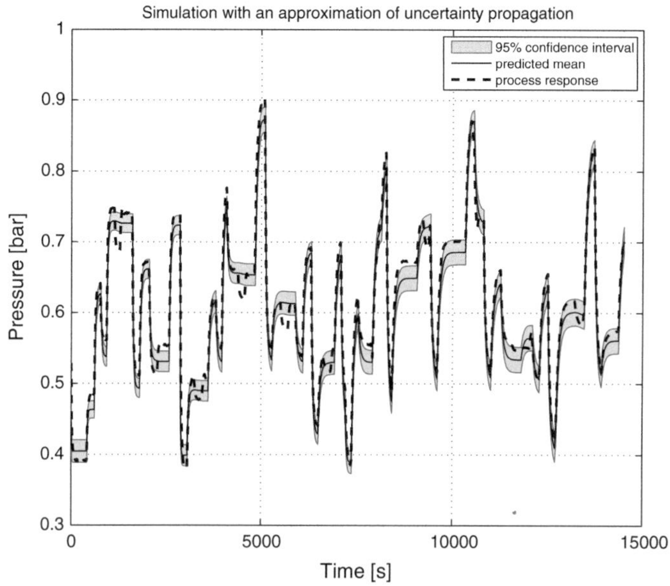

Fig. 6.5 Simulation results for the identified model on the validation input signal with the simulation with the analytical approximation of statistical moments used for the propagation of uncertainty

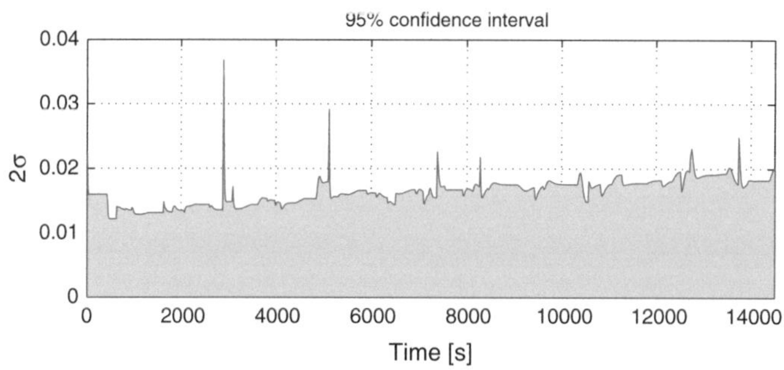

Fig. 6.6 95 % confidence interval for the response obtained with the simulation with the analytical approximation of statistical moments used for the propagation of uncertainty

- The modelling approach may show its advantages on processes where first-principle modelling is more difficult (e.g. biological and medical systems), but it is not the main purpose of this case study to demonstrate the benefits of the modelling approach. Instead, the emphasis is on the application and implementation of a Gaussian process model, which can be beneficial in process control and elsewhere.
- It can be argued that the plant can be modelled and controlled in other ways, but the choice to make the feasibility study on a familiar process was made deliberately so that the experience from other applied and differently obtained control algorithms could be utilised and comparisons could be made.

A moving-horizon minimisation problem of the special form [7]

$$\min_{\mathbf{u}} \; [r(k+P) - \hat{y}(k+P)]^2, \tag{6.19}$$

subject to

$$\mathrm{var}(\hat{y}(k+P)) \leq \sigma^2_{\max}, \tag{6.20}$$

is used in our case, where $\mathbf{u} = [u(k) \ldots u(k+P)]$ is the sequence of input-signal samples, P is the coincidence point (the point where a match between the output and the reference value is expected) and the inequality in Eq. (6.20) represents an arbitrarily selected constraint on the output variance $\sigma^2_{\max}$. It is possible to add hard constraints on other variables, but in our case they were not taken into account. The process model is a GP. Since the GP model predictions are probabilistic it is reasonable to use probabilistic constraints in the control algorithm like the constraint on the model prediction variance.

The variances are, in the case of the used stationary covariance function for the GP model, the sum of the variances that correspond to information about the regions where there are varying degrees of confidence in the model accuracy, depending upon the local density of the available identification data, and the variances of what is modelled as noise (Eq. 2.10). When the variances increase too much, a possible design option is that the response can be optimised with a constrained control. The expected consequence is that the control algorithm does not allow any excursion in the region where the accuracy of the model is below the prescribed value. This is a possible way to guarantee a safe operation based on the known accuracy of the model.

The optimisation algorithm, which is constrained nonlinear programming, is solved for each sample time over a prediction horizon of length P, for a series of moves that is equal to the control horizon. The optimisation problem is solved with the Matlab Optimization Toolbox routine for constrained nonlinear minimisation.

The reference trajectory, which defines the trajectory along which the plant should return to the set-point trajectory, is very important in defining the closed-loop behaviour of the controlled plant [7]. Often—and so it is in our case—the reference trajectory approaches the set-point exponentially from the current output value, with the time constant (T_{ref}) of the exponential defining the speed of the response. The current error is

$$e(k) = w(k) - y(k), \tag{6.21}$$

where $w(k)$ is the set-point trajectory and $y(k)$ is the current output value. The reference trajectory is chosen such that the error i steps later, if the output signal followed it exactly, would be

$$e(k+i) = e^{\frac{-iT_s}{T_{ref}}} e(k), \tag{6.22}$$

where T_s is the sampling interval. The reference trajectory is defined to be

$$r(k+i) = w(k+i) - e(k+i) = w(k+i) - e^{\frac{-iT_s}{T_{ref}}} e(k). \qquad (6.23)$$

In our case, the coincidence point was chosen to be eight samples, and the control horizon was one sample. The value of the coincidence point is determined iteratively as a compromise between the closed-loop performance and the real-time computation feasibility. This control algorithm is used for an experimental assessment of the GP model-based predictive control on a gas–liquid separation plant.

Three experiments are presented. The first is reference-tracking control for a square-wave set-point pressure signal with the level changing in the entire operating region. Nevertheless, the system operates within the region where the model is giving a good description of the system dynamics, because this is the region where the model was identified. The time constant of the reference trajectory is $T_{ref} = 150$ s. The closed-loop response and its detail are given in Figs. 6.7 and 6.8. The changing level and the manipulative signal are shown together with the model prediction variance in Fig. 6.9.

The closed-loop performances with the linear PID and gain-scheduling controller for the same process in a similar operation are described in [6] and could be used for a comparison.

It can be seen from Fig. 6.7 that the closed-loop performance is uniform, regardless of the level changes, and the predicted standard deviation from the model is, as a

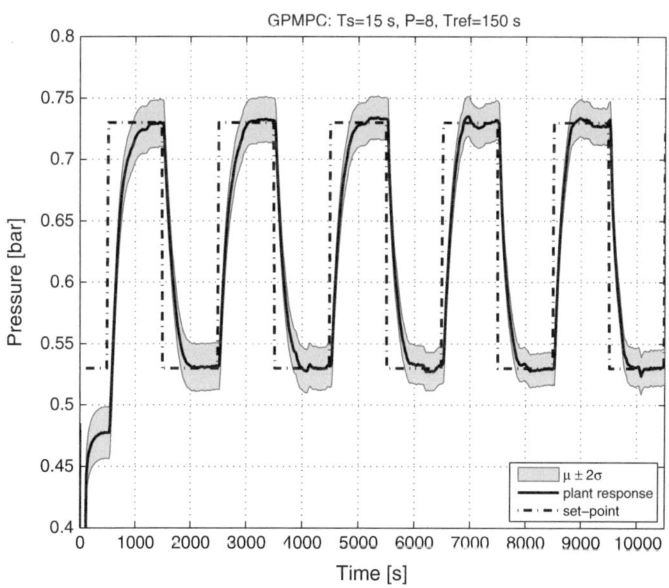

Fig. 6.7 Closed-loop pressure response for the changing liquid level

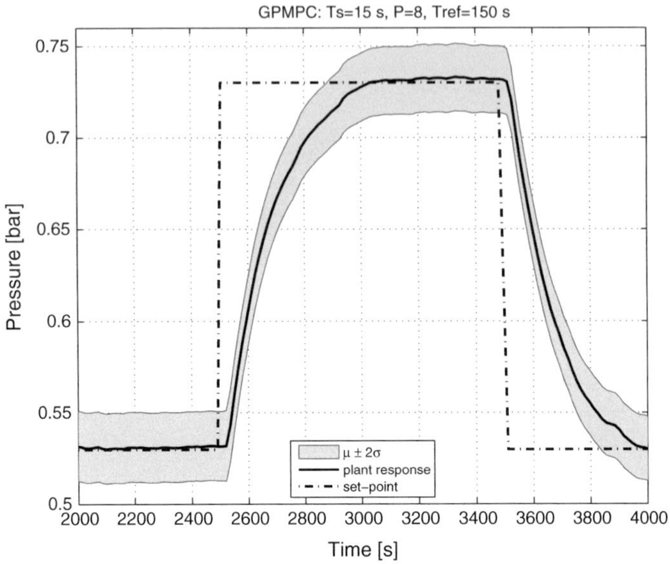

Fig. 6.8 A detail of the closed-loop pressure response for the changing liquid level

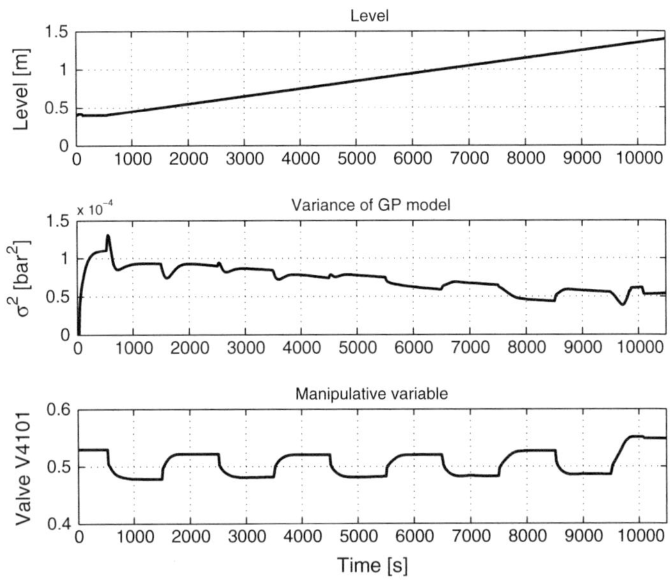

Fig. 6.9 Liquid level, variance of GP model and manipulative variable

result, low and difficult to distinguish from the response in Fig. 6.7. It appears as if the hard constraint on the variance ($\sigma^2_{\max} = 0.01$) does not play an important role. However, the situation is such that if the hard constraint is not taken into account, the optimisation algorithm gets stuck in a local minimum at a distance from the global one, away from well-modelled region where model mean predictions tend towards 0 because of the stationarity of the selected covariance function, and finds the input signal which drives the closed-loop system in the region where the model is very uncertain and the plant operation becomes hazardous. The hard constraint on the variance is actually used as an instrument to keep the closed-loop system within the operating region close to the global minimum. The global minimum or at least a minimum very close to the global one is then achieved by optimisation.

The next two experiments are conducted at the very edge of the region where the model was identified. This is the region where the accuracy of the model rapidly decreases. Both experiments are pursued at a constant low liquid level, and the pressure set-point changing step-wise from values where the model is accurate towards the values where the model is not accurate any more. The time constant of the reference trajectory is $T_{ref} = 400$ s.

In the second experiment, again the constraint of $\sigma^2_{\max} = 0.01$ is imposed on the control-variable optimisation algorithm. The value of the constraint is high enough to ensure that the predicted model variance does not reach that limit value, but, on the other hand, is tight enough to prevent the optimisation algorithm from getting stuck in a local minimum at a distance from the global one, and keeps the closed-loop system within a safe operating region, as was the case in the previous experiment. The closed-loop response is shown in Fig. 6.10. The level and manipulative signal are shown together with the model prediction variance in Fig. 6.11.

It can be seen from Figs. 6.10 and 6.11 that the standard deviation and variance of the model prediction increase rapidly when the GP model leaves the region where

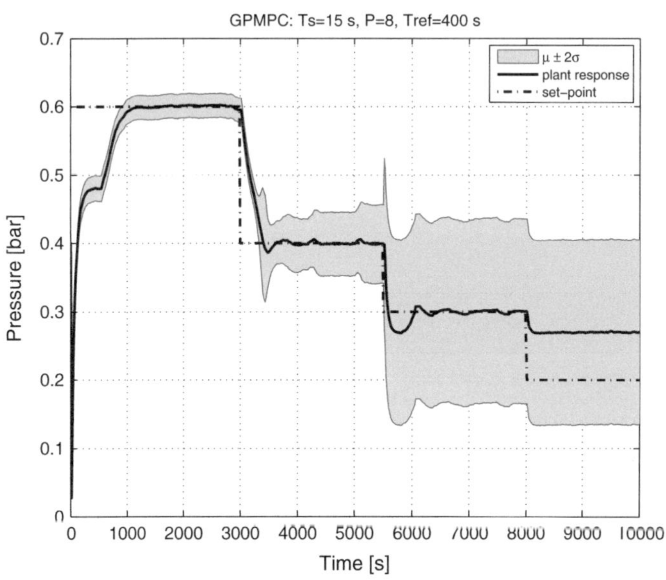

Fig. 6.10 Closed-loop pressure response at the border of the operating region

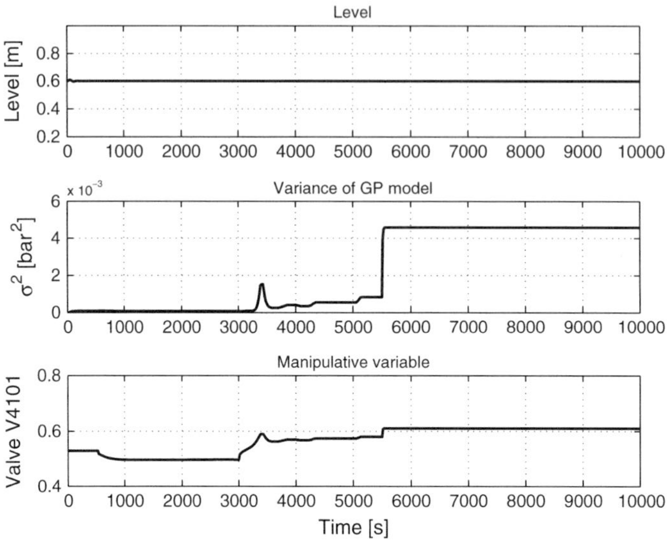

Fig. 6.11 Liquid level, variance of GP model and manipulative variable at the border of the operating region

the density of the identification data decreased due to the selection of a stationary covariance function for modelling. The mismatch between the model and the real process becomes so large that, firstly, the performance, determined by the set-point trajectory, is reduced (between 5500 and 8000 s) and later in the operation a large steady-state error occurs (beyond 8000 s). The latter means that the control algorithm not only does not follow the set-point anymore, but the uncontrollable plant operation might not be within the safety parameters. The value of the steady-state error in the region of the mismatch between the model and the process differs depending on the values of the process and the model input variables and can be obtained only by an experiment on the plant itself.

In the third experiment, the constraint of $\sigma^2_{\text{max}} = 0.00016$ is imposed on the control-variable optimisation algorithm. This constraint value allows the algorithm to operate only in the region where the process model is good enough to guarantee the performance of model-predictive control with a specified accuracy. If this constraint value is set even lower, the operating region of the plant becomes so constrained that the impaired closed-loop performance would not suit our demonstration case well.

The closed-loop response is shown in Fig. 6.12. The level and manipulative signal are shown together with the model prediction variance in Fig. 6.13.

Discussion

The value of σ^2_{max} is very important in the performance of the controller. The selection of σ^2_{max} depends on the compromise that the designer is ready to make between the performance and the dimensions of the operating region and is constrained by the available model, safety, and the stability of the closed-loop system. As the constraints

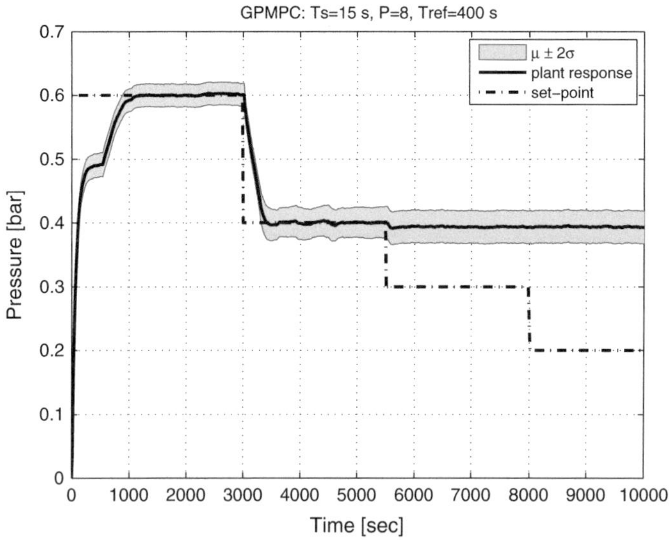

Fig. 6.12 Closed-loop pressure response at the border of the operating region with the variance constraint at 0.00016

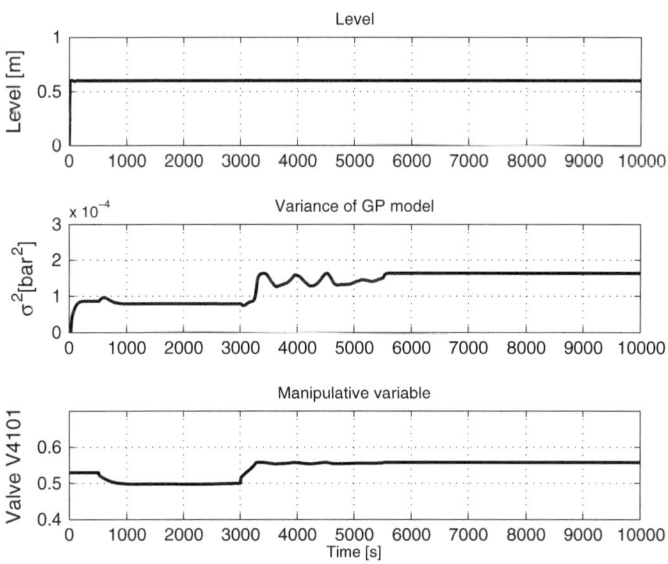

Fig. 6.13 Liquid level, variance of GP model and manipulative variable at the border of the operating region with the variance constraint at 0.00016

on variance are tightened, the performance more closely matches the specified one, and the region of operation is more contracted. However, it is important to stress that it is the model that defines the region of operation; $\sigma^2_{\max}$ only relaxes the borders of the region.

It is clear from Figs. 6.12 and 6.13 that the standard deviation and the variance of the model prediction increases when the identified GP model leaves the region where the density of the identification data decreases. In this region, the constraint comes into effect and the closed-loop system response now avoids the region with large variance, at the cost of not tracking the set-point, and therefore, an increase of the steady-state error. However, the safety of operations is ensured since the plant operates only within the known region.

Thus, the goal to design and implement a control algorithm that will guarantee operation only in the region of guaranteed performance is fulfilled with high probability.

6.2 Faulty Measurements Detection and Reconstruction in Urban Traffic

The content of this section is adopted in part from [8]. Soft sensors are models that, based on measurements of various variables provide the estimation of another variable. The expression soft sensor is mainly used in the field of the process industry, but the utility of mathematical modelling is widespread in engineering fields.

Surveys of the use of soft sensors that mainly focus on the process industry can be found in [9–11]. The approach is, however, not limited to process industries and similar ideas are utilised in other engineering fields, for example, in traffic engineering, e.g. [12, 13]. Soft sensors are used where their hardware counterparts are not available, are very costly or their installation is very costly.

Soft sensors can, in general, be divided into two different classes [9], i.e. model-driven and data-driven soft sensors. The model-driven family of soft sensors is most commonly based on theoretical or so-called first-principle models or on state estimators, e.g. extended Kalman filters. The data-driven family of soft sensors is based on data measured within the systems. The most common methods [9] for the development of data-driven soft sensors are PCA, partial least squares, artificial neural networks, neuro-fuzzy systems and support vector machines.

The most common applications of soft sensors are: online prediction, monitoring and process fault detection, sensor fault detection and reconstruction. Despite the relatively large number of soft sensor applications there are some issues that are still not sufficiently well addressed [9]. These are mainly related to measured data that need a lot of pre-processing due to: missing data, data outliers, drifting data, data co-linearity, different sampling rates and measurement delays. Even though the authors of [9] list these issues in context of process control, they are present to the certain extent everywhere that data is measured, and urban traffic is no exception.

System and Problem Description

The problem of detecting faulty measurements and its reconstruction, which is the focus of this case-study investigation has been partially addressed in the literature.

For example, papers [14–17] give surveys of the methods dealing with outlier detection and removal, while [18] describes faulty detection and signal reconstruction for dynamic systems with a lagged PCA. Outlier detection in the context of GP regression can be found in [8, 19]. The problem solutions can be found more often in the context of GP classification, e.g. [20, 21].

In urban traffic-control systems, different sensors are usually employed in order to collect relevant information about the traffic situation in the city. These sensors range from inductive loops over video-based systems to elaborate intelligent-traffic-systems installations providing toll collection and enforcement. Typical data collected by traffic-control systems include a count of the vehicles travelling over a detector, the occupancy of the detector, or the velocity of the vehicles.

Despite their higher installation costs, inductive loops are still the sensor of choice for providing vehicle counts and occupancy when installing signalised intersections. When properly designed, they last a long time and are virtually maintenance-free, compared to alternatives like video-based sensors. Nevertheless, the inductive loop itself, the connecting cables, or the interface board, which is a part of an intersection controller, may eventually fail, and sometimes the inductive-loop interface gradually de-tunes from the proper working point. When the detector stops working the hardware usually recognises the failure and reports it to the maintenance staff—however, it takes some time, e.g. from several days to several weeks, before the system is repaired.

In order to control an intersection in traffic-actuated mode, measurements provided by detectors are crucial. A typical solution in the case of recognised failure is to program the interface board of selected detectors to generate permanent requests in the case of a loop rupture. For example, the signal 'vehicle is present' is artificially generated. This provisional solution may have a severe impact on the throughput of the controlled system of intersections in the case that the detector is located on a lane experiencing low traffic and short green periods. In such a case, employing a soft sensor that is able to identify and reconstruct the measurements of the failing detector may significantly improve the traffic situation in the transient period before the detector is repaired.

A typical control scenario of an urban network similar to the one depicted in Fig. 6.14 would collect the counts of incoming vehicles on the border of the traffic-controlled area, use them to identify the current traffic demands, and finally set the signals at a particular intersection to the set-points, proving an optimum of some control criterion. In the case that some of the input sensors start to fail and become unreliable, the control system has no means to react and the traffic model used for the control of the urban network drifts off the current situation, resulting in incorrect control actions being applied.

This is the reason why we seek a simple solution to make the sensor more reliable and, due to hardware constraints, without the need to store large amount of historic data. Our aim is to devise a soft sensor for vehicle-count measurements, with the ability to evaluate the incoming data, online outlier and inconsistent data detection, and, if necessary, to reconstruct the sensor signal to its most likely mean value.

Fig. 6.14 Example of a traffic-controlled network (Zličin shopping centre, Prague) [22]

The measurements of traffic intensity (the number of vehicles per time period) and queue length are, from the microscopic view, measurements of discrete events. Only from the macroscopic view they may look like a signal that is noisy. The candidates for modelling such time series would be methods that are applicable and give acceptable results in the presence of substantial noise, give a prediction without a major delay or are quick enough that they can be used for online calculations and that the obtained prediction confidence interval is calculated with a very modest amount of calculation. The modelling of GP models is applied in our case.

The GP prior is put over the space of the functions meaning that every prediction that is made by the GP model has a Gaussian distribution. This implies that every prediction that is made has some most likely value and the less likely values are equally possible on both sides of the most likely value. This is, from our point of view, not unrealistic.

In our case a nonlinear, autoregressive GP model or GP-NAR model is considered, such that the current output value depends on previous output values up to a given lag n.

$$y(k) = f(y(k-1), y(k-2), \dots, y(k-n)) + \nu. \tag{6.24}$$

Autoregressive models [23] are often used when external inputs are not known, which is quite common in, e.g. econometrics, finances, environmental sciences, physics, etc.

Just for the sake of a comparison we also consider the GP-NAR model that can be used for periodic functions and uses regressors that are lagged for multiples of the signal's period T_p.

$$y(k) = f(y(k - T_p), y(k - 2T_p), \ldots, y(k - nT_p)) + \nu. \tag{6.25}$$

This model requires the storage of historic data for the used periods and is, because of the constrained hardware memory, not viable for our problem solution. Nevertheless, it is an intriguing solution for a model comparison.

GP-NAR models can be found in modelling the applications of building-energy consumption [24], electrical-power network load forecasting [25], mine gas emissions [26], landslide displacement [27], stock-market modelling [28, 29], etc. The modelling of autoregressive conditional heteroscedasticity models using GP models is elaborated in [28].

Modelling

The data samples studied in this paper are extracted from a collection of traffic-intensity measurements taken at the test site of the Czech traffic-control project in the Zličin area of Prague. This area contains mostly shopping centres and a bus terminal serving both long-haul buses and regular public transport.

The whole collection of measurements covers the period from December 10, 2007 to September 2, 2008 for all the detectors installed at the site. According to the apparent differences between the data, the data collection is divided into three subsets representing working days, i.e. Mondays to Fridays, weekends, i.e. Saturdays and Sundays, and holidays, i.e. work-free days. From this collection, two separate data sets of working detector data for each of the three types of traffic-intensity data are extracted for the GP model identification and validation—see the example, for typical working-days data in Fig. 6.15. This means that the final model is composed of three sub-models for each type of data. The identification data set consisted of data from two typical days and the validation set of data from three typical days. The identification data set is selected so that the training does not consume too many resources. The alternative is to use sparse or online learning methods, as shown in the case study of Sect. 6.3.

Yet another data set containing failing detector data from two consecutive weeks is used to test the failure identification and reconstruction capabilities of the model.

The selection of covariance function, model order and regressors is pursued among a fairly large number of combinations of covariance functions and regressors. Since the purpose of the model is reconstruction of the input signals to its most likely mean value, the model predictions are compared to the values of the validation signals that represent the most likely mean values.

These most likely mean values of the validation signals are obtained by smoothing the validation signals using GP model regression. The stationary covariance function,

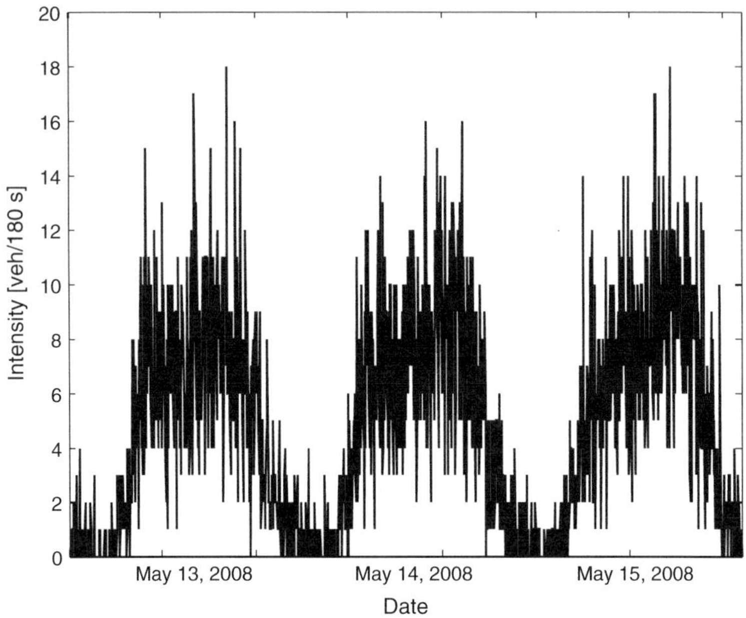

Fig. 6.15 The profile of vehicle arrival data for three consecutive working days, sampled with 180 s interval

i.e. the squared exponential covariance function described with Eq. (2.13) and the sample instants k representing time, are used as the kernel function and the regressor, respectively.

Among the large number of tested combinations, a selection of the best results is given in Table 6.2 for the model of working days, in Table 6.3 for the model of weekends and in Table 6.4 for the model of holidays. The tables show the average of 2-fold cross-validation results: the log marginal likelihood for selected identification data, and the performance measures SMSE (Eq. 2.54) and MSLL (Eq. 2.57) for the identification and validation data. Validation is pursued with respect to the mean target values of the preprocessed validation signal.

Linear covariance functions with regressors that represent the history of 39 min for weekend days, 45 min for working days and 57 min for holidays perform the best for the models of working days, weekends and holidays. The predictions for any kind of day can be obtained by switching among the three models according to the type of day.

An excessive amount of noise contributed to the relatively low quality of the finally selected model predictions. Note that a different selection of identification data from the available measurements ends with slightly different models, but the used selection is demonstrated in the continuation to be representative enough for our purposes.

Table 6.2 Model-order and covariance function selection for working-days

Cov. function	Order	Ident. data			Valid. data	
		ℓ	$\text{SMSE}_{\text{2-fold}}$	$\text{MSLL}_{\text{2-fold}}$	SMSE	MSLL
SE	7	−2128	0.3880	−0.4744	0.0614	−0.7816
SE + ARD	6	−2130	0.3901	−0.4712	0.0682	−0.7750
LIN + ARD	15	**−2099**	**0.3825**	**−0.4817**	**0.0521**	**−0.7979**
SE + ARD + LIN + ARD	4	−2146	0.3954	−0.4654	0.0843	−0.7511
NN	8	−2122	0.3888	−0.4752	0.0565	−0.7876
Matérn $d = \frac{3}{2}$	5	−2136	0.3934	−0.4678	0.0725	−0.7679

Validation data is evaluated with respect to the target mean values. Notation in the table: ℓ—log marginal likelihood for selected identification data, SMSE and MSLL—prediction performance measures, SE—square exponential covariance function, LIN—linear covariance function, NN—neural network covariance function, ARD—the variant of covariance function with ARD property

Table 6.3 Model-order and covariance function selection for the model of weekends

Cov. function	Order	Ident. data			Valid. data	
		ℓ	$\text{SMSE}_{\text{2-fold}}$	$\text{MSLL}_{\text{2-fold}}$	SMSE	MSLL
SE	5	−2099	0.3922	−0.4699	0.0962	−0.5560
SE + ARD	5	−2098	0.3920	−0.4706	0.1134	−0.5540
LIN + ARD	13	**−2070**	**0.3785**	**−0.4876**	**0.0613**	**−0.5815**
SE + ARD + LIN + ARD	3	−2127	0.4253	−0.4336	0.1650	−0.4934
NN	5	−2095	0.3925	−0.4691	0.1017	−0.5540
Matérn $d = \frac{3}{2}$	3	−2129	0.4180	−0.4407	0.1599	−0.4964

Validation data is evaluated with respect to the target mean values. Notation in the table: ℓ—log marginal likelihood for selected identification data, SMSE and MSLL—prediction performance measures, SE—square exponential covariance function, LIN—linear covariance function, NN—neural network covariance function, ARD—the variant of covariance function with ARD property

Table 6.4 Model-order and covariance function selection for holidays' model

Cov. function	Order	Ident. data			Valid. data	
		ℓ	$\text{SMSE}_{\text{2-fold}}$	$\text{MSLL}_{\text{2-fold}}$	SMSE	MSLL
SE	5	−2099	0.3922	−0.4699	0.1579	−0.2385
SE + ARD	5	−2098	0.3920	−0.4706	0.1882	−0.2466
LIN + ARD	19	**−2056**	**0.3783**	**−0.4884**	**0.0670**	**−0.2971**
SE + ARD + LIN + ARD	3	−2127	0.4253	−0.4336	0.2823	−0.1852
NN	5	−2095	0.3925	−0.4691	0.1501	−0.2400
Matérn $d = \frac{3}{2}$	3	−2111	0.4695	−0.3806	0.2800	−0.1790

Validation data is evaluated with respect to the target mean values. Notation in the table: ℓ—log marginal likelihood for selected identification data, SMSE and MSLL—prediction performance measures, SE—square exponential covariance function, LIN—linear covariance function, NN—neural network covariance function, ARD—the variant of covariance function with ARD property

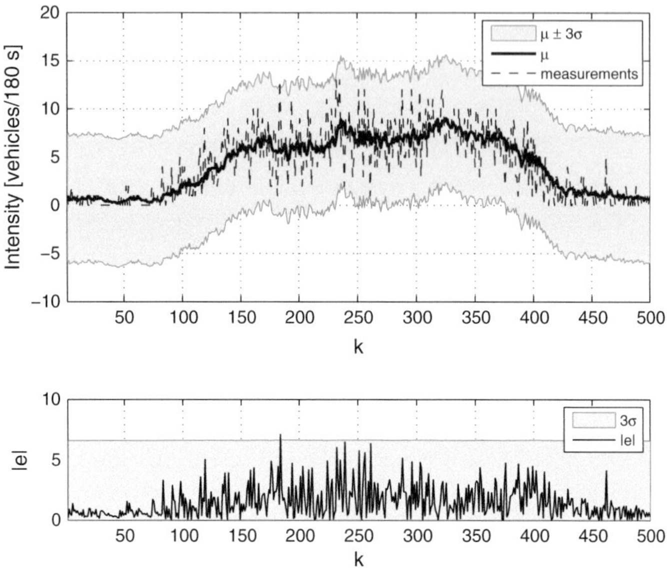

Fig. 6.16 One-step-ahead predictions on the selection of identification data representing one working day (*upper figure*) and residuals of predictions with 99 % interval (*lower figure*)

One-step-ahead predictions on estimation data are given in Fig. 6.16. It can be seen from Fig. 6.16 that the residuals are contained in the 3σ interval approximately corresponding to a 99 % confidence interval. The 3σ interval is selected for the recognition of failure in data, as we will see later.

One-step-ahead predictions on validation data that are taken from measurements on a different day in a different week are shown in Fig. 6.17.

Figure 6.18 shows how the values of the predictions deteriorate with a progressive number of steps in multistep-ahead prediction. These multi-step-ahead predictions are obtained with iterative, i.e. 'naive', computation, so that the predicted mean values are used to replace the missing values of delayed measurements in the regression vector progressively.

At this point, we would like to show briefly a possible alternative if the performance of the iterative multi-step-ahead predictions is not to our satisfaction and under the condition that there is enough storage memory available in the hardware where the identified model runs. This is the model described with Eq. (6.26). Since the mentioned condition about memory is not satisfied in our case, the rigorous selection of model order is not pursued. Three regressors are selected as samples delayed for a week, two and three weeks, respectively. The rationale behind the selection is that the three-week period is short enough not to be affected by seasonal changes and long enough to be robust with respect to the emerging daily changes. The identified model has regressors as follows:

$$y(k) = f(y(k - T_p), y(k - 2T_p), y(k - 3T_p)) + \nu. \tag{6.26}$$

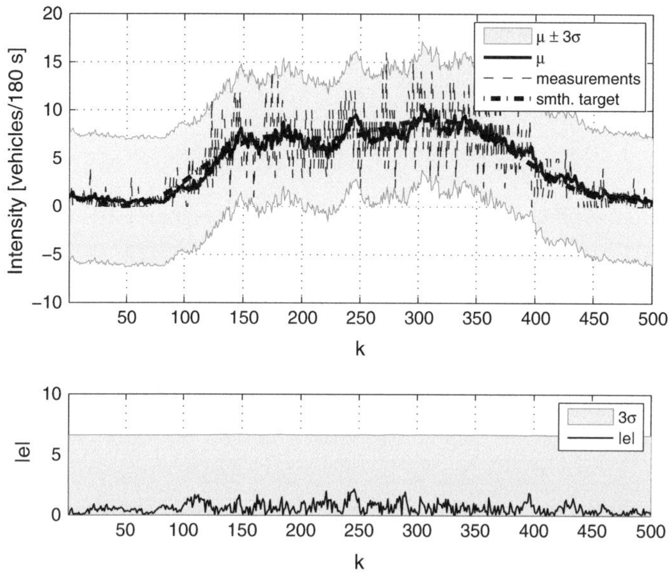

Fig. 6.17 One-step-ahead predictions on the selection of validation data representing one working day (*upper figure*) and residuals of predictions with 99 % interval (*lower figure*)

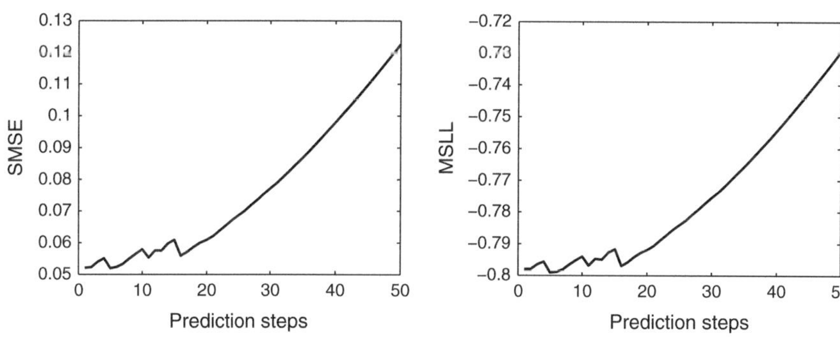

Fig. 6.18 The comparison of SMSE (*left figure*) and MSLL (*right figure*) for multistep-ahead predictions

The selection of the covariance function and the best model-identification results as well as the validation results are given in Table 6.5.

It is clear from Table 6.5 that the model with a covariance function that is a sum of the squared exponential and linear covariance functions, both with ARD properties, performs the best on the identification data. The validation results are not in favour of this model, but the differences are acceptable in our case.

Table 6.5 Alternative model for periodic responses: Covariance function selection for the working-days model

Cov. function	Ident. data			Valid. data	
	ℓ	SMSE	MSLL	SMSE	MSLL
SE	−1120	0.4055	−0.4525	**0.1120**	−0.6448
SE + ARD	−1125	0.3742	−0.4935	0.1471	−0.6286
LIN + ARD	−1133	0.4427	−0.4078	0.1775	−0.5676
SE + ARD + LIN + ARD	**−1113**	**0.3543**	**−0.5201**	0.1422	−0.6388
NN	−1114	0.4036	−0.4540	0.1123	**−0.6495**
Matérn $d = \frac{3}{2}$	−1118	0.3923	−0.4694	0.1125	**−0.6495**

Validation data is evaluated with regards to target mean values. Notation in the table: ℓ—log marginal likelihood for selected identification data, SMSE and MSLL—prediction performance measures for selected validation data, SE—square exponential covariance function, LIN—linear covariance function, NN—neural network covariance function, ARD—the variant of covariance function with ARD property

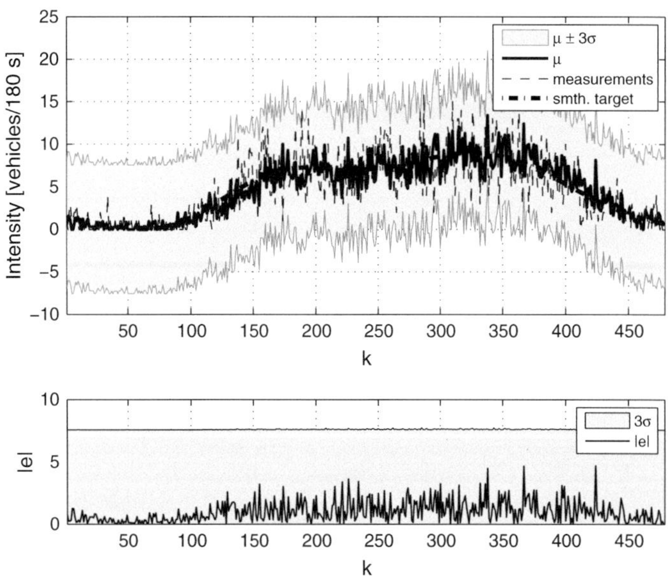

Fig. 6.19 Alternative model predictions on the selection of validation data representing one working day (*upper figure*) and residuals of predictions with 99 % interval (*lower figure*)

Predictions based on validation data that are taken from measurements on a different day in a different week are shown in Fig. 6.19. The visual inspection of the prediction responses does not show large differences in the model's input/output consistency with the primary GP-NAR model from Fig. 6.17. The mean value of GP-NAR model from Fig. 6.17 is more noisy because of the noise in regressors used for the model identification. This is the errors-in-variables problem. A possible way to overcome it, demonstrated for the squared exponential covariance function, is described in [30].

Using the model described with Eq. 6.26 we avoid the deterioration of the performance obtained with an iterative multistep-ahead prediction at the cost of keeping the necessary amount of data history. However, the presumed hardware constraints excluded the alternative model described with Eq. 6.26. In the continuation, the fulfilment of other requirements by the primary GP-NAR model is validated.

The model's purposiveness or usefulness tells us whether or not the obtained model satisfies its purpose, which means the model is validated when the problem that motivated the modelling exercise can be solved using the obtained model. In our case, the model has to give an output prediction where the standard deviation of the prediction will be used as a threshold of the measurement validity. In the case that the measurement is recognised as 'not valid' it is replaced with a prediction that might not be mirroring the actual value, but will be close enough so that the control system will continue its regular operation without exhibiting any extreme modes. Our model, even though not expressing high prediction accuracy, satisfies all the described requirements and is able to calculate the predictions quickly.

The threshold at which a measurement is characterised as not realistic is set at 3σ, i.e. at three times standard deviation of the output distribution given by the GP model prediction. This is the threshold frequently used in offline outliers detection algorithms, e.g. [16].

The proposed algorithm that runs in soft sensor and online in every time sample detects irregularities and tries to reconstruct the data is as follows:

Algorithm: FAILURE DETECTION(M)

---

repeat
  take the measurement M
  calculate the prediction $P = \mathcal{N}(\mu, \sigma^2)$
  if $|M - \mu| > 3\sigma$
    then $\begin{cases} M \leftarrow P \\ \text{indicate possible fault} \end{cases}$
until end of sample period

Figure 6.20 shows the detection of faulty data and their reconstruction by a model, i.e. soft sensor.

As can be seen from the proposed algorithm, the reconstruction is made as a replacement of the faulty measurement with model predictions. In the case that the fault is not just an outlier, but a longer period fault, then the model starts to predict the queue length until the operator detects and removes the fault.

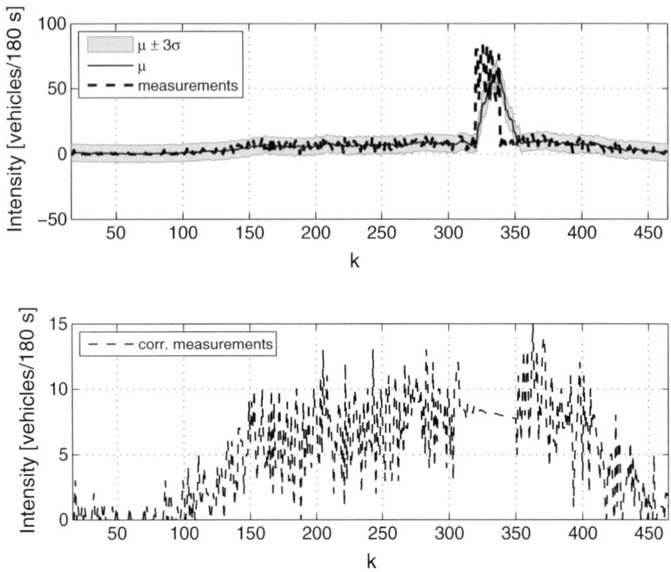

Fig. 6.20 Soft sensor applied on faulty data. *Upper graph* depicts measurements that are corrupted from the 320th sample until the 340th sample and predictions of soft sensor. *Bottom graph* depicts measurements with reconstructed samples of corrupted measurements

Discussion

This case study is about a possible application of a time-series autoregressive model as the soft sensor for faulty measurement detection and reconstruction in urban traffic. The proposed method for developing a computational model that serves as a soft sensor is based on a GP model. The resultant soft sensor detects single-measurement outliers as well as longer-lasting faults and will be used to replace the faulty measurements with a model prediction. In this way, it will increase the reliability of the hardware sensors that are used in an urban traffic-control system.

The main conclusions are as follows:

- GP models can be used as a method for modelling data when an excessive amount of noise is present.
- The obtained model that will serve as a soft sensor was validated as appropriate for fulfilling its task, i.e. outlier detection and the reconstruction of measurements.
- Soft sensors are also a useful technology for fault detection and signal reconstruction in traffic-control engineering.

The presented study is a feasibility study, the purpose of which is to show the potentials of the soft-sensor approach based on a GP model. The results for working days are mainly demonstrated, but the same valid conclusions can be drawn for the other two types of data, i.e. weekends and holidays.

6.3 Prediction of Ozone Concentration in the Air

The content of this section is adopted in part from [31]. Ozone (O_3) [32, 33] is a highly unstable and poisonous gas that can form and react in the atmosphere under the action of light and is present in two layers of the atmosphere. Ozone is a very specific air substance that is present throughout the whole of the Earth's atmosphere, from ground level to the top of the atmosphere. The stratospheric ozone prevents the harmful solar ultraviolet radiation from reaching the Earth's surface. However, in the tropospheric layer, which is at ground level, the ozone is an air pollutant that damages human respiratory system and the equilibrium of the ecosystem [34]. Overexposure to ozone can cause serious health problems in plants and people, and, thus, ozone pollution is a major problem in some regions of the world. It tends to increase during periods with high temperatures, intense sunlight, lots of air pollutants and stable air masses [32]. The ozone content changes in the troposphere, and the complexity of the processes defining these changes is the reason why the atmospheric ozone dynamics is the object of intensive research.

The most direct way to obtain accurate air-quality information is from measurements made at surface-monitoring stations across many countries. Fixed measurements of the hourly ozone concentrations in compliance with the European Directive on ambient air quality and cleaner air for Europe [35] give continuous information about the evolution of the surface ozone pollution at a large number of sites across Europe. In several European Union member states they are increasingly being supplemented by numerical model outputs delivered on a regional or local scale, in keeping with the European directive. The European standards that guarantee human-health protection are as follows: 'health protection level', 120 $\mu g/m^3$ eight hours mean concentration; 'informing the public level', 180 $\mu g/m^3$ one hour mean concentration; and 'warning the public level', 240 $\mu g/m^3$ one hour mean concentration. Therefore, predicting the ozone concentration and informing the population when the air-quality standards are not being met are important tasks.

The objective of this case study is to present a method for the prediction of the ozone concentration based on an online updated dynamic GP model obtained from measurement data, as described in Sect. 2.4.3, for the city of Burgas in Bulgaria. Ground-based measurements of the air quality have been carried out in places spread all over Bulgaria, similar to the situation in other countries in the world. These measurements are in the form of a series of simultaneous observations of the time evolution of the surface ozone concentrations for different meteorological conditions.

Ozone concentration can be modelled and forecasted using a variety of methods, and methods that describe the nonlinear dynamics from the available data are particularly useful. Thus, there are a number of methods for ozone-concentration prediction based on various modelling techniques, e.g. based on neural network NARX models (e.g. [36–38]), polynomial NARX models (e.g. [39]), fuzzy systems (e.g. [40, 41]), support vector machines (e.g. [42]), ARIMA stochastic models (e.g. [43]), and GP models (e.g. [37, 44, 45]). There are also methods that are based on a combination of some of the mentioned techniques, e.g. the approach in [46] combines the use of neural networks, support vector machines, and genetic algorithms.

Process and Problem Description

Burgas is in one of the regions in Bulgaria (Fig. 6.21) with the highest levels of ozone pollution in the air, which makes it important to obtain a prediction model for this region.

In [45], dynamic GP models of the first order based on measurements of the air-pollutant concentrations are identified and verified for one-step-ahead predictions of the ozone concentration in the air of Burgas. Furthermore, in [44] dynamic GP models of higher order are identified and verified by using measurements of both the air pollutants and the meteorological parameters. In both cases the GP models are trained offline using only a subset of the available data due to the high computational burden of modelling GP models. However, this limitation and, consequently, the quality of the GP models can be improved with sparse-matrix methods and online updating using the most recent measurements.

The purpose of this case study, which builds upon and extends the results of [31, 44, 45], is to illustrate a solution to the problem of ozone prediction that is operational throughout the year and is based on online learning of the model used for the prediction. The proposed solution is based on an online Gaussian process (OGP) [47] model that can also be utilised for few-hour-ahead predictions of the ozone concentration with the necessary amendments. For this purpose, the sparse OGP learning method is used in [48]. As the weather and its characteristics are constantly changing, the model should be updated and adjusted as well. This means it should not only update the model with information contained in the streaming

Fig. 6.21 Map of Bulgaria with Burgas

data, but it should concurrently optimise the hyperparameter values as well. The OGP method does not rely on selection heuristics. Usually, it requires several passes over the data and could exhibit some numerical instabilities, as reported in [49] and as we experienced ourselves. Therefore, what in our opinion is a more robust method is used in this case study for the online updating, i.e. evolving, of a GP model. Its performance is compared with offline trained GP models. The case study, adopted from [31], illustrates the method from Sect. 2.4, where the concept of the GP model evolving is described.

Modelling

The ozone concentration has a strong daily cycle. Figure 6.22 shows the daily cycle of the average hourly ozone concentrations for all seasons. It is clear that the average ozone concentration has a minimum at 5–8 h, depending on the season. After 6–8 h it increases rapidly, i.e. the formation and collection of ozone in the air starts after 6–8 h. The maximum is reached at 11–16 h. Therefore, if we are able to predict correctly the ozone concentration at that hourly interval, we will be able to predict the maximum hourly concentration for the day in all cases of high health risk.

We only have measurements from 2008, with some data gaps, available. Only measurements where complete data for days are available were used for the modelling. There are 194 of such days. This data set was then randomly re-sampled into the identification or training data set, which is about 30 % of the complete data set, and the validation data, which is about 70 % of the complete data set.

The identification data set is then further divided into ten subsets and a 10-fold cross-validation is used for the model selection.

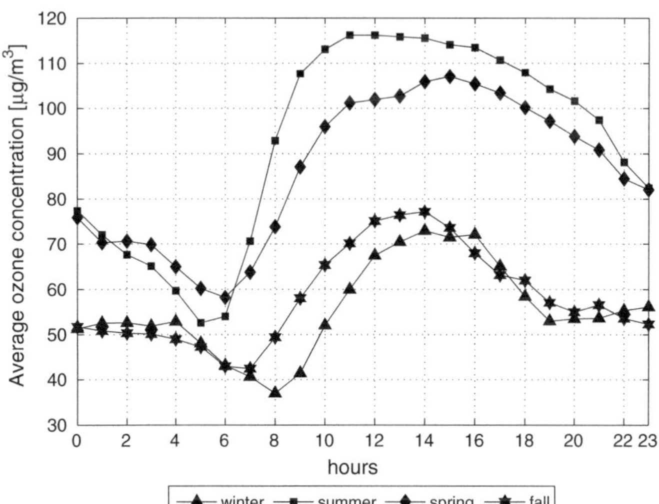

Fig. 6.22 Average hourly ozone concentrations for each season

Table 6.6 The log marginal likelihoods (ℓ), SMSE and MSLL measures for 10-fold cross-validation on the identification data set and on the validation data set for first-order models

Model	Ident. data			Valid. data	
	ℓ	$\text{SMSE}_{10\text{-fold}}$	$\text{MSLL}_{10\text{-fold}}$	SMSE	MSLL
A1	630.5	0.1876	−0.8333	0.1604	−0.9105
B1	754.2	**0.1437**	**−0.9620**	**0.1330**	**−1.0100**
C1	720.9	0.1617	−0.9066	0.1626	−0.9080
D1	**758.9**	0.1605	−0.9149	0.1365	−0.9969

Model Structure Investigation

Four first-order GP models, i.e. models containing the output variable with one delay as one of the regressors, for the prediction of the ozone concentration are identified and verified based on the preprocessed, available measurement data. In addition to the log-likelihoods ($\mathcal{L}$) of the identified models, the standardised mean-squared error (SMSE) and the mean standardised log loss (MSLL) computed for tenfold cross-validation on the training data and for the simulation with the validation data are given in Table 6.6.

Various combinations of regressors were investigated. Let us present the most representative models among them also from the physical aspect.

The four identified first-order GP models have the following regressors:

Model A1: The value of the ozone concentration in the previous hour:

$$c_{O_3}(k) = f_{\text{A1}}(c_{O_3}(k-1)) + \nu \tag{6.27}$$

where k is the current hour of the day and c_{O_3} is the concentration of ozone in the air.

Model B1: The values of the ozone concentration, the concentrations of the air pollutants, and the meteorological parameters in the previous hour:

$$\begin{aligned} c_{O_3}(k) = f_{\text{B1}}(&c_{O_3}(k-1), c_{NO_2}(k-1), c_{SO_2}(k-1), c_{C_6H_5OH}(k-1),\\ &c_{C_6H_6}(k-1), h(k-1), p(k-1), sr(k-1), temp(k-1),\\ &ws(k-1)) + \nu \end{aligned} \tag{6.28}$$

Here, c_{NO_2}, c_{SO_2}, $c_{C_6H_5OH}$, and $c_{C_6H_6}$ are the concentrations of nitrogen dioxide, sulphur dioxide, phenol, and benzene in the air, h is the air humidity, p is the air pressure, sr is the sun's radiation, $temp$ is the air temperature and ws is the wind speed.

Model C1: The values of the ozone concentration and the concentrations of the air pollutants in the previous hour:

$$\begin{aligned} c_{O_3}(k) = f_{\text{C1}}(&c_{O_3}(k-1), c_{NO_2}(k-1), c_{SO_2}(k-1),\\ &c_{C_6H_5OH}(k-1), c_{C_6H_6}(k-1)) + \nu \end{aligned} \tag{6.29}$$

Table 6.7 The log marginal likelihoods (ℓ), SMSE and MSLL measures for 10-fold cross-validation on the identification data set and on the validation data set for high-order models

Model	Ident. data			Valid. data	
	ℓ	$\text{SMSE}_{10\text{-fold}}$	$\text{MSLL}_{10\text{-fold}}$	SMSE	MSLL
A2	746.6	0.1663	−0.9006	0.1530	−0.9357
A3	650.8	0.1729	−0.766	0.1477	−0.9592
B2	**805.5**	**0.1405**	**−0.9862**	**0.1254**	**−1.0444**
C2	724.8	0.1642	−0.9076	0.1554	−0.9602
D2	769.3	0.1461	−0.9659	0.1255	−1.0417

Model D1: The values of the ozone concentration and the meteorological parameters in the previous hour:

$$c_{O_3}(k) = f_{\text{D1}}(c_{O_3}(k-1), h(k-1), p(k-1), sr(k-1), temp(k-1), ws(k-1)) + \nu \tag{6.30}$$

In Table 6.6, the best obtained values for all the performance measures are given in bold. It is clear that the best values are obtained with the model B1. This is expected from the physical point of view. We also investigated high-order models as well. Again, we present only the most representative models.

Four second-order GP models and one third-order GP model for the prediction of the ozone concentration are identified and validated. The training and the validation data are formed in the same way as for the first-order models. The performance measures are given in Table 6.7.

The presented identified high-order GP models have the following regressors:

Model A2: The values of the ozone concentration in the two previous hours

$$c_{O_3}(k) = f_{\text{A2}}(c_{O_3}(k-1), c_{O_3}(k-2)) + \nu. \tag{6.31}$$

Model A3: The values of the ozone concentration in the three previous hours

$$c_{O_3}(k) = f_{\text{A3}}(c_{O_3}(k-1), c_{O_3}(k-2), c_{O_3}(k-3)) + \nu. \tag{6.32}$$

Model B2: The values of the ozone concentration, the concentrations of the air pollutants, and the meteorological parameters in the two previous hours

$$\begin{aligned} c_{O_3}(k) = \; & f_{\text{B2}}(c_{O_3}(k-1), c_{NO_2}(k-1), c_{SO_2}(k-1), c_{C_6H_5OH}(k-1), \\ & c_{C_6H_6}(k-1), h(k-1), p(k-1), sr(k-1), temp(k-1), ws(k-1), \\ & c_{O_3}(k-2), c_{NO_2}(k-2), c_{SO_2}(k-2), c_{C_6H_5OH}(k-2), \\ & c_{C_6H_6}(k-2), h(k-2), p(k-2), sr(k-2), temp(k-2), ws(k-2)) + \nu. \end{aligned} \tag{6.33}$$

Model C2: The values of the ozone concentration and the concentrations of the air pollutants in the two previous hours

$$\begin{aligned} c_{O_3}(k) = f_{C2}(&c_{O_3}(k-1), c_{NO_2}(k-1), c_{SO_2}(k-1), c_{C_6H_5OH}(k-1),\\ &c_{C_6H_6}(k-1), c_{O_3}(k-2), c_{NO_2}(k-2), c_{SO_2}(k-2),\\ &c_{C_6H_5OH}(k-2), c_{C_6H_6}(k-2)) + \nu. \end{aligned} \tag{6.34}$$

Model D2: The values of the ozone concentration and the meteorological parameters in the two previous hours

$$\begin{aligned} c_{O_3}(k) = f_{D2}(&c_{O_3}(k-1), h(k-1), p(k-1), sr(k-1), temp(k-1),\\ &ws(k-1), c_{O_3}(k-2), h(k-2), p(k-2), sr(k-2),\\ &temp(k-2), ws(k-2)) + \nu. \end{aligned} \tag{6.35}$$

It is clear from Tables 6.6 and 6.7 that for the same model type, the higher-order models are more accurate than the first-order models (model A2 is more accurate than model A1, model B2 is more accurate than model B1, etc.).

It is clear from Table 6.7 that the best performance is obtained for the model B2. However, the results obtained with model D2, which is much simpler than model B2, are not that different. Other higher-order GP models were also tested, but they did not provide a significant improvement in the prediction quality. It might be argued that the increase in the complexity of model B2 and those of higher order not shown here is not worth the computational effort, in terms of the predictive performance of these models in comparison to model D2. Favouring simple models in the face of the complexity/performance trade-off model D2 is selected and used in the subsequent investigation.

All the models considered in this section are trained offline on the data of a one-year period only. As the weather is constantly changing, an alternative approach that updates with in-coming data is used as follows.

Modelling with Evolving GP Models

Data for the whole period of interest, in our case the year 2008, was available for training and validation. It should be noted that the same data was not used for the training and the validation, but the data was from the same period. Therefore, all the characteristics of the data from the whole period can be trained. In the continuation we use an alternative approach to training GP models that is needed when the training data is not available for the whole period of interest, and consequently, not all the characteristics of the period can be trained. In this case, the model should be constantly adapted in accordance with the new characteristics. Such an approach can be used in environments that are constantly changing. As the weather is constantly changing, this approach seems promising.

To assess the effectiveness of the proposed method we divided the data into four periods: astronomical winter, astronomical spring, astronomical summer and astronomical autumn. The astronomical winter lasts from December 21 to March

20, the astronomical spring lasts from March 21 to June 20, the summer lasts from June 21 to September 22 and the autumn lasts from September 23 to December 20. The division is like this because the data is only available for the year 2008 and it can be shown (Fig. 6.22) that the selected seasons have different characteristics.

From Fig. 6.22 it can be seen that the selected four groups can be further divided into two main groups. In the first group are the seasons spring and summer, which have a high-ozone concentration from 10 to 17 h, and in the second group are the seasons autumn and winter, which have a high-ozone concentration from 12 to 15 h. The first group has a much higher and longer high-ozone-concentration peak than the second group.

To demonstrate the effectiveness of the online training model in the case of changing characteristics we performed both an offline and online, without forgetting, training of the GP models on winter-season data and validated the models on summer-season data. It should be noted that as the basis for the online training, the offline trained model was used and then this was constantly adapted through the validation data, i.e. the whole year. Considering the previously made analysis of the regressors, the model structure D2 was used.

The predictions of the ozone concentration for one day (July 20, 2008) based on both models are shown in Fig. 6.23 and their performance measures are written in Table 6.8. This day was chosen as it is from the summer season and it has a harmful ozone concentration.

It is clear that the predictions of the ozone concentration based on the online trained model are much more accurate than the predictions based on the offline trained model, especially in the period between 10 and 18 h, when the ozone concentration usually reaches harmful levels.

To analyse the influence of the exponential forgetting on predictions of the ozone concentration, several evaluations of the online trained GP models with various

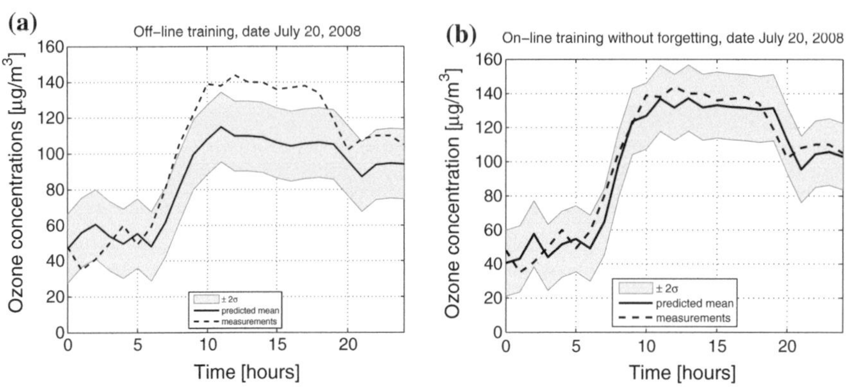

Fig. 6.23 The predicted mean value and 95 % confidence interval of the ozone concentration for July 20, 2008: **a** using offline training in the winter season; **b** using online training through the whole validation data, i.e. the whole year

Table 6.8 Performance measures of the ozone-concentration predictions for one day (July 20, 2008) based on an offline trained and an online trained model.

Model	SMSE	MSLL
Offline	0.3469	0.7130
Online	**0.0530**	**−1.4349**

Table 6.9 Performance measures of the ozone-concentration predictions for all the available data based on online trained models with forgetting factors λ: 1, 0.99995, 0.99985, 0.99975 and 0.99965

Model		SMSE	MSLL
Online	$\lambda = 1$	0.12082	−1.0293
	$\lambda = 0.99995$	0.12079	−1.03577
	$\boldsymbol{\lambda} = \mathbf{0.99985}$	**0.12019**	**−1.04904**
	$\lambda = 0.99975$	0.12033	−1.04857
	$\lambda = 0.99965$	0.12049	−1.04529

The best obtained results are in bold text

forgetting factors λ: 1 (without forgetting), 0.99995, 0.99985, 0.99975, 0.99965 and below were performed. The performance measures of the ozone-concentration predictions for all the validation data (spring, summer and autumn seasons) of each model are written in Table 6.9.

As the online trained model with forgetting factor $\lambda = 0.99985$ has minimal performance measures (Table 6.9), its predictions of the ozone concentration for two days are depicted in Fig. 6.24. It is clear that the predictions are very similar; therefore, the proposed approach seems to be a promising alternative for training GP models that predict the ozone concentration.

Figure 6.25 depicts the plot of predicted mean values versus the measured values for the whole of 2008 obtained from the online trained model.

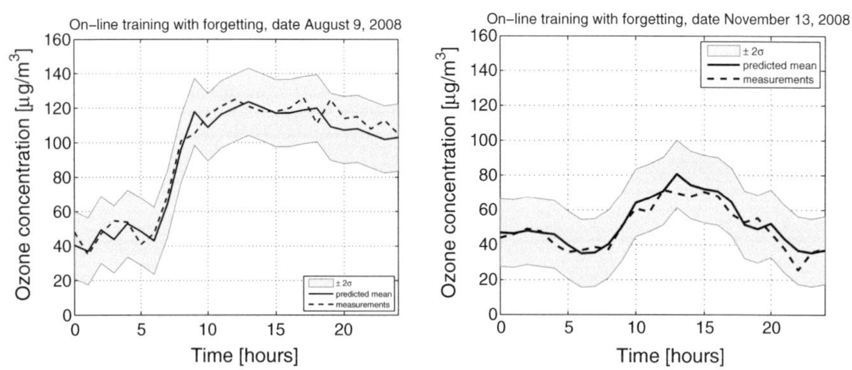

Fig. 6.24 The predicted mean value and 95 % confidence interval of the ozone concentration for August 9 and November 13, 2008

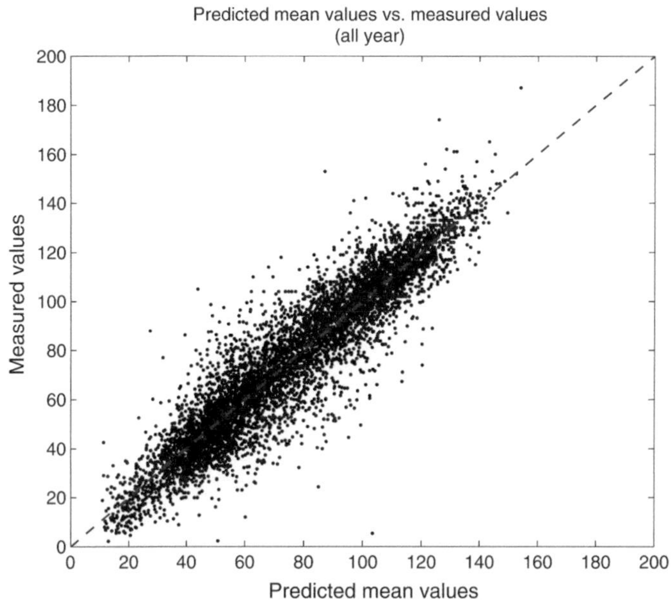

Fig. 6.25 Predicted mean values versus measured values for the year 2008

Discussion

In this case study, various first- and higher-order GP models for the prediction of the ozone concentration in the air of Burgas, Bulgaria are identified offline and compared. The results show that for the same model type, the higher-order models are more accurate than the first-order models. The best model according to the SMSE and MSLL performance measures used on the validation data is the second-order model, whose regressors are values of the ozone concentration and the meteorological parameters for the two previous hours.

Furthermore, as an alternative approach, an online updating (evolving) GP model is proposed. Such an approach is needed when the training data is not available for the whole period of interest and consequently not all the characteristics of the period can be trained or when the environment that is to be modelled is constantly changing. The proposed approach is evaluated as a second-order model with the model structure obtained in the first experiment with various forgetting factors. The analysis of the influence of the exponential forgetting for predictions of the ozone concentration shows that the forgetting factor $\lambda = 0.99985$ provides the most accurate predictions according to the used performance measures. The investigation shows that the predictions obtained with the proposed approach are very similar to the predictions obtained with a model that is trained offline on all-year data. Nevertheless, the online updated GP model is more suitable for changing environments and, consequently, gives better results for the prediction of the ozone concentration. Note that the investigation results are considered satisfactory for one-step-ahead predictions only. For longer horizons, the model structure has to be re-investigated.

References

1. Likar, B., Kocijan, J.: Predictive control of a gas-liquid separation plant based on a Gaussian process model. Comput. Chem. Eng. **31**(3), 142–152 (2007)
2. Kocijan, J., Likar, B.: Gas-liquid separator modelling and simulation with Gaussian-process models. Simul. Modell. Prac. Theory **16**(8), 910–922 (2008)
3. Vrančić, D., Juričić, D., Petrovčič, J.: Measurements and mathematical modelling of a semi-industrial liquid-gas separator for the purpose of fault diagnosis. Technical Report DP-7260, Jožef Stefan Institute, Ljubljana (1995). http://dsc.ijs.si/Damir.Vrancic/Files/dp7260.pdf
4. Girard, A.: Approximate methods for propagation of uncertainty with Gaussian process models. Ph.D. thesis, University of Glasgow, Glasgow (2004). http://citeseerx.ist.psu.edu/viewdoc/summary?doi=10.1.1.63.8313
5. Rasmussen, C.E., Williams, C.K.I.: Gaussian Processes for Machine Learning. MIT Press, Cambridge, MA (2006)
6. Kocijan, J., Žunič, G., Strmčnik, S., Vrančić, D.: Fuzzy gain-scheduling control of a gas-liquid separation plant implemented on a PLC. Int. J. Control **75**, 1082–1091 (2002)
7. Maciejowski, J.M.: Predictive control with constraints. Pearson Education Limited, Harlow (2002)
8. Kocijan, J., Přikryl, J.: Soft sensor for faulty measurements detection and reconstruction in urban traffic. In: Proceedings 15th IEEE Mediterranean Electromechanical Conference (MELECON), pp. 172–177. Valletta (2010)
9. Kadlec, P., Gabrys, B., Strand, S.: Data-driven soft sensors in the process industry. Comput. Chem. Eng. **33**, 795–814 (2009)
10. Gonzalez, G.D.: Soft sensors for processing plants. In: Proceedings of the second international conference on intelligent processing and manufacturing of materials, IPMM'99 (1999)
11. Fortuna, L.: Soft sensors for monitoring and control of industrial processes. Springer, Berlin (2007)
12. Adamski, A., Habdank-Wojewodzki, S.: Traffic congestion and incident detector realized by fuzzy discrete dynamic system. Arch. Transp. **17**(2), 5–13 (2004)
13. Zhu, W.B., Li, D.S., Lu, Y.: Real time speed measure while automobile braking on soft sensing technique. J. Phys.: Conf. Ser. **48**(1), 730–733 (2006)
14. Gupta, M., Gao, J., Aggarwal, C.C., Han, J.: Outlier detection for temporal data: a survey. IEEE Trans. Knowl. Data Eng. **26**(9), 2250–2267 (2014)
15. Hodge, V., Austin, J.: A survey of outlier detection methodologies. Artif. Intell. Rev. **22**(2), 85–126 (2004)
16. Lin, B., Recke, B., Knudsen, J., Jørgensen, S.B.: A systematic approach for soft sensor development. Comput. Chem. Eng. **31**(5), 419–425 (2007)
17. Menold, P.H., Pearson, R.K., Allgöwer, F.: Online outlier detection and removal. In: Proceedings of the 7th Mediterranean on control and automation (MED'99), pp. 1110–1133. Haifa (1999)
18. Lee, C., Choi, S., Lee, I.B.: Sensor fault identification based on time-lagged PCA in dynamic processes. Chemom. Intell. Lab. Syst. **70**(2), 165–178 (2004)
19. Osborne, M.A., Garnett, R., Swersky, K., de Freitas, N.: Prediction and fault detection of environmental signals with uncharacterised faults. In: 26th AAAI Conference on Artificial Intelligence (AAAI-12). Toronto (2012)
20. Gao, Y., Li, Y.: Improving Gaussian process classification with outlier detection, with applications in image classification. In: Kimmel, R.K.R., Sugimoto, A. (eds.) Computer Vision—ACCV 2010, Part IV, LNCS, vol. 6495, pp. 153–164. Springer, Berlin (2011)
21. Smith, M., Reece, S., Roberts, S., Rezek, I.: Online maritime abnormality detection using Gaussian processes and extreme value theory. In: IEEE 12th International Conference on Data Mining (ICDM), pp. 645–654. IEEE (2012)
22. Google maps. http://maps.google.com/
23. Shumway, R.H., Stoffer, D.S.: Time Series Analysis and Its Applications, With R Examples. Springer Texts in Statistics, 3rd edn. Springer, New York, NY (2011)

24. Noh, H.Y., Rajagopal, R.: Data-driven forecasting algorithms for building energy consumption. In: Lynch, J.P., Yun, C.B., Wang, K.W. (eds.) Proceedings SPIE 8692, Sensors and Smart Structures Technologies for Civil, Mechanical, and Aerospace Systems 2013, vol. 8692. San Diego, CA (2013). doi:10.1117/12.2009894
25. Lourenco, J., Santos, P.: Short-term load forecasting using a Gaussian process model: the influence of a derivative term in the input regressor. Intell. Decis. Technol. **6**(4), 273–281 (2012)
26. Dong, D.: Mine gas emission prediction based on Gaussian process model. Procedia Eng. **45**(1), 334–338 (2012)
27. Liu, Z., Xu, W., Shao, J.: Gauss process based approach for application on landslide displacement analysis and prediction. CMES—Comput. Model. Eng. Sci. **84**(2), 99–122 (2012)
28. Ou, P., Wang, H.: Modeling and forecasting stock market volatility by Gaussian processes based on GARCH, EGARCH and GJR models. In: Proceedings of the World Congress on Engineering 2011, WCE 2011, vol. 1, pp. 338–342 (2011)
29. Peltola, V., Honkela, A.: Variational inference and learning for non-linear state-space models with state-dependent observation noise. In: Proceedings of the 2010 IEEE International Workshop on Machine Learning for Signal Processing, MLSP 2010, pp. 190–195 (2010)
30. McHutchon, A., Rasmussen, C.E.: Gaussian process training with input noise. In: Shawe-Taylor, J., Zemel, R., Bartlett, P., Pereira, F., Weinberger, K. (eds.) Advances in Neural Information Processing Systems, vol. 24, pp. 1341–1349 (2011)
31. Petelin, D., Grancharova, A., Kocijan, J.: Evolving Gaussian process models for the prediction of ozone concentration in the air. Simul. Modell. Prac. Theory **33**(1), 68–80 (2013)
32. Hites, R.A.: Elements of environmental chemistry. Wiley - Interscience, Hoboken, NJ (2007)
33. Horvath, M., Bilitzky, L., Huttner, J.: Ozone. Elsevier, Amsterdam (1985)
34. Bytnerowicz, A., Omasa, K., Paoletti, E.: Integrated effects of air pollution and climate change on forests: a northern hemisphere perspective. Environ. Pollut. **147**(3), 438–445 (2006)
35. Directive 2008/50/EC of the European Parliament and of the Council of 21 May 2008 on ambient air quality and cleaner air for Europe. http://ec.europa.eu/environment/air/legis.htm
36. Al-Alawi, S.M., Abdul-Wahab, S.A., Bakheit, C.S.: Combining principal component regression and artificial neural-networks for more accurate predictions of ground-level ozone. Environ. Modell. Softw. **23**, 396–403 (2008)
37. Grašič, B., Mlakar, P., Božnar, M.: Ozone prediction based on neural networks and Gaussian processes. Nuovo cimento Soc. ital. fis., C Geophys. space phys. **29**(6), 651–661 (2006)
38. Solaiman, T.A., Coulibaly, P., Kanaroglou, P.: Ground-level ozone forecasting using data-driven methods. Air Qual. Atmos. Health **1**, 179–193 (2008)
39. Pisoni, E., Farina, M., Carnevale, C., Piroddi, L.: Forecasting peak air pollution levels using NARX models. Eng. Appl. Artif. Intell. **22**, 593–602 (2009)
40. Lin, Y., Cobourn, W.G.: Fuzzy system models combined with nonlinear regression for daily ground-level ozone predictions. Atmos. Environ. **41**, 3502–3513 (2007)
41. Nebot, A., Mugica, V., Escobet, A.: Ozone prediction based on meteorological variables: a fuzzy inductive reasoning approach. Atmos. Chem. Phys. Discuss. **8**, 12343–12370 (2008)
42. Chelani, A.B.: Prediction of daily maximum ground ozone concentration using support vector machine. Environ. Monit. Assess. **162**, 169–176 (2010)
43. Dueñas, C., Fernández, M.C., Cañete, S., Carretero, J., Liger, E.: Stochastic model to forecast ground-level ozone concentration at urban and rural areas. Chemosphere **61**, 1379–1389 (2005)
44. Grancharova, A., Kocijan, J., Krastev, A., Hristova, H.: High order Gaussian process models for prediction of ozone concentration in the air. In: Proceedings of the 7th EUROSIM Congress on Modelling and Simulation, EUROSIM'10 (2010)
45. Grancharova, A., Nedialkov, D., Kocijan, J., Hristova, H., Krastev, A.: Application of Gaussian processes to the prediction of ozone concentration in the air of Burgas. In: Proceedings of International Conference on Automatics and Informatics, pp. IV-17-IV-20 (2009)
46. Feng, Y., Zhang, W., Sun, D., Zhang, L.: Ozone concentration forecast method based on genetic algorithm optimized back propagation neural networks and support vector machine data classification. Atmos. Environ. **45**, 1979–1985 (2011)

47. Csató, L., Opper, M.: Sparse online Gaussian processes. Neural Comput. **14**(3), 641–668 (2002)
48. Petelin, D., Kocijan, J., Grancharova, A.: On-line Gaussian process model for the prediction of the ozone concentration in the air. Comptes Rendus de L'Academie Bulgare des Sciences **64**(2013), 117–124 (2011)
49. Seeger, M., Williams, C.K.I., Lawrence, N.D.: Fast forward selection to speed up sparse Gaussian process regression. In: Ninth International Workshop on Artificial Intelligence and Statistics, Society for Artificial Intelligence and Statistics (2003)

Appendix A
Mathematical Preliminaries

Random Variable

Roughly speaking, a *random variable* is a variable with a value that is subject to variations due to chance. It does not have a single value, rather it has a set of possible different values, each with an associated probability. A random variable x is defined as a function that assigns real numbers to samples

$$x : \Omega \rightarrow \mathbb{R} \tag{A.1}$$

where Ω is a sample space. The random variable maps the possible outcomes of an experiment, rather than describing the actual outcome. See mathematical references, e.g., [1–3], for a correct mathematical introduction to random variables.

There is a difference between probabilities for discrete variables and probability densities for continuous variables. The term 'probability' is used throughout the book to refer to both.

Mean and Variance

The *mean* value of the random variable x is the average of all its possible values χ, denoted by $E(x)$ and is defined as

$$E(x) = \int_{-\infty}^{\infty} \chi p(\chi) d\chi, \tag{A.2}$$

where $p(\cdot)$ is the probability density function of a random variable x.

The *variance* value of the random variable x is a measure of how much variability is in the values of x around the mean value of x and is defined as the average value of the squared difference between x and its mean

$$\mathrm{var}(x) = E((x - E(x))^2). \tag{A.3}$$

J. Kocijan, *Modelling and Control of Dynamic Systems Using Gaussian Process Models*, Advances in Industrial Control,
DOI 10.1007/978-3-319-21021-6

The *standard deviation* of the random variable x is defined as the positive square root of the variance

$$\sigma_x = \sigma(x) = +\sqrt{\mathrm{var}(x)}. \tag{A.4}$$

The *covariance* of the two random variables x_i and x_j expresses the extent to which two random variables change together

$$\mathrm{cov}(x_i, x_j) = E((x_i - E(x_i))(x_j - E(x_j))). \tag{A.5}$$

When the covariance of two random variables is zero, they are uncorrelated. When the covariance of two random variables is positive, their values mainly vary together with the same trend. If the values of two variables mainly vary in an opposing trend, the covariance is negative.

Stochastic Process

A vector $\mathbf{x} \in \mathbb{R}^N$ is called a *random vector* if all its components are random variables. The mean of a random vector is the vector of the mean values of its components

$$E(\mathbf{x}) = \begin{bmatrix} E(x_1) \\ E(x_2) \\ \vdots \\ E(x_N) \end{bmatrix}. \tag{A.6}$$

The variance of a random vector is an $N \times N$ covariance matrix composed of the covariances between the components of a random vector

$$\begin{aligned} \mathrm{var}(\mathbf{x}) &= \mathbf{\Sigma}(\mathbf{x}) = E((\mathbf{x} - E(\mathbf{x}))(\mathbf{x} - E(\mathbf{x}))^{\mathrm{T}}) \\ &= \begin{bmatrix} \mathrm{cov}(x_1, x_1) & \mathrm{cov}(x_1, x_2) & \cdots & \mathrm{cov}(x_1, x_N) \\ \mathrm{cov}(x_2, x_1) & \mathrm{cov}(x_2, x_2) & \cdots & \mathrm{cov}(x_2, x_N) \\ \vdots & \vdots & \vdots & \vdots \\ \mathrm{cov}(x_N, x_1) & \mathrm{cov}(x_N, x_2) & \cdots & \mathrm{cov}(x_N, x_N) \end{bmatrix}. \end{aligned} \tag{A.7}$$

The covariance of two random vectors $\mathbf{x}_i \in \mathbb{R}^N$ and $\mathbf{x}_j \in \mathbb{R}^M$ is an $N \times M$ covariance matrix composed of the covariances between the components of two random vectors

$$\begin{aligned} \mathrm{cov}(\mathbf{x}_i, \mathbf{x}_j) &= \mathbf{\Sigma}(\mathbf{x}_i, \mathbf{x}_j) = E((\mathbf{x}_i - E(\mathbf{x}_i))(\mathbf{x}_j - E(\mathbf{x}_j))^{\mathrm{T}}) \\ &= \begin{bmatrix} \mathrm{cov}(x_{i1}, x_{j1}) & \mathrm{cov}(x_{i1}, x_{i2}) & \cdots & \mathrm{cov}(x_{i1}, x_{jM}) \\ \mathrm{cov}(x_{i2}, x_{j1}) & \mathrm{cov}(x_{i2}, x_{j2}) & \cdots & \mathrm{cov}(x_{i2}, x_{jM}) \\ \vdots & \vdots & \vdots & \vdots \\ \mathrm{cov}(x_{iN}, x_{j1}) & \mathrm{cov}(x_{iN}, x_{j2}) & \cdots & \mathrm{cov}(x_{iN}, x_{jM}) \end{bmatrix}. \end{aligned} \tag{A.8}$$

The *stochastic process* is a collection or family of random variables. In our case these are random variables whose values change frequently with respect to values of some independent variable.

A stochastic process may also be seen as a random function with each function value being a random variable.

Gaussian Probability Density Function

A Gaussian or normal probability density function for a random variable x is defined as

$$p(x) = \mathcal{N}(\mu, \sigma^2) = \frac{1}{\sqrt{2\pi}\sigma} e^{-\frac{1}{2}\frac{(x-\mu)^2}{\sigma^2}}, \tag{A.9}$$

where μ stands for $E(x)$ and σ^2 for $\mathrm{var}(x)$.

A Gaussian or normal probability density function for a random vector $\mathbf{x}$ is defined as

$$p(\mathbf{x}) = \mathcal{N}(\boldsymbol{\mu}, \boldsymbol{\Sigma}) = \frac{1}{(2\pi)^{\frac{n}{2}} |\boldsymbol{\Sigma}|^{\frac{1}{2}}} e^{-\frac{1}{2}(\mathbf{x}-\boldsymbol{\mu})^{\mathrm{T}} \boldsymbol{\Sigma}^{-1} (\mathbf{x}-\boldsymbol{\mu})}, \tag{A.10}$$

where $\boldsymbol{\mu}$ stands for $E(\mathbf{x})$ and $\boldsymbol{\Sigma}$ for $\mathrm{var}(\mathbf{x})$.

Conditional Probability, Bayes' Theorem, Marginal Probability

The conditional probability function for the random variables a and b is

$$p(a|b) = \frac{p(a, b)}{p(b)} \tag{A.11}$$

where $p(a, b)$ is the joint probability and $p(b)$ is the probability of the random variable b. If a and b are independent, then $p(a)$ and the conditional $p(a|b)$ are equal.

Equation (A.11) can be transformed using the rule $p(a, b) = p(b, a)$ in

$$p(a|b) = p(b|a)p(a) = p(a|b)p(b). \tag{A.12}$$

Using this equation we obtain Bayes' theorem

$$p(b|a) = \frac{p(a|b)p(b)}{p(a)} \tag{A.13}$$

where $p(b|a)$ is the posterior probability distribution or posterior, $p(a|b)$ is the likelihood, $p(b)$ is the prior probability distribution or prior and $p(a)$ is called the marginal probability or evidence of a and can be obtained as

$$p(a) = \int p(a, b)db. \tag{A.14}$$

The integral is replaced by a sum if the variable b is not continuous but has discrete values.

References

1. Fine, T.L.: Probability and Probabilistic Reasoning for Electrical Engineering. Prentice Hall, Englewood Cliffs, NJ (2005)
2. Johnson, R.A., Miller, I., Freund, J.E.: Probability and Statistics for Engineers, 8th edn. Pearson, Boston, MA (2011)
3. Rohatgi, V.K.: An Introduction to Probability Theory and Mathematical Statistics. Wiley, New York, NY (1976)

Appendix B
Predictions

Predictions at Deterministic Inputs

Predictions for a Single Output

The mean value and the variance for the prediction with the Gaussian distribution $y^* \in \mathbb{R}$ from the vector of deterministic input values $\mathbf{z}^* \in \mathbb{R}^D$ [1]:

$$\begin{aligned} \mu^* = E(y^*) &= \mathbf{k}^{\mathrm{T}}(\mathbf{z}^*)\mathbf{K}^{-1}\mathbf{y} \\ &= \mathbf{k}^{\mathrm{T}}(\mathbf{z}^*)\boldsymbol{\beta} = \sum_{i=1}^{N} \beta_i C(\mathbf{z}_i, \mathbf{z}^*), \end{aligned} \tag{B.1}$$

$$\sigma^{2*} = \mathrm{var}(y^*) = \kappa(\mathbf{z}^*) - \mathbf{k}^{\mathrm{T}}(\mathbf{z}^*)\mathbf{K}^{-1}\mathbf{k}(\mathbf{z}^*), \tag{B.2}$$

where $\boldsymbol{\beta} = \mathbf{K}^{-1}\mathbf{y}$, $\mathbf{K} = \boldsymbol{\Sigma}_f + \sigma_n^2\mathbf{I} \in \mathbb{R}^{N\times N}$, $\mathbf{k}(\mathbf{z}^*) = [C(\mathbf{z}_1, \mathbf{z}^*), \ldots, C(\mathbf{z}_N, \mathbf{z}^*)]^{\mathrm{T}} \in \mathbb{R}^N$, and $\kappa(\mathbf{z}^*) = C(\mathbf{z}^*, \mathbf{z}^*) \in \mathbb{R}$.

Predictions for Multiple Conditionally Independent Outputs

The mean value and the variance for the prediction with the joint Gaussian distribution $\mathbf{y}^* \in \mathbb{R}^E$ from the vector of deterministic input values $\mathbf{z}^* \in \mathbb{R}^D$ [2]:

$$\begin{aligned} \boldsymbol{\mu}^* = E(\mathbf{y}^*) &= \left[\mu_1^* \ldots \mu_E^*\right] \\ &= \left[E(y_1^*) \ldots E(y_E^*)\right], \end{aligned} \tag{B.3}$$

$$\boldsymbol{\Sigma}^* = \begin{bmatrix} \sigma_1^{2*} & 0 & \ldots & 0 \\ 0 & \sigma_2^{2*} & \ldots & 0 \\ \vdots & \vdots & \ddots & \vdots \\ 0 & 0 & \ldots & \sigma_E^{2*} \end{bmatrix}, \tag{B.4}$$

J. Kocijan, *Modelling and Control of Dynamic Systems Using Gaussian Process Models*, Advances in Industrial Control,
DOI 10.1007/978-3-319-21021-6

Predictions at Stochastic Inputs with the Gaussian Distribution

Predictions for a Single Output with Approximate Moments

The mean value and the variance for the predictions with the Gaussian distribution $y^* \in \mathbb{R}$ from the vector of stochastic inputs $\mathbf{z}^* \sim \mathcal{N}(\boldsymbol{\mu}_{\mathbf{z}^*}, \boldsymbol{\Sigma}_{\mathbf{z}^*}) \in \mathbb{R}^D$ for the approximation of moments with a Taylor expansion [3]:

$$\mu^* = E(y^*) \approx \mathbf{k}(\boldsymbol{\mu}_{\mathbf{z}^*})^{\mathrm{T}}\mathbf{K}^{-1}\mathbf{y} \tag{B.5}$$

$$\begin{aligned}\sigma^{2*} = \mathrm{var}(y^*) &\approx \sigma^2(\boldsymbol{\mu}_{\mathbf{z}^*}) + \frac{1}{2}\mathrm{tr}\left(\left.\frac{\partial^2\sigma^2(\mathbf{z}^*)}{\partial\mathbf{z}^*\partial\mathbf{z}^{*\mathrm{T}}}\right|_{\mathbf{z}^*=\boldsymbol{\mu}_{\mathbf{z}^*}} \boldsymbol{\Sigma}_{\mathbf{z}^*}\right)\\ &+ \left.\frac{\partial\mu(\mathbf{z}^*)}{\partial\mathbf{z}^*}\right|_{\mathbf{z}^*=\boldsymbol{\mu}_{\mathbf{z}^*}}^{\mathrm{T}} \boldsymbol{\Sigma}_{\mathbf{z}^*} \left.\frac{\partial\mu(\mathbf{z}^*)}{\partial\mathbf{z}^*}\right|_{\mathbf{z}^*=\boldsymbol{\mu}_{\mathbf{z}^*}},\end{aligned} \tag{B.6}$$

where $\sigma^2(\boldsymbol{\mu}_{\mathbf{z}^*})$ denotes the variance of the Gaussian predictive distribution in the case when there are no uncertain input values. The partial derivatives in Eq. (B.6) are calculated by components of dimension [4]

$$\frac{\partial\mu(\mathbf{z}^*)}{\partial z_d^*} = \frac{\partial\mathbf{k}(\mathbf{z}^*)}{\partial z_d^*}\mathbf{K}^{-1}\mathbf{y}, \tag{B.7}$$

$$\frac{\partial^2\sigma^2(\mathbf{z}^*)}{\partial z_d^*\partial z_e^*} = \frac{\partial^2\kappa(\mathbf{z}^*)}{\partial z_d^*\partial z_e^*} - 2\frac{\partial\mathbf{k}(\mathbf{z}^*)^{\mathrm{T}}}{\partial z_d^*}\mathbf{K}^{-1}\frac{\partial\mathbf{k}(\mathbf{z}^*)}{\partial z_e^*} - 2\frac{\partial^2\mathbf{k}(\mathbf{z}^*)^{\mathrm{T}}}{\partial z_d^*\partial z_e^*}\mathbf{K}^{-1}\mathbf{k}(\mathbf{z}^*), \tag{B.8}$$

according to the input dimensions $d, e = 1, \dots, D$ and then evaluated at $\mathbf{z}^* = \boldsymbol{\mu}_{\mathbf{z}^*}$.

For the **squared exponential covariance function** described with Eq. (2.14) and denoted as C_f these partial derivatives are calculated by components of dimension [4, 5]:

$$\frac{\partial C_f(\mathbf{z}, \mathbf{z}_i^*)}{\partial z_{di}^*} = -l^{-2d}(z_{di}^* - z_{dj}^*)C(\mathbf{z}_i^*, \mathbf{z}_j^*) = -\frac{\partial C_f(\mathbf{z}, \mathbf{z}_j^*)}{\partial z_{dj}^*}, \tag{B.9}$$

$$\frac{\partial^2\sigma^2(\mathbf{z}_i^*)}{\partial z_{di}^*\partial z_{ei}} = (-l^{-2d}\delta_{de} + l^{-2d}(z_{di}^* - z_{dj}^*)l^{-2e}(z_{ei}^* - z_{ej}^*))C_f(\mathbf{z}_i^*, \mathbf{z}_j^*) \tag{B.10}$$

and

$$\frac{\partial^2\sigma^2(\mathbf{z}_i^*)}{\partial z_{di}^*\partial z_{ej}} = (l^{-2d}\delta_{de} - l^{-2d}(z_{di}^* - z_{dj}^*)l^{-2e}(z_{ei}^* - z_{ej}^*))C_f(\mathbf{z}_i^*, \mathbf{z}_j^*) \tag{B.11}$$

where l^{-2d}; $d = 1, \dots, D$ are hyperparameters from Eq. (2.14) and $\delta_{de} = 1$ if $d = e$ and $\delta_{de} - 0$ otherwise.

Predictions for a Single Output with Exact Moments

Squared Exponential Covariance Function

The mean value and the variance of the predictions which is the Gaussian distribution $y^* \in \mathbb{R}$ from the vector of stochastic inputs $\mathbf{z}^* \sim \mathcal{N}(\boldsymbol{\mu}_{\mathbf{z}^*}, \boldsymbol{\Sigma}_{\mathbf{z}^*}) \in \mathbb{R}^D$ for the squared exponential covariance function of Eq. (2.14) and the noise covariance function of Eq. (2.11) for the approximation with exact moments [2, 5]:

$$\mu^* = E(y^*) = \boldsymbol{\beta}^{\mathrm{T}}\mathbf{q}, \tag{B.12}$$

where the elements of vector $\mathbf{q}$ are

$$q_i = \sigma_f^2 |\boldsymbol{\Sigma}_{\mathbf{z}^*}\boldsymbol{\Lambda}^{-1} + \mathbf{I}|^{-\frac{1}{2}} \exp\left(-\frac{1}{2}(\mathbf{z}_i - \boldsymbol{\mu}_{\mathbf{z}^*})^{\mathrm{T}}(\boldsymbol{\Sigma}_{\mathbf{z}^*} + \boldsymbol{\Lambda})^{-1}(\mathbf{z}_i - \boldsymbol{\mu}_{\mathbf{z}^*})\right), \tag{B.13}$$

where σ_f^2 and $\boldsymbol{\Lambda}^{-1}$ are hyperparameters of the squared exponential covariance function, as defined in Eq. (2.14). q_i is an expectation of $C_f(\mathbf{z}_i, \mathbf{z}^*)$ with respect to the probability distribution of $\mathbf{z}^*$.

$$\sigma^{2*} = \mathrm{var}(y^*) = \boldsymbol{\beta}^{\mathrm{T}}\tilde{\mathbf{Q}}\boldsymbol{\beta} - \mu^{*2} + \sigma_f^2 - \mathrm{tr}\left((\boldsymbol{\Sigma}_f + \sigma_n^2\mathbf{I})^{-1}\tilde{\mathbf{Q}}\right) + \sigma_n^2, \tag{B.14}$$

where the elements of $\tilde{\mathbf{Q}} \in \mathbb{R}^{N \times N}$ are given by

$$\begin{aligned}\tilde{Q}_{ij} = &\frac{C(\mathbf{z}_i, \boldsymbol{\mu}_{\mathbf{z}^*})C(\mathbf{z}_j, \boldsymbol{\mu}_{\mathbf{z}^*})}{|2\boldsymbol{\Sigma}_{\mathbf{z}^*}\boldsymbol{\Lambda}^{-1} + \mathbf{I}|^{-\frac{1}{2}}}\\ &\exp\left(-\frac{1}{2}(\tilde{\zeta}_{ij} - \boldsymbol{\mu}_{\mathbf{z}^*})^{\mathrm{T}}(\boldsymbol{\Sigma}_{\mathbf{z}^*} + \frac{1}{2}\boldsymbol{\Lambda}^2)^{-1}\boldsymbol{\Sigma}_{\mathbf{z}^*}\boldsymbol{\Lambda}^{-1}(\tilde{\zeta}_{ij} - \boldsymbol{\mu}_{\mathbf{z}^*})\right),\end{aligned} \tag{B.15}$$

with $\tilde{\zeta}_{ij} = \frac{1}{2}(\mathbf{z}_i + \mathbf{z}_j)$.

Linear Covariance Function

The mean value and the variance for the predictions with the Gaussian distribution $y^* \in \mathbb{R}$ from the vector of stochastic inputs $\mathbf{z}^* \sim \mathcal{N}(\boldsymbol{\mu}_{\mathbf{z}^*}, \boldsymbol{\Sigma}_{\mathbf{z}^*}) \in \mathbb{R}^D$ for the linear covariance function of Eq. (2.22) and the noise covariance function of Eq. (2.11) [5, 6]:

$$\mu^* = E(y^*) = \boldsymbol{\mu}_{\mathbf{z}^*}(\boldsymbol{\Lambda}^{-1}\mathbf{Z})(\mathbf{Z}^{\mathrm{T}}\boldsymbol{\Lambda}^{-1}\mathbf{Z} + \sigma_n^2\mathbf{I})^{-1}\mathbf{y}, \tag{B.16}$$

where $\mathbf{Z} = [\mathbf{z}_1, \mathbf{z}_2, \ldots \mathbf{z}_N] \in \mathbb{R}^{D \times N}$ and $\boldsymbol{\Lambda}^{-1}$ is a matrix of hyperparameters of the linear covariance function as defined in Eq. (2.22).

$$\sigma^{2*} = \mathrm{var}(y^*) = \boldsymbol{\mu}_{\mathbf{z}^*}\boldsymbol{\alpha}\boldsymbol{\mu}_{\mathbf{z}^*}^{\mathrm{T}} + \mathrm{tr}[\boldsymbol{\alpha}\boldsymbol{\Sigma}_{\mathbf{z}^*}] + \mathrm{tr}[\boldsymbol{\gamma}\boldsymbol{\Sigma}_{\mathbf{z}^*}] + \sigma_n^2, \tag{B.17}$$

where

$$\boldsymbol{\alpha} = \boldsymbol{\Lambda}^{-1} - (\boldsymbol{\Lambda}^{-1}\mathbf{Z})(\mathbf{Z}^{\mathrm{T}}\boldsymbol{\Lambda}^{-1}\mathbf{Z} + \sigma_n^2\mathbf{I})^{-1}(\mathbf{Z}^{\mathrm{T}}\boldsymbol{\Lambda}^{-1})$$

and

$$\gamma = (\boldsymbol{\Lambda}^{-1}\mathbf{Z})(\mathbf{Z}^{\mathrm{T}}\boldsymbol{\Lambda}^{-1}\mathbf{Z} + \sigma_n^2\mathbf{I})^{-1}\mathbf{y}\mathbf{y}^{\mathrm{T}}((\mathbf{Z}^{\mathrm{T}}\boldsymbol{\Lambda}^{-1}\mathbf{Z} + \sigma_n^2\mathbf{I})^{-1})^{\mathrm{T}}(\mathbf{Z}^{\mathrm{T}}\boldsymbol{\Lambda}^{-1}).$$

Predictions for Multiple Outputs with Exact Moments

The mean value and the variance of the prediction which is the joint Gaussian distribution $\mathbf{y}^* \in \mathbb{R}^E$ from the vector of stochastic inputs $\mathbf{z}^* \sim \mathcal{N}(\boldsymbol{\mu}_{\mathbf{z}^*}, \boldsymbol{\Sigma}_{\mathbf{z}^*}) \in \mathbb{R}^D$ using the **squared exponential covariance function** of Eq. (2.14) together with the noise covariance function of Eq. (2.11) and the approximation with exact moments [2, 7]:

$$\boldsymbol{\mu}^* = E(\mathbf{y}^*) = \left[\, \boldsymbol{\beta}_1^{\mathrm{T}}\mathbf{q}_1 \ldots \boldsymbol{\beta}_E^{\mathrm{T}}\mathbf{q}_E \,\right], \tag{B.18}$$

where the vectors $\mathbf{q}_i;\ i = 1, \ldots, E$ are given by Eq. (B.13).

$$\boldsymbol{\Sigma}^* = \begin{bmatrix} \mathrm{var}(y_1^*) & \ldots & \mathrm{cov}(y_1^*, y_E^*) \\ \vdots & \ddots & \vdots \\ \mathrm{cov}(y_E^*, y_1^*) & \ldots & \mathrm{var}(y_E^*) \end{bmatrix}. \tag{B.19}$$

The cross-covariances in Eq. (B.19) are given by

$$\mathrm{cov}(y_a^*, y_b^*) = E(y_a^* y_b^*) - \mu_a^*\mu_b^*;\ a, b \in \{1, \ldots, E\} \tag{B.20}$$

which leads to [2]:

$$\mathrm{cov}(y_a^*, y_b^*) = \begin{cases} \boldsymbol{\beta}_a^{\mathrm{T}}\mathbf{Q}\boldsymbol{\beta}_b - E(y_a^*)E(y_b^*), & a \neq b \\ \boldsymbol{\beta}_a^{\mathrm{T}}\mathbf{Q}\boldsymbol{\beta}_a - E(y_a^*)E(y_a^*) + \sigma_{f_a}^2 - \mathrm{tr}\left((\boldsymbol{\Sigma}_{f_a} + \sigma_n^2\mathbf{I})^{-1}\mathbf{Q}\right) + \sigma_n^2, & a = b \end{cases} \tag{B.21}$$

where the elements of $\mathbf{Q}$ are

$$Q_{ij} = \frac{C_a(\mathbf{z}_i, \boldsymbol{\mu}_{\mathbf{z}^*})C_b(\mathbf{z}_j, \boldsymbol{\mu}_{\mathbf{z}^*})}{|\boldsymbol{\Sigma}_{\mathbf{z}^*}(\boldsymbol{\Lambda}_a^{-1} + \boldsymbol{\Lambda}_b^{-1}) + \mathbf{I}|^{\frac{1}{2}}} \exp\left(\frac{1}{2}(\boldsymbol{\zeta}_{ij}^{\mathrm{T}}(\boldsymbol{\Sigma}_{\mathbf{z}^*}(\boldsymbol{\Lambda}_a^{-1} + \boldsymbol{\Lambda}_b^{-1}) + \mathbf{I})^{-1}\boldsymbol{\Sigma}_{\mathbf{z}^*}\boldsymbol{\zeta}_{ij})\right), \tag{B.22}$$

with $\boldsymbol{\zeta}_{ij} = \boldsymbol{\Lambda}_a^{-1}(\mathbf{z}_i - \boldsymbol{\mu}_{\mathbf{z}^*}) + \boldsymbol{\Lambda}_b^{-1}(\mathbf{z}_j - \boldsymbol{\mu}_{\mathbf{z}^*})$.

Cross-Covariances Between Stochastic Inputs and Outputs

Cross-covariances between the stochastic inputs and the stochastic outputs are [2]:

$$\mathrm{cov}(\mathbf{z}^*, \mathbf{y}^*) = E(\mathbf{z}^*\mathbf{y}^{*\mathrm{T}}) - \boldsymbol{\mu}_{\mathbf{z}^*}\boldsymbol{\mu}^{*\mathrm{T}}. \tag{B.23}$$

For each output dimension $a = 1, \ldots, E$ we compute $\mathrm{cov}(\mathbf{z}^*, y_a^*)$ as

$$\mathrm{cov}(\mathbf{z}^*, y_a^*) = \sum_{i=1}^{N} \beta_{a_i} q_{a_i} \boldsymbol{\Sigma}_{\mathbf{z}^*}(\boldsymbol{\Sigma}_{\mathbf{z}^*} + \boldsymbol{\Lambda}_a)^{-1}(\mathbf{z}_i - \boldsymbol{\mu}_{\mathbf{z}^*}). \tag{B.24}$$

References

1. Rasmussen, C.E., Williams, C.K.I.: Gaussian Processes for Machine Learning. MIT Press, Cambridge, MA (2006)
2. Deisenroth, M.P.: Efficient reinforcement learning using Gaussian processes. Ph.D. thesis, Karlsruhe Institute of Technology, Karlsruhe (2010)
3. Girard, A., Rasmussen, C.E., Candela, J.Q., Murray-Smith, R.: Gaussian process priors with uncertain inputs—application to multiple-step ahead time series forecasting. In: Becker, S., Thrun, S., Obermayer, K. (eds.) Advances in Neural Information Processing Systems, vol. 15, pp. 542–552. MIT Press, Cambridge, MA (2003)
4. Girard, A., Rasmussen, C., Murray-Smith, R.: Gaussian process priors with uncertain inputs: Multiple-step ahead prediction. Technical Report DCS TR-2002-119, University of Glasgow, Glasgow (2002)
5. Girard, A.: Approximate methods for propagation of uncertainty with Gaussian process models. Ph.D. thesis, University of Glasgow, Glasgow (2004). http://citeseerx.ist.psu.edu/viewdoc/summary?doi=10.1.1.63.8313
6. Kocijan, J., Girard, A., Leith, D.J.: Incorporating linear local models in Gaussian process model. Technical Report DP-8895, Jozef Stefan Institute, Ljubljana (2003)
7. Deisenroth, M.P., Turner, R.D., Huber, M.F., Hanebeck, U.D., Rasmussen, C.E.: Robust filtering and smoothing with Gaussian processes. IEEE Trans. Autom. Control **57**(7), 1865–1871 (2012). doi:10.1109/TAC.2011.2179426

Appendix C
Matlab Code

Implementations of some of the presented algorithms in Matlab® are available for download at the website:

https://github.com/Dynamic-Systems-and-GP/GPdyn.git or **http://extras.springer.com**

The programs are short stand-alone implementations and not part of a larger package. They are based on GPML package of Carl E. Rasmussen and Chris K.I. Williams that is available on the website of book [1]. The GPML package is provided together with the provided code.

Reference

1. Rasmussen, C.E., Williams, C.K.I.: Processes for Machine Learning. MIT Press, Cambridge, MA (2006)

J. Kocijan, *Modelling and Control of Dynamic Systems Using Gaussian Process Models*, Advances in Industrial Control,
DOI 10.1007/978-3-319-21021-6

Index

J. Kocijan, *Modelling and Control of Dynamic Systems Using Gaussian Process Models*, Advances in Industrial Control,
DOI 10.1007/978-3-319-21021-6

MIX
Papier aus verantwortungsvollen Quellen
Paper from responsible sources
FSC® C105338

If you have any concerns about our products,
you can contact us on
ProductSafety@springernature.com

In case Publisher is established outside the EU,
the EU authorized representative is:
Springer Nature Customer Service Center GmbH
Europaplatz 3, 69115 Heidelberg, Germany

Printed by Libri Plureos GmbH
in Hamburg, Germany